The Concise Oxford Dictionary of Mathematics

Second Edition

CHRISTOPHER CLAPHAM

Oxford New York

OXFORD UNIVERSITY PRESS

OXFORD
UNIVERSITY PRESS

Great Clarendon Street, Oxford OX2 6DP

Oxford University Press is a department of the University of Oxford.
It furthers the University's objective of excellence in research, scholarship,
and education by publishing worldwide in

Oxford New York

Auckland Bangkok Buenos Aires Cape Town Chennai
Dar es Salaam Delhi Hong Kong Istanbul Karachi Kolkata
Kuala Lumpur Madrid Melbourne Mexico City Mumbai Nairobi
São Paulo Shanghai Taipei Tokyo Toronto

Oxford is a registered trade mark of Oxford University Press
in the UK and in certain other countries

Published in the United States
by Oxford University Press Inc., New York

First edition 1990
Second edition 1996

British Library Cataloguing in Publication Data

Data available

Library of Congress Cataloging in Publication Data

Clapham, Christopher.
The concise Oxford dictionary of mathematics / Christopher
Clapham.—2nd ed.
p. cm.—(Oxford paperback reference)
Rev. ed. of: A concise dictionary of mathematics. 1990.
1. Mathematics—Dictionaries. I. Clapham, Christopher. Concise
dictionary of mathematics. II. Title.
QA5.C53 1996 510'.3—dc20 95-37085

ISBN 0-19-280041-8

20 19

Printed in Great Britain by
Clays Ltd, St Ives plc

OXFORD PAPERBACK REFERENCE

The Concise Oxford Dictionary of **Mathematics**

Christopher Clapham was until 1993 Senior Lecturer in Mathematics at the University of Aberdeen and has also taught at universities in Nigeria, Lesotho, and Malawi. He is the author of *Introduction to Abstract Algebra* and *Introduction to Mathematical Analysis*. He lives in Exeter.

Contents

Contributors

C. Chatfield, BSc, PhD
R. Cheal, BSc
J. B. Gavin, BSc, MSc
University of Bath

J. R. Pulham, BSc, PhD
University of Aberdeen

D. P. Thomas, BSc, PhD
University of Dundee

Preface

This dictionary is intended to be a reference book that gives reliable definitions or clear and precise explanations of mathematical terms. The level is such that it will suit, among others, sixth-form pupils, college students and first-year university students who are taking mathematics as one of their courses. Such students will be able to look up any term they may meet and be led on to other entries by following up cross-references or by browsing more generally.

The concepts and terminology of all those topics that feature in pure and applied mathematics and statistics courses at this level today are covered. There are also entries on mathematicians of the past and important mathematics of more general interest. Computing is not included. The reader's attention is drawn to the appendices which give useful tables for ready reference.

Some entries give a straight definition in an opening phrase. Others give the definition in the form of a complete sentence, sometimes following an explanation of the context. In this case, the keyword appears again in bold type at the point where it is defined. Other keywords in bold type may also appear if this is the most appropriate context in which to define or explain them. Italic is used to indicate words with their own entry, to which cross-reference can be made if required.

This edition is more than half as large again as the first edition. A significant change has been the inclusion of entries covering applied mathematics and statistics. In these areas, I am very much indebted to the contributors, whose names are given on page v. I am most grateful to these colleagues for their specialist advice and drafting work. They are not, however, to be held responsible for the final form of the entries on their subjects. There has also been a considerable increase in the number of short biographies, so that all the major names are included. Other additional entries have greatly increased the comprehensiveness of the dictionary.

The text has benefited from the comments of colleagues who have read different parts of it. Even though the names of all of them will not be given, I should like to acknowledge here their help and express my thanks.

Christopher Clapham

Abel, Niels Henrik (1802–1829) Norwegian mathematician who, at the age of 19, proved that the general equation of degree greater than 4 cannot be solved algebraically. In other words, there can be no formula for the roots of such an equation similar to the familiar formula for a quadratic equation. He was also responsible for fundamental developments in the theory of algebraic functions. He died in some poverty at the age of 26, just a few days before he would have received a letter announcing his appointment to a professorship in Berlin.

abelian group Suppose that G is a *group* with the operation $\circ$. Then G is **abelian** if the operation $\circ$ is commutative; that is, if, for all elements a and b in G, $a \circ b = b \circ a$.

abscissa The x-coordinate in a Cartesian coordinate system in the plane.

absolute error See *error.*

absolute value For any real number a, the **absolute value** (also called the *modulus*) of a, denoted by $|a|$, is a itself if $a \geq 0$, and $-a$ if $a < 0$. Thus $|a|$ is positive except when $a = 0$. The following properties hold:

(i) $|ab| = |a||b|$.
(ii) $|a + b| \leq |a| + |b|$.
(iii) $|a - b| \geq ||a| - |b||$.
(iv) For $a > 0$, $|x| \leq a$ if and only if $-a \leq x < a$.

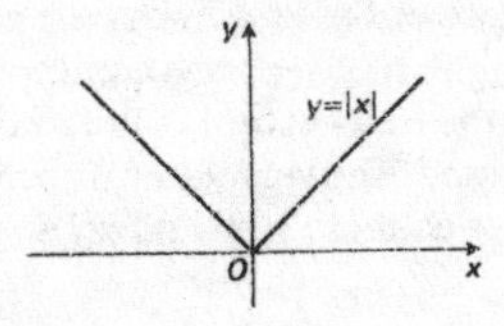

absorbing state See *random walk.*

absorption laws For all sets A and B (subsets of some *universal set*), $A \cap (A \cup B) = A$ and $A \cup (A \cap B) = A$. These are the **absorption laws.**

abstract algebra The area of mathematics concerned with algebraic structures, such as *groups, rings* and *fields,* involving sets of elements with particular operations satisfying certain axioms. The purpose is to derive, from the set of axioms, general results that are then applicable to any particular example of the algebraic structure in question. The theory of certain algebraic structures is highly developed; in particular, the theory of vector spaces is so extensive that its study, known as *linear algebra*, would probably no longer be classified as abstract algebra.

acceleration Suppose that a particle is moving in a straight line, with a point O on the line taken as origin and one direction taken as positive. Let x be the *displacement* of the particle at time t. The **acceleration** of the particle is equal to $\ddot{x}$ or d^2x/dt^2, the *rate of change* of the *velocity* with respect to t. If the velocity is positive (that is, if the particle is moving in the positive direction), the acceleration is positive when the particle is speeding up and negative when it is slowing down. However, if the velocity is negative, a positive acceleration means that the particle is slowing down and a negative acceleration means that it is speeding up.

In the preceding paragraph, a common convention has been followed, in which the unit vector $\mathbf{i}$ in the positive direction along the line has been suppressed. Acceleration is in fact a vector quantity, and in the one-dimensional case above it is equal to $\ddot{x}\mathbf{i}$.

When the motion is in two or three dimensions, vectors are used explicitly. The acceleration $\mathbf{a}$ of a particle is a vector equal to the rate of change of the velocity $\mathbf{v}$ with respect to t. Thus $\mathbf{a} = d\mathbf{v}/dt$. If the particle has *position vector* $\mathbf{r}$, then $\mathbf{a} = d^2\mathbf{r}/dt^2 = \ddot{\mathbf{r}}$. When Cartesian coordinates are used, $\mathbf{r} = x\mathbf{i} + y\mathbf{j} + z\mathbf{k}$, and then $\ddot{\mathbf{r}} = \ddot{x}\mathbf{i} + \ddot{y}\mathbf{j} + \ddot{z}\mathbf{k}$.

Acceleration has the dimensions LT^{-2}, and the SI unit of measurement is the metre per second per second, abbreviated to '$\mathrm{m\,s}^{-2}$'.

acceleration–time graph A graph that shows acceleration plotted against time for a particle moving in a straight line. Let $v(t)$ and $a(t)$ be the velocity and acceleration, respectively, of the particle at time t. The acceleration–time graph is the graph $y = a(t)$, where the t-axis is horizontal and the y-axis is vertical with the positive direction upwards. With the convention that any area below the horizontal axis is negative, the area under the graph between $t = t_1$ and $t = t_2$ is equal to $v(t_2) - v(t_1)$. (Here a common convention has been followed, in which the unit vector $\mathbf{i}$ in the positive direction along the line has been suppressed. The velocity and acceleration of the particle are in fact vector quantities equal to $v(t)\mathbf{i}$ and $a(t)\mathbf{i}$, respectively.)

acceptance region See *hypothesis testing*.

acute angle An angle that is less than a *right angle*. An **acute-angled** triangle is one all of whose angles are acute.

addition (of complex numbers) Let the complex numbers z_1 and z_2, where $z_1 = a + bi$ and $z_2 = c + di$, be represented by the points P_1 and P_2 in the *complex plane*. Then $z_1 + z_2 = (a + c) + (b + d)i$, and $z_1 + z_2$ is represented in the complex plane by the point Q such that OP_1QP_2 is a parallelogram; that is, such that $\overrightarrow{OQ} = \overrightarrow{OP_1} + \overrightarrow{OP_2}$. Thus, if the complex number z is associated with the *directed line-segment* $\overrightarrow{OP}$, where P

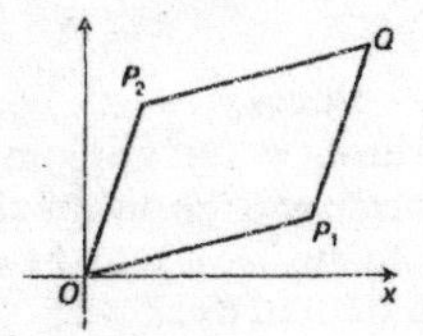

represents z, then the addition of complex numbers corresponds exactly to the addition of the directed line-segments.

addition (of directed line-segments) See *addition* (of vectors).

addition (of matrices) Let **A** and **B** be $m \times n$ matrices, with $\mathbf{A} = [a_{ij}]$ and $\mathbf{B} = [b_{ij}]$. The operation of **addition** is defined by taking the **sum** $\mathbf{A} + \mathbf{B}$ to be the $m \times n$ matrix **C**, where $\mathbf{C} = [c_{ij}]$ and $c_{ij} = a_{ij} + b_{ij}$. The sum $\mathbf{A} + \mathbf{B}$ is not defined if **A** and **B** are not of the same order. This operation + of addition on the set of all $m \times n$ matrices is *associative* and *commutative*.

addition (of vectors) Given vectors **a** and **b**, let $\overrightarrow{OA}$ and $\overrightarrow{OB}$ be *directed line-segments* that represent **a** and **b**, with the same initial point O. The sum of $\overrightarrow{OA}$ and $\overrightarrow{OB}$ is the directed line-segment $\overrightarrow{OC}$, where $OACB$ is a parallelogram, and the **sum** $\mathbf{a} + \mathbf{b}$ is defined to be the vector **c** represented by $\overrightarrow{OC}$. This is called the **parallelogram law**. Alternatively, the sum of vectors **a** and **b** can be defined by representing **a** by a directed line-segment $\overrightarrow{OP}$ and **b** by $\overrightarrow{PQ}$, where the final point of the first directed line-segment is the initial point of the second. Then $\mathbf{a} + \mathbf{b}$ is the vector represented by $\overrightarrow{OQ}$. This is called the **triangle law**. Addition of vectors has the following properties, which hold for all **a**, **b** and **c**:

(i) $\mathbf{a} + \mathbf{b} = \mathbf{b} + \mathbf{a}$, the commutative law.
(ii) $\mathbf{a} + (\mathbf{b} + \mathbf{c}) = (\mathbf{a} + \mathbf{b}) + \mathbf{c}$, the associative law.
(iii) $\mathbf{a} + \mathbf{0} = \mathbf{0} + \mathbf{a} = \mathbf{a}$, where $\mathbf{0}$ is the zero vector.
(iv) $\mathbf{a} + (-\mathbf{a}) = (-\mathbf{a}) + \mathbf{a} = \mathbf{0}$, where $-\mathbf{a}$ is the negative of **a**.

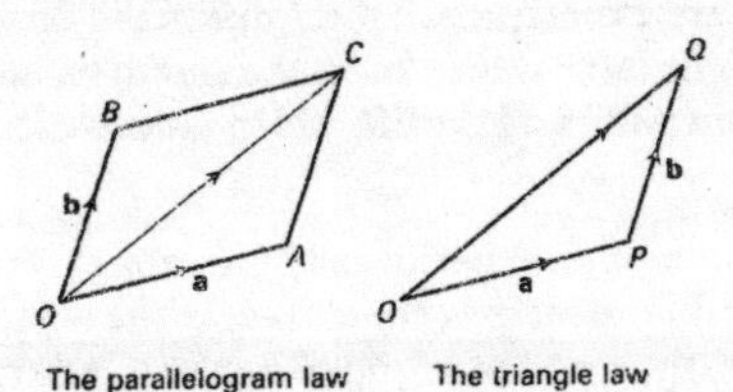

The parallelogram law The triangle law

addition modulo *n* See *modulo n, addition* and *multiplication*.

additive group A *group* with the operation +, called addition, may be called an **additive group**. The operation in a group is normally denoted by addition only if it is *commutative*, so an additive group is usually *abelian*.

additive inverse See *inverse element*.

adjacency matrix For a *simple graph G*, with n vertices $v_1, v_2, \ldots, v_n$, the **adjacency matrix A** is the $n \times n$ matrix $[a_{ij}]$ with $a_{ij} = 1$, if v_i is joined to v_j, and $a_{ij} = 0$, otherwise. The matrix **A** is *symmetric* and the diagonal entries are zero. The number of ones in any row (or column) is equal to the *degree* of the corresponding vertex. An example of a graph and its adjacency matrix **A** is shown below.

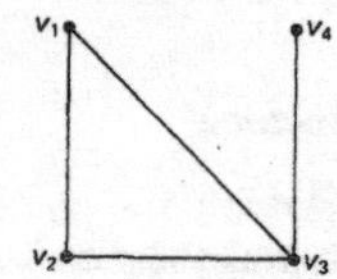

$$\mathbf{A} = \begin{bmatrix} 0 & 1 & 1 & 0 \\ 1 & 0 & 1 & 0 \\ 1 & 1 & 0 & 1 \\ 0 & 0 & 1 & 0 \end{bmatrix}$$

adjoint The **adjoint** of a square matrix **A**, denoted by adj **A**, is the transpose of the matrix of cofactors of **A**. For $\mathbf{A} = [a_{ij}]$, let A_{ij} denote the *cofactor* of the entry a_{ij}. Then the matrix of cofactors is the matrix $[A_{ij}]$ and $\text{adj}\,\mathbf{A} = [A_{ij}]^T$. For example, a 3×3 matrix **A** and its adjoint can be written

$$\mathbf{A} = \begin{bmatrix} a_{11} & a_{12} & a_{13} \\ a_{21} & a_{22} & a_{23} \\ a_{31} & a_{32} & a_{33} \end{bmatrix}, \qquad \text{adj}\,\mathbf{A} = \begin{bmatrix} A_{11} & A_{21} & A_{31} \\ A_{12} & A_{22} & A_{32} \\ A_{13} & A_{23} & A_{33} \end{bmatrix}.$$

In the 2×2 case, a matrix **A** and its adjoint have the form

$$\mathbf{A} = \begin{bmatrix} a & b \\ c & d \end{bmatrix}, \qquad \text{adj}\,\mathbf{A} = \begin{bmatrix} d & -b \\ -c & a \end{bmatrix}.$$

The adjoint is important because it can be used to find the *inverse* of a matrix. From the properties of cofactors, it can be shown that $\mathbf{A}\,\text{adj}\,\mathbf{A} = (\det \mathbf{A})\mathbf{I}$. It follows that, when $\det \mathbf{A} \neq 0$, the inverse of **A** is $(1/\det \mathbf{A})\,\text{adj}\,\mathbf{A}$.

adjugate = *adjoint*.

aerodynamic drag A body moving through the air, such as an aeroplane flying in the Earth's atmosphere, experiences a force due to the flow of air over the surface of the body. The force is the sum of the **aerodynamic drag**, which is tangential to the flight path, and the **lift**, which is normal to the flight path.

air resistance The resistance to motion experienced by an object moving through the air caused by the flow of air over the surface of the object. It is a force that affects, for example, the speed of a drop of rain or of a parachutist falling towards the Earth's surface. As well as depending on the nature of the object, air resistance depends on the speed of the object. Possible *mathematical models* are to assume that the magnitude of the air resistance is proportional to the speed or to the square of the speed.

Algebra, Fundamental Theorem of See *Fundamental Theorem of Algebra*.

algebra of sets The set of all subsets of a *universal set* E is closed under the binary operations $\cup$ (*union*) and $\cap$ (*intersection*) and the unary operation $'$ (*complementation*). The following are some of the properties, or laws, that hold for subsets A, B and C of E:

(i) $A \cup (B \cup C) = (A \cup B) \cup C$ and $A \cap (B \cap C) = (A \cap B) \cap C$, the associative properties.
(ii) $A \cup B = B \cup A$ and $A \cap B = B \cap A$, the commutative properties.
(iii) $A \cup \emptyset = A$ and $A \cap \emptyset = \emptyset$, where $\emptyset$ is the *empty set*.
(iv) $A \cup E = E$ and $A \cap E = A$.
(v) $A \cup A = A$ and $A \cap A = A$.
(vi) $A \cap (B \cup C) = (A \cap B) \cup (A \cap C)$ and $A \cup (B \cap C) = (A \cup B) \cap (A \cup C)$, the distributive properties.
(vii) $A \cup A' = E$ and $A \cap A' = \emptyset$.
(viii) $E' = \emptyset$ and $\emptyset' = E$.
(ix) $(A')' = A$.
(x) $(A \cup B)' = A' \cap B'$ and $(A \cap B)' = A' \cup B'$, De Morgan's laws.

The application of these laws to subsets of E is known as the **algebra of sets**. Despite some similarities with the algebra of numbers, there are important and striking differences.

algebraic number A real number that is the root of a *polynomial equation* with integer coefficients. All *rational numbers* are algebraic, since a/b is the root of the equation $bx - a = 0$. Some *irrational numbers* are algebraic; for example, $\sqrt{2}$ is the root of the equation $x^2 - 2 = 0$. An irrational number that is not algebraic (such as π) is called a *transcendental number*.

algebraic structure The term used to describe an abstract concept defined as consisting of certain elements with operations satisfying given axioms. Thus, a *group* or a *ring* or a *field* is an algebraic structure. The purpose of the definition is to recognize similarities that appear in different contexts within mathematics and to encapsulate these by means of a set of axioms.

algorithm A precisely described routine procedure that can be applied and systematically followed through to a conclusion.

al-Khwārizmī See under K.

alternate angles See *transversal*.

alternative hypothesis See *hypothesis testing*.

altitude A line through one vertex of a triangle and perpendicular to the opposite side. The three altitudes of a triangle are concurrent at the *orthocentre*.

amicable numbers A pair of numbers with the property that each is equal to the sum of the positive divisors of the other. (For the purposes of this definition, a number is not included as one of its own divisors.) For example, 220 and 284 are amicable numbers because the positive divisors of 220 are 1, 2, 4, 5, 10, 11, 20, 22, 44, 55 and 110, whose sum is 284, and the positive divisors of 284 are 1, 2, 4, 71 and 142, whose sum is 220.

These numbers, known to the Pythagoreans, were used as symbols of friendship. The amicable numbers 17 296 and 18 416 were found by Fermat, and a list of 64 pairs was produced by Euler. In 1867, a sixteen-year-old Italian boy found the second smallest pair, 1184 and 1210, overlooked by Euler. More than 600 pairs are now known. It has not been shown whether or not there are infinitely many pairs of amicable numbers.

amplitude Suppose that $x = A\sin(\omega t + \alpha)$, where A (> 0), ω and α are constants. This may, for example, give the displacement x of a particle, moving in a straight line, at time t. The particle is thus oscillating about the origin O. The constant A is the **amplitude**, and gives the maximum distance in each direction from O that the particle attains.

The term may also be used in the case of *damped oscillations* to mean the corresponding coefficient, even though it is not constant. For example, if $x = 5e^{-2t}\sin 3t$, the oscillations are said to have amplitude $5e^{-2t}$, which tends to zero as t tends to infinity.

analysis The area of mathematics generally taken to include those topics that involve the use of limiting processes. Thus *differential calculus* and *integral calculus* certainly come under this heading. Besides these, there are other topics, such as the summation of infinite series, which involve 'infinite' processes of this sort. The *Binomial Theorem*, a theorem of algebra, leads on into analysis when the index is no longer a positive integer, and the study of sine and cosine, which begins as trigonometry, becomes analysis when the power series for the functions are derived. The term 'analysis' has also come to be used to indicate a rather more rigorous approach to the topics of calculus, and to the foundations of the real number system.

analysis of variance A general procedure for partitioning the overall variability in a set of data into components due to specified causes and random variation. It involves calculating such quantities as the 'between-groups sum of squares' and the 'residual sum of squares', and dividing by the *degrees of freedom* to give so-called 'mean squares'. The results are usually presented in an **ANOVA** table, the name being derived from the opening letters of the words 'analysis of variance'. Such a table provides a concise summary from which the influence of the *explanatory variables* can be estimated and hypotheses can be tested, usually by means of *F-tests*.

anchor ring = *torus*.

and See *conjunction*.

angle (between lines in space) Given two lines in space, let $\mathbf{u}_1$ and $\mathbf{u}_2$ be vectors with directions along the lines. Then the **angle** between the lines, even if they do not meet, is equal to the angle between the vectors $\mathbf{u}_1$ and $\mathbf{u}_2$ (see *angle* (between vectors)), with the directions of $\mathbf{u}_1$ and $\mathbf{u}_2$ chosen so that the angle θ satisfies $0 \le \theta \le \pi/2$ (θ in radians), or $0 \le \theta \le 90$ (θ in degrees). If l_1, m_1, n_1 and l_2, m_2, n_2 are direction ratios for directions along the lines, the angle θ between the lines is given by

$$\cos\theta = \frac{|l_1 l_2 + m_1 m_2 + n_1 n_2|}{\sqrt{l_1^2 + m_1^2 + n_1^2}\sqrt{l_2^2 + m_2^2 + n_2^2}}.$$

angle (between lines in the plane) In coordinate geometry of the plane, the angle α between two lines with gradients m_1 and m_2 is given by

$$\tan\alpha = \frac{m_1 - m_2}{1 + m_1 m_2}.$$

This is obtained from the formula for $\tan(A - B)$. In the special cases when $m_1 m_2 = -1$ or when m_1 or m_2 is infinite, it has to be interpreted appropriately.

angle (between planes) Given two planes, let $\mathbf{n}_1$ and $\mathbf{n}_2$ be vectors *normal* to the two planes. Then a method of obtaining the **angle** between the planes is to take the angle between $\mathbf{n}_1$ and $\mathbf{n}_2$ (see *angle* (between vectors)), with the directions of $\mathbf{n}_1$ and $\mathbf{n}_2$ chosen so that the angle θ satisfies $0 \le \theta \le \pi/2$ (θ in radians), or $0 \le \theta \le 90$ (θ in degrees).

angle (between vectors) Given vectors **a** and **b**, let $\overrightarrow{OA}$ and $\overrightarrow{OB}$ be *directed line-segments* representing **a** and **b**. Then the **angle** θ between the vectors **a** and **b** is the angle $\angle AOB$, where θ is taken to satisfy $0 \le \theta \le \pi$ (θ in radians), or $0 \le \theta \le 180$ (θ in degrees). It is given by

$$\cos\theta = \frac{\mathbf{a}\,.\,\mathbf{b}}{|\mathbf{a}||\mathbf{b}|}.$$

angle of friction The angle λ such that $\tan\lambda = \mu_s$, where μ_s is the *coefficient of static friction*. Consider a block resting on a horizontal plane, as shown in the figure. In the limiting case when the block is about to move to the right on account of an applied force of magnitude P, $N = mg$, $P = F$ and $F = \mu_s N$. Then the *contact force*, whose components are N and F, makes an angle λ with the vertical.

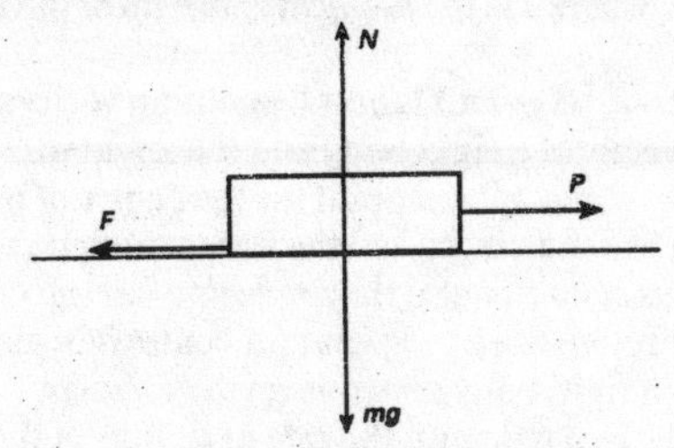

angle of inclination See *inclined plane.*

angle of projection The angle that the direction in which a particle is projected makes with the horizontal. Thus it is the angle that the initial velocity makes with the horizontal.

angular acceleration Suppose that the particle P is moving in the plane, in a circle with centre at the origin O and radius r_0. Let (r_0, θ) be the polar coordinates of P. At an elementary level, the **angular acceleration** may be defined to be $\ddot{\theta}$.

At a more advanced level, the **angular acceleration** $\boldsymbol{\alpha}$ of the particle P is the vector defined by $\boldsymbol{\alpha} = \dot{\boldsymbol{\omega}}$, where $\boldsymbol{\omega}$ is the *angular velocity.* Let $\mathbf{i}$ and $\mathbf{j}$ be unit vectors in the directions of the positive x- and y-axes and let $\mathbf{k} = \mathbf{i} \times \mathbf{j}$. Then, in the case above of a particle moving along a circular path, $\boldsymbol{\omega} = \dot{\theta}\mathbf{k}$ and $\boldsymbol{\alpha} = \ddot{\theta}\mathbf{k}$. If $\mathbf{r}$, $\mathbf{v}$ and $\mathbf{a}$ are the position vector, velocity and acceleration of P, then

$$\mathbf{r} = r_0\mathbf{e}_r, \qquad \mathbf{v} = \dot{\mathbf{r}} = r_0\dot{\theta}\mathbf{e}_\theta, \qquad \mathbf{a} = \ddot{\mathbf{r}} = -r_0\dot{\theta}^2\mathbf{e}_r + r_0\ddot{\theta}\mathbf{e}_\theta,$$

where $\mathbf{e}_r = \mathbf{i}\cos\theta + \mathbf{j}\sin\theta$ and $\mathbf{e}_\theta = -\mathbf{i}\sin\theta + \mathbf{j}\cos\theta$ (see *circular motion*). Using the fact that $\mathbf{v} = \boldsymbol{\omega} \times \mathbf{r}$, it follows that the acceleration $\mathbf{a}$ is given by $\mathbf{a} = \boldsymbol{\alpha} \times \mathbf{r} + \boldsymbol{\omega} \times (\boldsymbol{\omega} \times \mathbf{r})$.

angular frequency The constant ω in the equation $\ddot{x} = -\omega^2 x$ for *simple harmonic motion.* In certain respects ωt, where t is the time, acts like an angle. The angular frequency ω is usually measured in radians per second. The *frequency* of the oscillations is equal to $\omega/2\pi$.

angular measure There are two principal ways of measuring angles: by using *degrees*, in more elementary work, and by using *radians*, essential in more advanced work.

angular momentum Suppose that the particle P of mass m has position vector $\mathbf{r}$ and is moving with velocity $\mathbf{v}$. Then the **angular momentum** $\mathbf{L}$ of P about the point A with position vector $\mathbf{r}_A$ is the vector defined by $\mathbf{L} = (\mathbf{r} - \mathbf{r}_A) \times m\mathbf{v}$. It is the *moment* of the *linear momentum* about the point A. See also *conservation of angular momentum.*

Consider a rigid body rotating with angular velocity $\boldsymbol{\omega}$ about a fixed axis, and let $\mathbf{L}$ be the angular momentum of the rigid body about a point on the fixed axis. Then $\mathbf{L} = I\boldsymbol{\omega}$, where I is the *moment of inertia* of the rigid body about the fixed axis.

To consider the general case, let $\boldsymbol{\omega}$ and $\mathbf{L}$ now be column vectors representing the angular velocity of a rigid body and the angular momentum of the rigid body about a fixed point (or the centre of mass). Then $\mathbf{L} = \mathbf{I}\boldsymbol{\omega}$, where $\mathbf{I}$ is a 3×3 matrix, called the **inertia matrix**, whose elements involve the *moments of inertia* and the *products of inertia* of the rigid body relative to axes through the fixed point (or centre of mass).

The rotational motion of a rigid body depends on the angular momentum of the rigid body. In particular, the rate of change of the angular momentum about a fixed point (or centre of mass) equals the sum of the moments of the forces acting on the rigid body about the fixed point (or centre of mass).

angular speed The magnitude of the *angular velocity.*

angular velocity Suppose that the particle P is moving in the plane, in a circle with centre at the origin O and radius r_0. Let (r_0, θ) be the polar coordinates of P. At an elementary level, the **angular velocity** may be defined to be $\dot{\theta}$.

At a more advanced level, the **angular velocity** $\boldsymbol{\omega}$ of the particle P is the vector defined by $\boldsymbol{\omega} = \dot{\theta}\mathbf{k}$, where $\mathbf{i}$ and $\mathbf{j}$ are unit vectors in the directions of the positive x- and y-axes, and $\mathbf{k} = \mathbf{i} \times \mathbf{j}$. If $\mathbf{r}$ and $\mathbf{v}$ are the position vector and velocity of P, then

$$\mathbf{r} = r_0\mathbf{e}_r, \qquad \mathbf{v} = \dot{\mathbf{r}} = r_0\dot{\theta}\mathbf{e}_\theta,$$

where $\mathbf{e}_r = \mathbf{i}\cos\theta + \mathbf{j}\sin\theta$ and $\mathbf{e}_\theta = -\mathbf{i}\sin\theta + \mathbf{j}\cos\theta$ (see *circular motion*). By using the fact that $\mathbf{k} = \mathbf{e}_r \times \mathbf{e}_\theta$, it follows that the velocity $\mathbf{v}$ is given by $\mathbf{v} = \boldsymbol{\omega} \times \mathbf{r}$.

Consider a rigid body rotating about a fixed axis, and take coordinate axes so that the z-axis is along the fixed axis. Let (r_0, θ) be the polar coordinates of some point of the rigid body, not on the axis, lying in the plane $z = 0$. Then the angular velocity $\boldsymbol{\omega}$ of the rigid body is defined by $\boldsymbol{\omega} = \dot{\theta}\mathbf{k}$.

In general, for a rigid body that is rotating, such as a top spinning about a fixed point, the rigid body possesses an angular velocity $\boldsymbol{\omega}$ whose magnitude and direction depend on time.

annulus (plural: annuli) The region between two concentric circles. If the circles have radii r and $r + w$, the area of the annulus is equal to $\pi(r+w)^2 - \pi r^2$, which equals $w \times 2\pi(r + \frac{1}{2}w)$. It is therefore the same as the area of a rectangle of width w and length equal to the circumference of the circle midway in size between the two original circles.

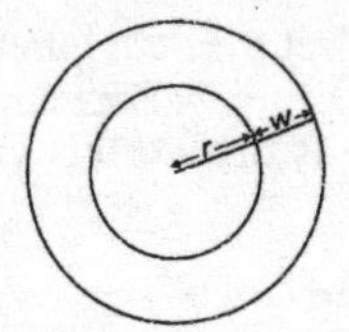

ANOVA See *analysis of variance.*

antiderivative Given a *real function* f, any function ϕ such that $\phi'(x) = f(x)$, for all x (in the domain of f), is an **antiderivative** of f. If ϕ_1 and ϕ_2 are both antiderivatives of a *continuous function* f, then $\phi_1(x)$ and $\phi_2(x)$ differ by a constant. In that case, the notation

$$\int f(x)\,dx$$

may be used for an antiderivative of *f*, with the understanding that an arbitrary constant can be added to any antiderivative. Thus,

$$\int f(x)\,dx + c,$$

where c is an arbitrary constant, is an expression that gives all the antiderivatives.

antilogarithm The **antilogarithm** of x, denoted by antilog x, is the number whose *logarithm* is equal to x. For example, suppose that common logarithm tables are used to calculate 2.75×3.12. Then, approximately, $\log 2.75 = 0.4393$ and $\log 3.12 = 0.4942$ and $0.4393 + 0.4942 = 0.9335$. Now antilog 0.9335 is required and, from tables, the answer 8.58 is obtained. Now that logarithm tables have been superseded by calculators, the term 'antilog' is little used. If y is the number whose logarithm is x, then $\log_a y = x$. This is equivalent to $y = a^x$ (from the definition of logarithm). So, if base a is being used, $\text{antilog}_a\, x$ is identical with a^x; for common logarithms, $\text{antilog}_{10}\, x$ is just 10^x, and this notation is preferable.

antipodal points Two points on a sphere that are at opposite ends of a diameter.

antiprism Normally, a convex *polyhedron* with two 'end' faces that are congruent regular polygons lying in parallel planes in such a way that, with each vertex of one polygon joined by an edge to two vertices of the other polygon, the remaining faces are isosceles triangles. The term could be used for a polyhedron of a similar sort in which the end faces are not regular and the triangular faces are not isosceles, in which case the first definition would be said to give a right-regular antiprism. If the end faces are regular and the triangular faces are equilateral, the antiprism is a *semi-regular polygon*.

antisymmetric matrix = *skew-symmetric matrix*.

antisymmetric relation A binary relation $\sim$ on a set S is **antisymmetric** if, for all a and b in S, whenever $a \sim b$ and $b \sim a$, then $a = b$. For example, the relation $\leq$ on the set of integers is antisymmetric. (Compare this with the definition of an *asymmetric* relation.)

apex (plural: apices) See *base* (of a triangle) and *pyramid*.

aphelion See *apse*.

apogee See *apse*.

Apollonius of Perga (about 262–190 BC) Greek mathematician whose most famous work *The Conics* was, until modern times, the definitive work on the *conic sections*: the ellipse, parabola and hyperbola. He proposed the idea of epicyclic motion for the planets. Euclid, Archimedes and Apollonius were pre-eminent in the period covering the third century BC known as the Golden Age of Greek mathematics.

Apollonius' circle Given two points A and B in the plane and a constant k, the locus of all points P such that $AP/PB = k$ is a circle. A circle obtained like this is an **Apollonius' circle.** Taking $k = 1$ gives a straight line, so either this value must be excluded or, in this context, a straight line must be considered to be a special case of a circle. In the figure, $k = 2$.

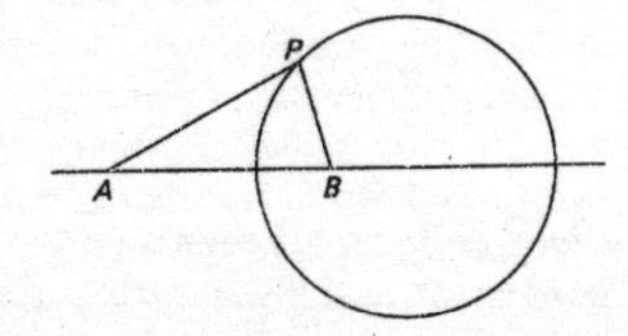

approximation When two quantities X and x are approximately equal, written $X \approx x$, one of them may be used in suitable circumstances in place of, or as an **approximation** for, the other. For example, $\pi \approx \frac{22}{7}$ and $\sqrt{2} \approx 1.414$.

apse A point in an *orbit* at which the body is moving in a direction perpendicular to the radius vector. In an elliptical orbit in which the centre of attraction is at one focus there are two apses, the points at which the body is at its nearest and its furthest from the centre of attraction. When the centre of attraction is the Sun, the points at which the body is nearest and furthest are the **perihelion** and the **aphelion.** When the centre of attraction is the Earth, they are the **perigee** and the **apogee.**

apsis (plural: apses) = *apse.*

Arabic numeral See *numeral.*

arc (of a curve) The part of a curve between two given points on the curve. If A and B are two points on a circle, there are two arcs AB. When A and B are not at opposite ends of a diameter, it is possible to distinguish between the longer and shorter arcs by referring to the **major arc** AB and the **minor arc** AB.

arc (of a digraph) See *digraph.*

arccos, arccosec, arccot, arcsin, arcsec, arctan See *inverse trigonometric function.*

arccosh, arccosech, arccoth, arcsinh, arcsech, arctanh See *inverse hyperbolic function.*

arc length Let $y = f(x)$ be the graph of a function f such that f' is *continuous* on $[a, b]$. The length of the arc, or **arc length,** of the curve $y = f(x)$ between $x = a$ and $x = b$ equals

$$\int_a^b \sqrt{1+(f'(x))^2}\,dx.$$

PARAMETRIC FORM: For the curve $x = x(t), y = y(t)$ $(t \in [\alpha, \beta])$, the arc length equals

$$\int_\alpha^\beta \sqrt{\left(\frac{dx}{dt}\right)^2 + \left(\frac{dy}{dt}\right)^2}\,dt.$$

POLAR FORM: For the curve $r = r(\theta)$ $(\alpha \le \theta \le \beta)$, the arc length equals

$$\int_\alpha^\beta \sqrt{r^2 + \left(\frac{dr}{d\theta}\right)^2}\,d\theta.$$

Archimedean solid A convex *polyhedron* is called **semi-regular** if the faces are regular polygons, though not all congruent, and if the vertices are all alike, in the sense that the different kinds of face are arranged in the same order around each vertex. Right-regular *prisms* with square side faces and (right-regular) *antiprisms* whose side faces are equilateral triangles are semi-regular. Apart from these, there are thirteen semi-regular polyhedra, known as the **Archimedean solids**. These include the *truncated tetrahedron*, the *truncated cube*, the *cuboctahedron* and the *icosidodecahedron*.

Archimedean spiral A curve whose equation in polar coordinates is $r = a\theta$, where a (> 0) is a constant. In the figure, $OA = a\pi$, $OB = 2a\pi$, and so $OB = 2OA$.

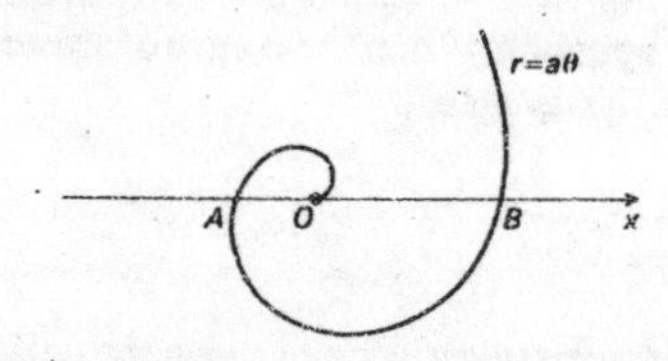

Archimedes (287–212 BC) Greek mathematician who can be rated one of the greatest of all time. He made considerable contributions to geometry, discovering methods of finding, for example, the surface area and the volume of a sphere and the area of a segment of a parabola. His work on hydrostatics and equilibrium was also fundamental. His most fascinating work, *The Method*, was rediscovered as recently as 1906. He may or may not have shouted 'Eureka' and run naked through the streets, but he was certainly murdered by a Roman soldier, an event that marks the end of an era in mathematics.

area of a surface of revolution Let $y = f(x)$ be the graph of a function f such that f' is *continuous* on $[a, b]$ and $f(x) \ge 0$ for all x in $[a, b]$. The **area of the surface** obtained by rotating, through one revolution about the x-axis, the arc of the curve $y = f(x)$ between $x = a$ and $x = b$, equals

$$2\pi \int_a^b y\sqrt{1+\left(\frac{dy}{dx}\right)^2}\, dx, \quad \text{or} \quad 2\pi \int_a^b f(x)\sqrt{1+(f'(x))^2}dx.$$

PARAMETRIC FORM: For the curve $x = x(t), y = y(t)$ $(t \in [\alpha, \beta])$, the surface area equals

$$2\pi \int_\alpha^\beta y\sqrt{\left(\frac{dx}{dt}\right)^2+\left(\frac{dy}{dt}\right)^2}\, dt.$$

POLAR FORM: For the curve $r = r(\theta)$ $(\alpha \leq \theta \leq \beta)$, the surface area equals

$$2\pi \int_\alpha^\beta r \sin\theta \sqrt{r^2+\left(\frac{dr}{d\theta}\right)^2}\, d\theta.$$

area under a curve Suppose that the curve $y = f(x)$ lies above the x-axis, so that $f(x) \geq 0$ for all x in $[a, b]$. The **area under the curve**, that is, the area of the region bounded by the curve, the x-axis and the lines $x = a$ and $x = b$, equals

$$\int_a^b f(x)\, dx.$$

The definition of *integral* is made precisely in order to achieve this result.

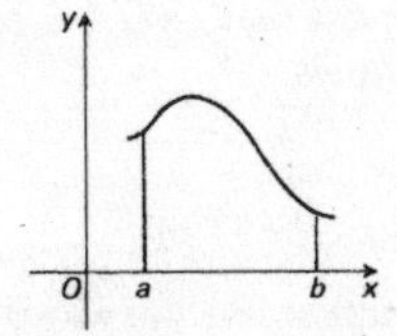

If $f(x) \leq 0$ for all x in $[a, b]$, the integral above is negative. However, it is still the case that its absolute value is equal to the area of the region bounded by the curve, the x-axis and the lines $x = a$ and $x = b$. If $y = f(x)$ crosses the x-axis, appropriate results hold. For example, if the regions A and B are as shown in the figure below, then

$$\text{area of region } A = \int_a^b f(x)\, dx, \quad \text{area of region } B = -\int_b^c f(x)\, dx.$$

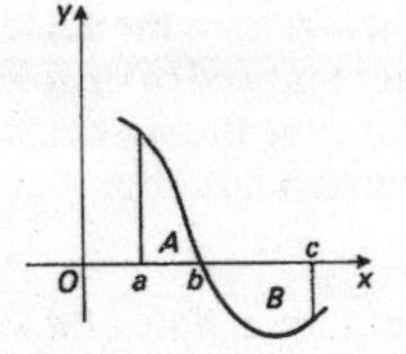

It follows that

$$\int_a^c f(x)\,dx = \int_a^b f(x)\,dx + \int_b^c f(x)\,dx$$
$$= \text{area of region } A - \text{area of region } B.$$

Similarly, to find the area of the region bounded by a suitable curve, the y-axis, and lines $y = c$ and $y = d$, an equation $x = g(y)$ for the curve must be found. Then the required area equals

$$\int_c^d g(y)\,dy,$$

assuming that the curve is to the right of the y-axis, so that $g(y) \geq 0$ for all y in $[c, d]$. As before, the value of the integral is negative if $g(y) \leq 0$.

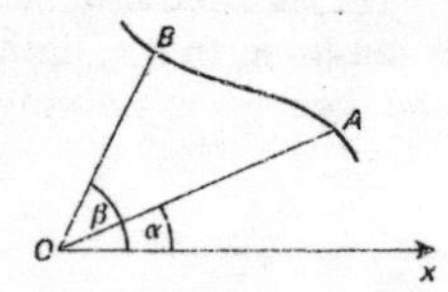

POLAR AREAS: If a curve has an equation $r = r(\theta)$ in polar coordinates, there is an integral that gives the area of the region bounded by an arc AB of the curve and the two radial lines OA and OB. Suppose that $\angle xOA = \alpha$ and $\angle xOB = \beta$. The area of the region described equals

$$\int_\alpha^\beta \tfrac{1}{2} r^2\,d\theta.$$

Argand, Jean Robert (1768–1822) Swiss-born mathematician who was one of several people, including Gauss, who invented a geometrical representation for complex numbers. This explains the name *Argand diagram.*

Argand diagram Another (less preferable) name for the *complex plane.*

argument Suppose that the *complex number* z is represented by the point P in the *complex plane*. The **argument** of z, denoted by arg z, is the angle θ (in radians) that OP makes with the positive real axis Ox, with the angle given a positive sense anticlockwise from Ox. As with polar coordinates, the angle θ may be taken so that $0 \leq \arg z < 2\pi$. Usually, however, the angle θ is chosen so that $-\pi < \arg z \leq \pi$. Sometimes, arg z is used to denote any of the values $\theta + 2n\pi$, where n is an integer. In that case, the particular value that lies in a certain interval, specified or understood, such as $[0, 2\pi)$ or $(-\pi, \pi]$, is called the **principal value** of arg z.

Aristarchus of Samos (about 270 BC) Greek astronomer, noted for being the first to affirm that the Earth rotates and travels around the Sun. He treated astronomy mathematically and used geometrical methods to calculate the relative sizes of the Sun and Moon and their relative distances from the Earth.

Arithmetic, Fundamental Theorem of See *Fundamental Theorem of Arithmetic.*

arithmetic mean See *mean.*

arithmetic sequence A finite or infinite sequence of terms $a_1, a_2, a_3, \ldots$ with a **common difference** d, so that $a_2 - a_1 = d, a_3 - a_2 = d$, and so on. The first term is usually denoted by a. For example, 2, 5, 8, 11, ... is the arithmetic sequence with $a = 2, d = 3$. In such an arithmetic sequence, the n-th term a_n is given by $a_n = a + (n - 1)d$.

arithmetic series A series $a_1 + a_2 + a_3 + \cdots$ (which may be finite or infinite) in which the terms form an *arithmetic sequence.* Thus the terms have a common difference d, with $a_k - a_{k-1} = d$ for all $k \geq 2$. If the first term equals a, then $a_k = a + (k - 1)d$. Let s_n be the sum of the first n terms of an arithmetic series, so that

$$s_n = a + (a + d) + (a + 2d) + \cdots + (a + (n - 1)d),$$

and let the last term here, $a + (n - 1)d$, be denoted by b. Then s_n is given by the formula $s_n = \frac{1}{2}n(a + b)$. The particular case in which $a = 1$ and $d = 1$ gives the sum of the first n natural numbers:

$$\sum_{r=1}^{n} r = 1 + 2 + \cdots + n = \tfrac{1}{2}n(n + 1).$$

arrangement See *permutation.*

Āryabhata (about 476–550) Indian mathematician, author of one of the oldest Indian mathematical texts. Written in verse, the *Āryabhatīya* is a summary of miscellaneous rules for calculation and mensuration. It deals with, for example, the areas of certain plane figures, values for π, and the summation of *arithmetic series.* Also included is the equivalent of a table of sines, based on the half-chord rather than the whole *chord* of the Greeks.

assignment problem A problem in which things of one type are to be matched with the same number of things of another type in a way that is, in a specified sense, the best possible. For example, when n workers are to be assigned to n jobs, it may be possible to specify the value v_{ij} to the company, measured in suitable units, if the i-th worker is assigned to the j-th job. The values v_{ij} may be displayed as the entries of an $n \times n$ matrix. By introducing suitable variables, the problem of assigning workers to jobs in such a way as to maximize the total value to the company can be formulated as a *linear programming* problem.

associative The *binary operation* $\circ$ on a set S is **associative** if, for all a, b and c in S, $(a \circ b) \circ c = a \circ (b \circ c)$.

astroid A *hypocycloid* in which the radius of the rolling circle is a quarter of the radius of the fixed circle. It has *parametric equations* $x = a\cos^3 t$, $y = a\sin^3 t$, where a is the radius of the fixed circle.

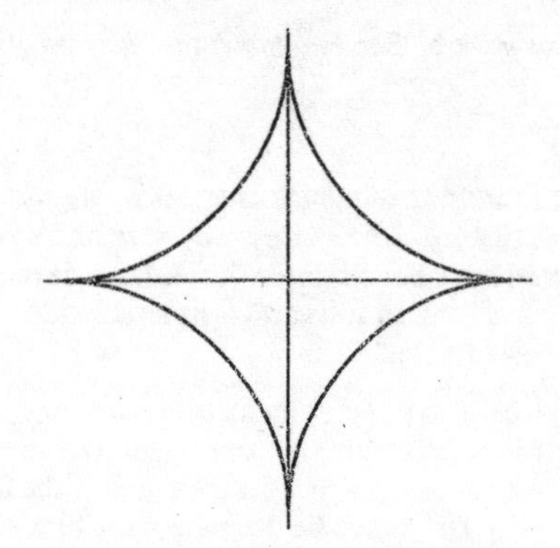

asymmetric A plane figure is **asymmetric** if it is neither symmetrical about a line nor symmetrical about a point.

asymmetric relation A binary relation $\sim$ on a set S is **asymmetric** if, for all a and b in S, whenever $a \sim b$, then $b \sim a$ does not hold. For example, the relation $<$ on the set of integers is asymmetric. (Compare this with the definition of an *antisymmetric* relation.)

asymptote A line l is an **asymptote** to a curve if the distance from a point P to the line l tends to zero as P tends to infinity along some unbounded part of the curve. Consider the following examples:

$$\text{(i) } y = \frac{x+3}{(x+2)(x-1)}, \quad \text{(ii) } y = \frac{3x^2}{x^2+x+1}, \quad \text{(iii) } y = \frac{x^3}{x^2+x+1}.$$

Example (i) has $x = -2$ and $x = 1$ as vertical asymptotes, and $y = 0$ as a horizontal asymptote. Example (ii) has no vertical asymptotes, and $y = 3$ as a horizontal asymptote. To investigate example (iii), it can be rewritten as

$$y = x - 1 + \frac{1}{x^2+x+1}.$$

Then it can be seen that $y = x - 1$ is a **slant asymptote**, that is, an asymptote that is neither vertical nor horizontal.

atmospheric pressure The *pressure* at a point in the atmosphere due to the gravitational force acting on the air. This pressure depends on position and time and is measured by a barometer. The standard atmospheric pressure at sea level is taken to be 101 325 pascals, which is the definition of 1 atmosphere. The variations from place to place are comparatively small, but are the main cause of the wind patterns of the Earth. In general, the atmospheric pressure decreases with height.

atto- Prefix used with *SI units* to denote multiplication by 10^{-18}.

augmented matrix For a given set of m linear equations in n unknowns $x_1, x_2, \ldots x_n$,

$$\begin{aligned} a_{11}x_1 + a_{12}x_2 + \cdots + a_{1n}x_n &= b_1, \\ a_{21}x_1 + a_{22}x_2 + \cdots + a_{2n}x_n &= b_2, \\ &\vdots \\ a_{m1}x_1 + a_{m2}x_2 + \cdots + a_{mn}x_n &= b_m, \end{aligned}$$

the **augmented matrix** is the matrix

$$\begin{bmatrix} a_{11} & a_{12} & \cdots & a_{1n} & b_1 \\ a_{21} & a_{22} & \cdots & a_{2n} & b_2 \\ \vdots & \vdots & \ddots & \vdots & \vdots \\ a_{m1} & a_{m2} & \cdots & a_{mn} & b_m \end{bmatrix}$$

obtained by adjoining to the matrix of coefficients an extra column of entries taken from the right-hand sides of the equations. The solutions of a set of linear equations may be investigated by transforming the augmented matrix to *echelon form* or reduced echelon form by elementary row operations. See *Gaussian elimination* and *Gauss–Jordan elimination*.

auxiliary equation See *linear differential equation with constant coefficients*.

axial plane One of the planes containing two of the coordinate axes in a 3-dimensional Cartesian coordinate system. For example, one of the axial planes is the yz-plane, or (y, z)-plane, containing the y-axis and the z-axis, and it has equation $x = 0$.

axiom A statement whose truth is either to be taken as self-evident or to be assumed. Certain areas of mathematics involve choosing a set of axioms and discovering what results can be derived from them, providing proofs for the theorems that are obtained.

axis (plural: axes) See *coordinates* (in the plane) and *coordinates* (in 3-dimensional space).

axis (of a cone) See *cone*.

axis (of a cylinder) See *cylinder*.

axis (of a parabola) See *parabola*.

B

Babbage, Charles (1792–1871) British mathematician and inventor of mechanical calculators. His 'analytical engine' was designed to perform mathematical operations mechanically using a number of features essential in the design of today's computers but, partly through lack of funds, the project was not completed.

back-substitution Suppose that a set of linear equations is in *echelon form*. Then the last equation can be solved for the first unknown appearing in it, any other unknowns being set equal to *parameters* taking arbitrary values. This solution can be substituted into the previous equation, which can then likewise be solved for the first unknown appearing in it. The process that continues in this way is **back-substitution**.

Banach, Stefan (1892–1945) Polish mathematician who was a major contributor to the subject known as functional analysis. Much subsequent work was inspired by his exposition of the theory in a paper of 1932.

bar chart A diagram representing the *frequency distribution* for nominal or discrete data with comparatively few possible values. It consists of a sequence of bars, or rectangles, corresponding to the possible values, and the length of each is proportional to the frequency. The bars have equal widths and are usually not touching. The figure shows the kinds of vehicles recorded in a small traffic survey.

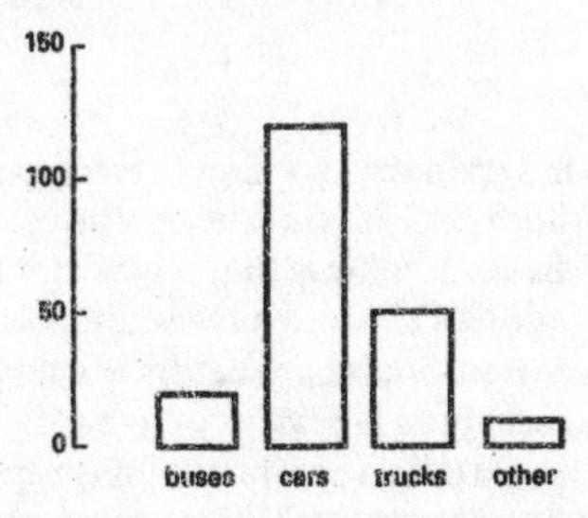

Barrow, Isaac (1630–1677) English mathematician whose method of finding tangents, published in 1670, is essentially that now used in differential calculus. He may have been the first to appreciate that the problems of finding tangents and areas under curves are inversely related. When he resigned his chair at Cambridge, he was succeeded by Newton, on Barrow's recommendation.

base (of an exponential function) See *exponential function to base a*.

base (of an isosceles triangle) See *isosceles triangle.*

base (of logarithms) See *logarithms.*

base (of natural logarithms) See *e.*

base (of a pyramid) See *pyramid.*

base (for representation of numbers) The integer represented as 4703 in standard decimal notation is written in this way because

$$4703 = (4 \times 10^3) + (7 \times 10^2) + (0 \times 10) + 3.$$

The same integer can be written in terms of powers of 8 as follows:

$$4703 = (1 \times 8^4) + (1 \times 8^3) + (1 \times 8^2) + (3 \times 8) + 7.$$

The expression on the right-hand side is abbreviated to $(11137)_8$ and is the **representation** of this number to **base** 8. In general, if g is an integer greater than 1, any positive integer a can be written uniquely as

$$a = c_n g^n + c_{n-1} g^{n-1} + \cdots + c_1 g + c_0,$$

where each c_i is a non-negative integer less than g. This is the **representation** of a to **base** g, and is abbreviated to $(c_n c_{n-1} \ldots c_1 c_0)_g$. Real numbers, not just integers, can also be written to any base, by using figures after a 'decimal' point, just as familiar *decimal* representations of real numbers are written to base 10. See also *binary representation*, *decimal* and *hexadecimal.*

base (of a triangle) It may be convenient to consider one side of a triangle to be the **base** of the triangle. The vertex opposite the base may then be called the **apex** of the triangle, and the distance from the apex to the base the **height** of the triangle.

base angles See *isosceles triangle.*

base unit See *SI units.*

basis (plural: bases) A set S of *vectors* is a **spanning set** if any vector can be written as a linear combination of those in S. If, in addition, the vectors in S are *linearly independent*, then S is a **basis**. It follows that any vector can be written *uniquely* as a linear combination of those in a basis. In 3-dimensional space, any set of three non-coplanar vectors $\mathbf{u}, \mathbf{v}, \mathbf{w}$ is a basis, since any vector $\mathbf{p}$ can be written uniquely as $\mathbf{p} = x\mathbf{u} + y\mathbf{v} + z\mathbf{w}$. In 2-dimensional space, any set of 2 non-parallel vectors has this property and so is a basis. Any one of the vectors in a set currently being taken as a basis may be called a **basis vector**.

basis vector See *basis.*

Bayes, Thomas (1702–1761) English mathematician remembered for his work in probability, which led to a method of statistical inference based on the theorem that bears his name. His paper on the subject was not published until 1763, after his death.

Bayes' Theorem The following theorem for calculating a posterior probability (see *prior probability*):

THEOREM: Let $A_1, A_2, \ldots, A_k$ be mutually exclusive events whose union is the whole sample space of an experiment and let B be an event with $\Pr(B) \neq 0$. Then

$$\Pr(A_i \mid B) = \frac{\Pr(B \mid A_i)\Pr(A_i)}{\Pr(B \mid A_1)\Pr(A_1) + \cdots + \Pr(B \mid A_k)\Pr(A_k)}.$$

For example, let A_1 be the event of tossing a double-headed coin, and A_2 the event of tossing a normal coin. Suppose that one of the coins is chosen at random so that $\Pr(A_1) = \frac{1}{2}$ and $\Pr(A_2) = \frac{1}{2}$. Let B be the event of obtaining 'heads'. Then $\Pr(B \mid A_1) = 1$ and $\Pr(B \mid A_2) = \frac{1}{2}$. So

$$\Pr(A_1 \mid B) = \frac{\Pr(B \mid A_1)\Pr(A_1)}{\Pr(B \mid A_1)\Pr(A_1) + \Pr(B \mid A_2)\Pr(A_2)}$$

$$= \frac{1 \times \frac{1}{2}}{1 \times \frac{1}{2} + \frac{1}{2} \times \frac{1}{2}} = \frac{2}{3}.$$

This says that, given that 'heads' was obtained, the probability that it was the double-headed coin that was tossed is 2/3.

Here, $\Pr(A_i)$ is a prior probability and $\Pr(A_i \mid B)$ is a posterior probability.

bearing The direction of the course upon which a ship is set or the direction in which an object is sighted may be specified by giving its **bearing**, the angle that the direction makes with north. The angle is measured in degrees in a clockwise direction from north. For example, north-east has a bearing of 45°, and west has a bearing of 270°.

beats Suppose that a body capable of oscillating freely with *angular frequency* ω is subject to an oscillatory applied force with angular frequency Ω. When ω and Ω are nearly equal, the motion appears to consist of oscillations with an amplitude that varies comparatively slowly. The **beats** occur as the amplitude achieves its maximum value. The effect can be heard when two notes whose frequencies are close together are sounded at the same time. Musical instruments can be tuned by listening for beats.

belongs to If x is an element of a set S, then x **belongs to** S and this is written $x \in S$. Naturally, $x \notin S$ means that x does not belong to S.

Bernoulli distribution The discrete probability *distribution* whose *probability mass function* is given by $\Pr(X = 0) = 1 - p$ and $\Pr(X = 1) = p$. It is the *binomial distribution* $\mathrm{B}(1, p)$.

Bernoulli family A family from Basle that produced a stream of significant mathematicians, some of them very important indeed. The best known are the brothers Jacques (or James or Jakob) and Jean (or John or Johann), and Jean's son Daniel. **Jacques Bernoulli** (1654–1705) did much work on the newly developed calculus, but is chiefly remembered for his contributions to probability theory: the *Ars conjectandi* was published, after his death, in 1713. The work of **Jean Bernoulli** (1667–1748) was more definitely within calculus: he discovered *l'Hôpital's rule* (see under H) and proposed the *brachistochrone* problem. He was one of the founders of the *calculus of variations.* In the next generation, **Daniel Bernoulli** (1700–1782) was the member of the family whose mathematical work was mainly in hydrodynamics.

Bernoulli trial One of a sequence of independent experiments, each of which has an outcome considered to be success or failure, all with the same probability p of success. The number of successes in a sequence of Bernoulli trials has a *binomial distribution.* The number of experiments required to achieve the first success has a *geometric distribution.*

Bessel, Friedrich Wilhelm (1784–1846) German astronomer and mathematician who made a major contribution to mathematics in the development of what are now called Bessel functions. These functions, which satisfy certain differential equations, are probably the most commonly occurring functions in physics and engineering after the elementary functions.

best unbiased estimator See *estimator.*

Bhāskara (1114–1185) Eminent Indian mathematician who continued in the tradition of *Brahmagupta*, making corrections and filling in many gaps in the earlier work. He solved examples of *Pell's equation* and grappled with the problem of division by zero.

biased estimator See *estimator.*

bijection A *one-to-one onto mapping*, that is, a mapping that is both *injective* and *surjective.*

bijective mapping A mapping that is *injective* (that is, *one-to-one*) and *surjective* (that is, *onto*).

bilateral symmetry See *symmetrical about a line.*

billion In Britain, a million million (10^{12}); in the United States, a thousand million (10^9). The American usage is increasing in Britain, and consequently use of the word can lead to ambiguity.

bimodal A frequency distribution is said to be **bimodal** if it shows two clear peaks.

binary code A **binary code** of **length** n is a set of *binary words* of length n which are called the **codewords.** For example, to send a message conveying the information as to whether a direction is north, south, east or west, a code of length 3 might be chosen, in which, say, 000 means 'north', 110 means 'south', 011 means 'east' and 101 means 'west'.

binary operation A **binary operation** $\circ$ on a set S is a rule that associates with any elements a and b of S an element denoted by $a \circ b$. If, for all a and b, the element $a \circ b$ also belongs to S, S is said to be **closed under** the operation $\circ$. This is often taken to be implied in saying that $\circ$ is a binary operation on S.

binary relation A formal definition of a **binary relation** on a set S is as a subset R of the *Cartesian product* $S \times S$. Thus, it can be said that, for a given *ordered pair* (a, b), either $(a, b) \in R$ or $(a, b) \notin R$. However, it is more natural to denote a relation by a symbol such as $\sim$ placed between the elements a and b, where $\sim$ stands for the words 'is related (in some way) to'. Familiar examples are normally written in this way: '$<$' is a binary relation on the set of integers; '$\subseteq$' is a binary relation on the set of subsets of some set E; and 'is perpendicular to' is a binary relation on the set of straight lines in the plane. The letter 'R' may be used in this way, '$a\,R\,b$' meaning that 'a is related to b'. If this notation is used, the set $\{(a, b) \mid (a, b) \in S \times S$ and $a\,R\,b\}$ may be called the *graph* of R. For any relation $\sim$, a corresponding relation $\not\sim$ can be defined that holds whenever $\sim$ does not hold.

binary representation The representation of a number to *base* 2. It uses just the two **binary digits** 0 and 1, and this is the reason why it is important in computing. For example, $37 = (100101)_2$, since

$$37 = (1 \times 2^5) + (0 \times 2^4) + (0 \times 2^3) + (1 \times 2^2) + (0 \times 2) + 1.$$

Real numbers, not just integers, can also be written in binary notation, by using binary digits after a 'decimal' point, just as familiar *decimal* representations of real numbers are written to base 10. For example, the number $\frac{1}{10}$, which equals 0.1 in decimal notation, has, in binary notation, a recurring representation: $(0.0001100110011 \ldots)_2$.

binary tree See *tree*.

binary word A **binary word** of **length** n is a string of n binary digits, or bits. For example, there are 8 binary words of length 3, namely, 000, 100, 010, 001, 110, 101, 011 and 111.

binomial coefficient The number, denoted by $\binom{n}{r}$, where n is a positive integer and r is an integer such that $0 \le r \le n$, defined by the formula

$$\binom{n}{r} = \frac{n(n-1)\ldots(n-r+1)}{1 \times 2 \times \cdots \times r} = \frac{n!}{r!(n-r)!}.$$

Since, by convention, $0! = 1$, we have $\binom{n}{0} = \binom{n}{n} = 1$. These numbers are called **binomial coefficients** because they occur as coefficients in the *Binomial Theorem*. They are sometimes denoted by ${}^nC_r$; this arose as the notation for the number of ways of selecting r objects out of n (see *selection*), but this number can be shown to be equal to the expression given above for the binomial coefficient. The numbers have the following properties:

(i) $\binom{n}{r}$ is an integer (this is not obvious from the definition).

(ii) $\binom{n}{n-r} = \binom{n}{r}$.

(iii) $\binom{n+1}{r} = \binom{n}{r-1} + \binom{n}{r}$.

(iv) $\binom{n}{0} + \binom{n}{1} + \binom{n}{2} + \cdots + \binom{n}{n} = 2^n$.

It is instructive to see the binomial coefficients laid out in the form of *Pascal's triangle.*

binomial distribution The discrete probability *distribution* for the number of successes when n independent experiments are carried out, each with the same probability p of success. The *probability mass function* is given by $\Pr(X = r) = {}^nC_r p^r(1-p)^{n-r}$, for $r = 0, 1, 2, \ldots, n$. This distribution is denoted by $\mathrm{B}(n, p)$ and has mean np and variance $np(1-p)$.

binomial series (or **expansion**) The series

$$1 + \frac{\alpha}{1!}x + \frac{\alpha(\alpha-1)}{2!}x^2 + \cdots + \frac{\alpha(\alpha-1)\ldots(\alpha-n+1)}{n!}x^n + \cdots,$$

being the *Maclaurin series* for the function $(1+x)^\alpha$. In general, it is valid for $-1 < x < 1$. If α is a non-negative integer, the expansion is a finite series and so is a polynomial, and then it is equal to $(1+x)^\alpha$ for all x.

Binomial Theorem The formulae $(x+y)^2 = x^2 + 2xy + y^2$ and $(x+y)^3 = x^3 + 3x^2y + 3xy^2 + y^3$ are used in elementary algebra. The **Binomial Theorem** gives an expansion like this for $(x+y)^n$, where n is any positive integer:

THEOREM: For all positive integers n,

$$(x+y)^n = \sum_{r=0}^{n} \binom{n}{r} x^{n-r} y^r$$

$$= x^n + \binom{n}{1}x^{n-1}y + \binom{n}{2}x^{n-2}y^2 + \cdots + \binom{n}{r}x^{n-r}y^r + \cdots + y^n,$$

where $\binom{n}{r} = \dfrac{n!}{r!(n-r)!}$ (see *binomial coefficient*).

The following is a special case of the Binomial Theorem. It can also be seen as a special case of the *binomial series* when the series is finite:

THEOREM: For all positive integers n,

$$(1+x)^n = \sum_{r=0}^{n} \binom{n}{r} x^r$$
$$= 1 + \binom{n}{1} x + \binom{n}{2} x^2 + \cdots + \binom{n}{r} x^r + \cdots + x^n.$$

bipartite graph A *graph* in which the vertices can be divided into two sets V_1 and V_2, so that no two vertices in V_1 are joined and no two vertices in V_2 are joined. The **complete bipartite graph** $K_{m,n}$ is the bipartite graph with m vertices in V_1 and n vertices in V_2, with every vertex in V_1 joined to every vertex in V_2.

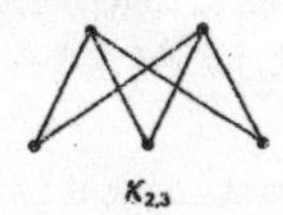

$K_{2,3}$

bisection method A numerical method for finding a root of an equation $f(x) = 0$. If values a and b are found such that $f(a)$ and $f(b)$ have opposite signs and f is *continuous* on the interval $[a, b]$, then (by the *Intermediate Value Theorem*) the equation has a root in (a, b). The method is to bisect the interval and replace it by either one half or the other, thereby closing in on the root.

Let $c = \frac{1}{2}(a + b)$. Calculate $f(c)$. If $f(c)$ has the same sign as $f(a)$, then take c as a new value for a; if not (so that $f(c)$ has the same sign as $f(b)$), take c as a new value for b. (If it should happen that $f(c) = 0$, then c is a root and the aim of finding a root has been achieved.) Repeat this whole process until the length of the interval $[a, b]$ is less than 2ϵ, where ϵ is specified in advance. The midpoint of the interval can then be taken as an approximation to the root, and the error will be less than ϵ.

bisector The line that divides an angle into two equal angles. See also *internal bisector*, *external bisector* and *perpendicular bisector*.

bit A binary digit; that is, 0 or 1.

bivariate Relating to two random variables. See *joint cumulative distribution function*, *joint distribution*, *joint probability density function* and *joint probability mass function*.

body An object in the real world idealized in a mathematical model as a *particle*, a *rigid body* or an *elastic* body, for example.

Bolyai, János (1802–1860) Hungarian mathematician who, in a work published in 1832 but probably dating from 1823, announced his discovery of *non-Euclidean geometry*. His work was independent of Lobachevsky. He had persisted with the problem, while serving as an army officer, despite the warnings of his father, an eminent mathematician who had himself spent many years on it without success. Later, János was disheartened by lack of recognition.

Bombelli, Rafael (1526–1572/3) Italian mathematician who, in his book *l'Algebra*, seems to have been the first to go some way in working with complex numbers, when the square roots of negative numbers occurred in the solution of cubic equations.

Boole, George (1815–1864) British mathematician who was one of the founding fathers of mathematical logic. His major work, published in 1854, is his *Investigation of the Laws of Thought*. The kind of symbolic argument that he developed led to the study of so-called Boolean algebras, which are of current significance in computing and algebra. His work, together with that of De Morgan and others, helped to pave the way for the development of modern formal algebra.

bound Let S be a non-empty subset of **R**. The real number b is said to be an **upper bound** for S if b is greater than or equal to every element of S. If S has an upper bound, then S is **bounded above**. Moreover, b is a **supremum** (or **least upper bound**) of S if b is an upper bound for S and no upper bound for S is less than b; this is written $b = \sup S$. For example, if $S = \{0.9, 0.99, 0.999, \ldots\}$ then $\sup S = 1$. Similarly, the real number c is a **lower bound** for S if c is less than or equal to every element of S. If S has a lower bound, then S is **bounded below**. Moreover, c is an **infimum** (or **greatest lower bound**) of S if c is a lower bound for S and no lower bound for S is greater than c; this is written $c = \inf S$. A set is **bounded** if it is bounded above and below.

It is a non-elementary result about the real numbers that any non-empty set that is bounded above has a supremum, and any non-empty set that is bounded below has an infimum.

bounded function A real *function* f, defined on a domain S, is **bounded** (on S) if there is a number M such that, for all x in S, $|f(x)| < M$. The fact that, if f is *continuous* on a closed interval $[a, b]$ then it is bounded on $[a, b]$, is a property for which a rigorous proof is not elementary (see *continuous function*).

bounded sequence The sequence $a_1, a_2, a_3, \ldots$ is **bounded** if there is a number M such that, for all n, $|a_n| < M$.

bounded set See *bound*.

Bourbaki, Nicolas The pseudonym used by a group of mathematicians, of changing membership, mostly French, who since 1939 have been publishing volumes intended to build into an encyclopaedic survey of pure mathematics, the *Éléments de mathématique*. Its influence is variously described as profound or baleful, but is undoubtedly extensive. Bourbaki has been the standard-bearer for what might be called the Structuralist School of modern mathematics.

box plot A diagram constructed from a set of numerical data showing a box that indicates the middle 50% of the ranked observations together with lines, sometimes called 'whiskers', showing the maximum and minimum observations in the sample. The median is marked on the box by a line. Box plots are particularly useful for comparing several samples. The figure shows box plots for three samples, each of size 20, drawn uniformly from the set of integers from 1 to 100.

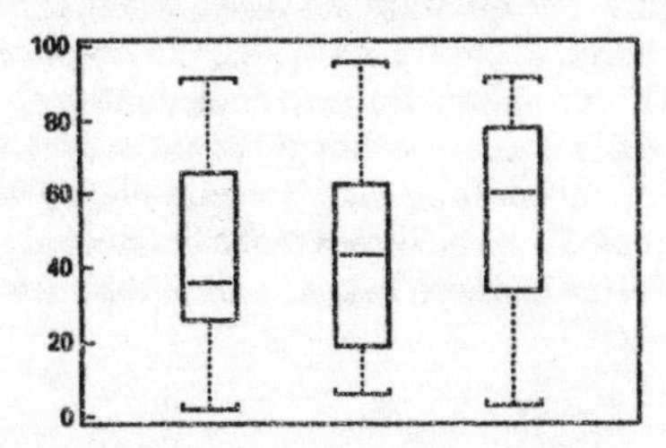

brachistochrone Suppose that A and B are points in a vertical plane, where B is lower than A but not vertically below A. Imagine a particle starting from rest at A and travelling along a curve from A to B under the force of gravity. The curve with the property that the particle reaches B as soon as possible is called the **brachistochrone** (from the Greek for 'shortest time'). The straight line from A to B does not give the shortest time. The required curve is a *cycloid*, vertical at A and horizontal at B. The problem was posed in 1696 by Jean Bernoulli and his solution, together with others by Newton, Lèibniz and Jacques Bernoulli, was published the following year.

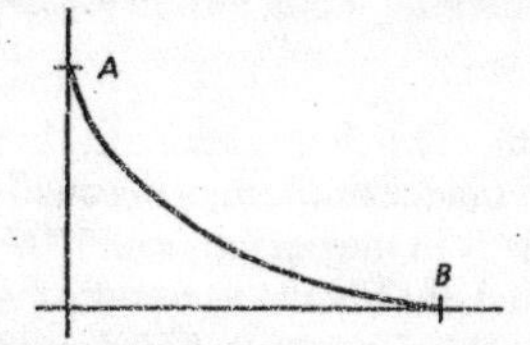

Brahmagupta (about 598–665) Indian astronomer and mathematician whose text on astronomy includes some notable mathematics for its own sake: the areas of quadrilaterals and the solution of certain *Diophantine equations*, for example. Here, the systematic use of negative numbers and zero occurs for probably the first time.

branch (of a hyperbola) The two separate parts of a *hyperbola* are called the two **branches**.

break-even point The point at which revenue begins to exceed cost. If one graph is drawn to show total revenue plotted against the number of items made and sold and another graph is drawn with the same axes to show total costs, the two graphs normally intersect at the break-even point. To the left of the break-even point, costs exceed revenue and the company runs at a loss while, to the right, revenue exceeds costs and the company runs at a profit.

bridges of Königsberg In the early eighteenth century, there were seven bridges in the town of Königsberg (or Kaliningrad). They crossed the different branches of the River Pregel (or Pregolya), as shown in diagrammatic form in the figure. The question was asked whether it was possible, from some starting point, to cross each bridge exactly once and return to the starting point. This prompted Euler to consider the problem in more generality and to publish what can be thought of as the first research paper in *graph* theory. The original question asked, essentially, whether the graph shown is an *Eulerian graph*. It can be shown that a connected graph is Eulerian if and only if every vertex has even degree, and so the answer is that it is not.

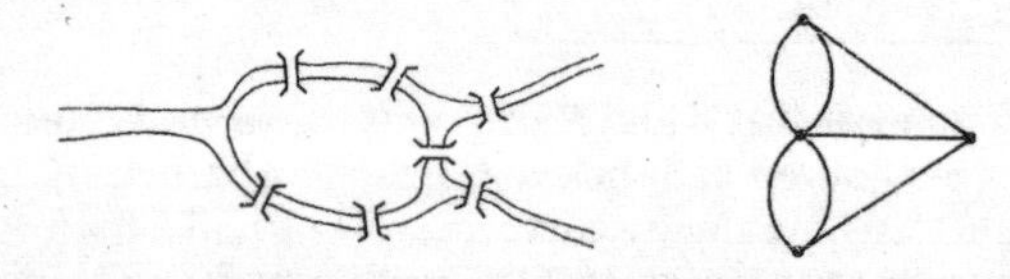

Briggs, Henry (1561–1630) English mathematician who was responsible for the introduction of common logarithms (base 10), at one time called Briggsian logarithms. Following the publication of tables of logarithms by Napier, Briggs consulted him and proposed an alternative definition using base 10. In 1617, the year of Napier's death, Briggs published his logarithms of the first 1000 numbers and, in 1624, tables including 30 000 logarithms to 14 decimal places.

Brouwer, Luitzen Egbertus Jan (1881–1966) Dutch mathematician, considered by many to be the founder of modern topology because of the significant theorems that he proved, mostly in the period from 1909 to 1913. He is also certainly the founder of the doctrine known as intuitionism, rejecting proofs which make use of the *principle of the excluded middle.*

Buffon's needle Suppose that a needle of length l is dropped at random onto a set of parallel lines a distance d apart, where $l < d$. The probability that the needle lands crossing one of the lines can be shown to equal $2l/\pi d$. The experiment in which this is repeated many times to estimate the value of π is called **Buffon's needle**. It was proposed by Georges Louis Leclerc, Comte de Buffon (1707–1788).

butterfly effect See *chaos.*

C

C See *complex number.*

calculus See *differential calculus, integral calculus* and *Fundamental Theorem of Calculus.*

calculus of variations A development of calculus concerned with problems in which a function is to be determined such that some related definite integral achieves a maximum or minimum value. Examples of its application are to the *brachistochrone* problem and to the problem of finding *geodesics.*

cancellation laws Let $\circ$ be a *binary operation* on a set S. The **cancellation laws** are said to hold if, for all a, b and c in S,

(i) if $a \circ b = a \circ c$, then $b = c$,
(ii) if $b \circ a = c \circ a$, then $b = c$.

It can be shown, for example, that in a *group* the cancellation laws hold.

canonical form See *quadric.*

Cantor, Georg (Ferdinand Ludwig Philipp) (1845–1918) Mathematician responsible for the establishment of set theory and for profound developments in the notion of the infinite. He was born in St Petersburg, but spent most of his life at the University of Halle in Germany. In 1873, he showed that the set of rational numbers is denumerable. He also showed that the set of real numbers is not. Later he fully developed his theory of infinite sets and so-called transfinite numbers. The latter part of his life was clouded by repeated mental illness.

Cantor set Take the closed interval $[0, 1]$. Remove the open interval that forms the middle third, that is, the open interval $(\frac{1}{3}, \frac{2}{3})$. From each of the remaining intervals again remove the open interval that forms the middle third. The **Cantor set** is the set that remains when this process is continued indefinitely. It consists of those real numbers whose *ternary representation* $(0.d_1d_2d_3\ldots)_3$ has each ternary digit d_i equal to either 0 or 2.

cap The operation $\cap$ (see *intersection*) is read by some as '**cap**', this being derived from the shape of the symbol, and in contrast to *cup* $\cup$.

Cardano, Girolamo (1501–1576) Italian physician and mathematician, whose *Ars magna* contained the first published solutions of the general cubic equation and the general quartic equation. Even though these were due to Tartaglia and Cardano's assistant Ludovico Ferrari respectively, Cardano was an outstanding mathematician of the time in the fields of algebra and trigonometry.

cardinality For a finite set A, the **cardinality** of A, denoted by $n(A)$, is the number of elements in A. The notation $\#(A)$ or $|A|$ is also used. For subsets A, B and C of some universal set E,

(i) $n(A \cup B) = n(A) + n(B) - n(A \cap B)$,
(ii) $n(A \cup B \cup C) = n(A) + n(B) + n(C) - n(A \cap B)$
$\qquad -n(A \cap C) - n(B \cap C) + n(A \cap B \cap C)$.

cardioid The curve traced out by a point on the circumference of a circle rolling round another circle of the same radius. Its equation, in which a is the radius of each circle, may be taken in polar coordinates as $r = 2a(1 + \cos\theta)$ $(-\pi < \theta \leq \pi)$. In the figure, $OA = 4a$ and $OB = 2a$.

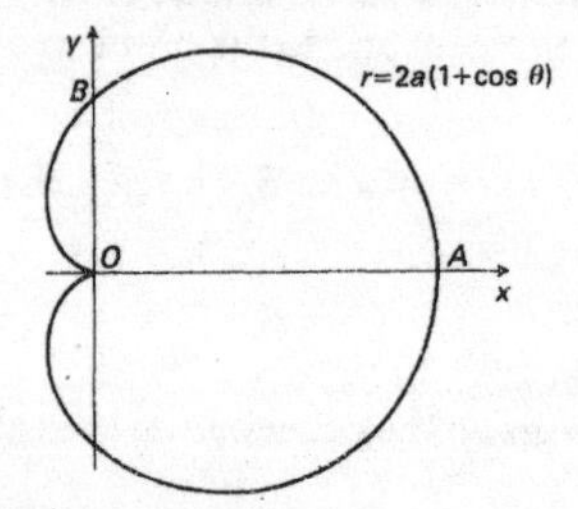

Carmichael number = *pseudo-prime.*

Cartesian coordinates See *coordinates* (in the plane) and *coordinates* (in 3-dimensional space).

Cartesian product The **Cartesian product** $A \times B$, of sets A and B, is the set of all *ordered pairs* (a, b), where $a \in A$ and $b \in B$. In some cases it may be possible to give a pictorial representation of $A \times B$ by taking two perpendicular axes and displaying the elements of A along one axis and the elements of B along the other axis; the ordered pair (a, b) is represented by the point with, as it were, those coordinates. In particular, if A and B are subsets of **R** and are intervals, this gives a pictorial representation as shown in the figure. Similarly, the Cartesian product $A \times B \times C$ of sets A, B and C can be defined as the set of all ordered triples (a, b, c), where $a \in A$, $b \in B$ and $c \in C$. More generally, for sets $A_1, A_2, \ldots, A_n$, the Cartesian product $A_1 \times A_2 \times \cdots \times A_n$ can be defined in a similar way.

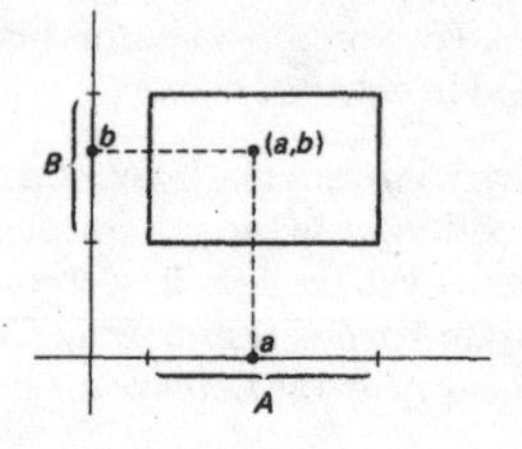

catastrophe theory A theory providing simple models for the behaviour of the equilibria of complex potential systems under variation of the potential. It was developed into a theory of biological morphogenesis by René Thom (1923–). It obtained its name because the transition from one stable state to another is often very rapid. It has been proposed as a model for many situations involving a rapid change of behaviour such as occurs when an attacking animal turns to flight or when a stock market crashes.

catenary The curve in which an ideal flexible heavy rope or chain of uniform density hangs between two points. With suitable axes, the equation of the curve is $y = c\cosh(x/c)$ (see *hyperbolic function*).

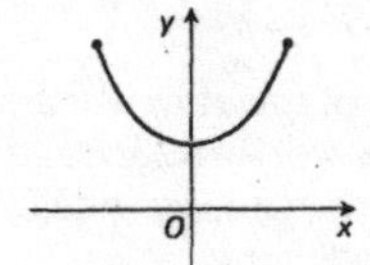

Cauchy, Augustin-Louis (1789–1857) One of the most important mathematicians of the early nineteenth century and a dominating figure in French mathematics. His work ranged over vast areas of mathematics, in almost 800 papers, but he is chiefly remembered as one of the founders of rigorous mathematical analysis. Using the definition of *limit* as it is now known, he developed sound definitions of continuity and convergence. He was also a pioneer in the theory of functions of a complex variable.

Cavalieri, Bonaventura (1598–1647) Italian mathematician known for his method of 'indivisibles' for calculating areas and volumes. In his method, an area is thought of as composed of lines and a volume as composed of areas. Here can be seen the beginnings of the ideas of integral calculus. The following theorem of his is typical of the approach: Two solids have the same volume if they have equal altitudes and if sections parallel to and at the same distance from the base have equal areas.

Cayley, Arthur (1821–1895) British mathematician who contributed greatly to the resurgence of pure mathematics in Britain in the nineteenth century. He published over 900 papers on many aspects of geometry and algebra. He conceived and developed the theory of *matrices*, and was one of the first to study abstract *groups*.

Cayley–Hamilton Theorem The *characteristic polynomial* $p(\lambda)$ of an $n \times n$ matrix $\mathbf{A}$ is defined by $p(\lambda) = \det(\mathbf{A} - \lambda\mathbf{I})$. The following result about the characteristic polynomial is called the **Cayley–Hamilton Theorem**:

THEOREM: If the characteristic polynomial $p(\lambda)$ of an $n \times n$ matrix $\mathbf{A}$ is written

$$p(\lambda) = (-1)^n(\lambda^n + b_{n-1}\lambda^{n-1} + \cdots + b_1\lambda + b_0),$$

then $\mathbf{A}^n + b_{n-1}\mathbf{A}^{n-1} + \cdots + b_1\mathbf{A} + b_0\mathbf{I} = \mathbf{O}$.

c.d.f. = *cumulative distribution function.*

centi- Prefix used with *SI units* to denote multiplication by 10^{-2}.

central conic A *conic* with a *centre of symmetry*, and thus an *ellipse* or a *hyperbola*. The conic with equation $ax^2 + 2hxy + by^2 + 2gx + 2fy + c = 0$ is central if and only if $ab \neq h^2$.

central force A force acting on a particle, directed towards or away from a fixed point. The fixed point O may be called the centre of the *field of force*. A central force $\mathbf{F}$ on a particle P is given by $\mathbf{F} = f(r)\mathbf{r}$, where $\mathbf{r}$ is the position vector of P and $r = |\mathbf{r}|$.

Examples are the *gravitational force* $\mathbf{F} = -(GMm/r^3)\mathbf{r}$, and the force $\mathbf{F} = -(k(r-l)/r)\mathbf{r}$ due to an elastic string (see *Hooke's law*).

Central Limit Theorem A fundamental theorem of statistics which says that the distribution of the mean of a sequence of *random variables* tends to a *normal distribution* as the number in the sequence increases indefinitely. It has more general forms, but one version is the following:

THEOREM: Let $X_1, X_2, X_3, \ldots$ be a sequence of independent, identically distributed random variables with mean μ and finite variance σ^2. Let

$$\overline{X}_n = \frac{X_1 + X_2 + \cdots + X_n}{n}, \qquad Z_n = \frac{\sqrt{n}(\overline{X}_n - \mu)}{\sigma}.$$

Then, as n increases indefinitely, the distribution of Z_n tends to the standard normal distribution.

It implies, in particular, that if a reasonably large number of samples are selected from any population with finite variance, then the mean of the observations can be assumed to have a normal distribution.

central quadric A non-*degenerate quadric* with a *centre of symmetry*, and thus an *ellipsoid*, a *hyperboloid of one sheet* or a *hyperboloid of two sheets*.

centre See *circle*, *ellipse* and *hyperbola*.

centre of gravity When a system of particles or a rigid body with a total mass m experiences a *uniform gravitational force*, the total effect on the system or body as a whole is equivalent to a single force acting at the **centre of gravity**. This point coincides with the *centre of mass*. The centre of gravity moves in the same way as a single particle of mass m would move under the uniform gravitational force.

When the gravitational force is given by the *inverse square law of gravitation*, the same is not true. However, when a particle experiences such a gravitational force due to a rigid body having spherical symmetry and total mass M, the total force on the particle is the same as that due to a single particle of mass M at the centre of the spherical body.

centre of mass Suppose that particles $P_1, \ldots, P_n$, with corresponding masses $m_1, \ldots, m_n$, have position vectors $\mathbf{r}_1, \ldots, \mathbf{r}_n$, respectively. The **centre of mass** (or **centroid**) is the point with position vector $\mathbf{r}_C$, given by

$$m\mathbf{r}_C = \sum_{i=1}^{n} m_i\mathbf{r}_i, \quad \text{where} \quad m = \sum_{i=1}^{n} m_i,$$

m being the total mass of the particles.

Now consider a rod of length l whose density at a point a distance x from one end is $\rho(x)$. Then the centre of mass is at the point a distance x_C from the end, given by

$$mx_C = \int_0^l \rho(x)x\,dx, \quad \text{where} \quad m = \int_0^l \rho(x)\,dx,$$

m being the mass of the rod.

For a lamina and for a 3-dimensional rigid body, the corresponding definitions involve double and triple integrals. In vector form, the position vector $\mathbf{r}_C$ of the centre of mass is given by

$$m\mathbf{r}_C = \int_V \rho(x)x\,dx, \quad \text{where} \quad m = \int_V \rho(\mathbf{r})\,dV,$$

where $\rho(\mathbf{r})$ is the density at the point with position vector $\mathbf{r}$, V is the region occupied by the body and m is the total mass of the body.

centre of symmetry See *symmetrical about a point.*

centrifugal force See *fictitious force.*

centripetal force Suppose that a particle P of mass m is moving with constant speed v in a circular path, with centre at the origin O and radius r_0. Let P have polar coordinates (r_0, θ). (See *circular motion.*) Then $v = r_0\dot{\theta}$ and the acceleration of the particle is in the direction towards O and has magnitude $r_0\dot{\theta}^2$. It follows that, if P is acted on by a force **F**, this force is in the direction towards O and has magnitude mv^2/r_0. It is called the **centripetal force.**

For example, when an unpowered satellite orbits the Earth at a constant speed, the centripetal force on the satellite is the gravitational force between the satellite and the Earth.

centroid = *centre of mass.* See also *centroid* (of a triangle).

centroid (of a triangle) The geometrical definition of the **centroid** G of a triangle ABC is as the point at which the *medians* of the triangle are concurrent. It is, in fact, 'two-thirds of the way down each median', so that,

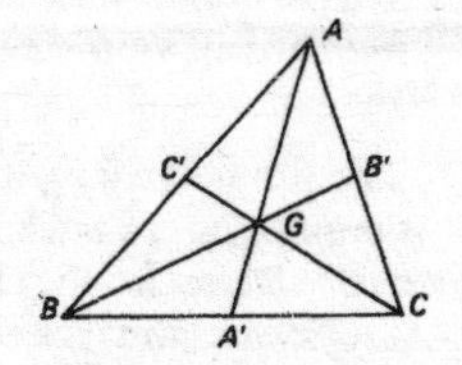

for example, if A' is the midpoint of BC, then $AG = 2GA'$. This is indeed the point at which a triangular *lamina* of uniform density has its *centre of mass*. It is also the centre of mass of three particles of equal mass situated at the vertices of the triangle. If A, B and C are points in the plane with Cartesian coordinates (x_1, y_1), (x_2, y_2) and (x_3, y_3), then G has coordinates

$$\left(\tfrac{1}{3}(x_1 + x_2 + x_3),\ \tfrac{1}{3}(y_1 + y_2 + y_3)\right).$$

For points A, B and C in 3-dimensional space with Cartesian coordinates (x_1, y_1, z_1), (x_2, y_2, z_2) and (x_3, y_3, z_3), there is no change in the definition of the centroid G and it has coordinates

$$((\tfrac{1}{3}(x_1 + x_2 + x_3),\ \tfrac{1}{3}(y_1 + y_2 + y_3),\ \tfrac{1}{3}(z_1 + z_2 + z_3)).$$

If A, B and C have *position vectors* $\mathbf{a}$, $\mathbf{b}$ and $\mathbf{c}$, then G has position vector $\frac{1}{3}(\mathbf{a} + \mathbf{b} + \mathbf{c})$.

Ceva's Theorem The following theorem, due to Giovanni Ceva (1648–1734) and published in 1678:

THEOREM: Let L, M and N be points on the sides BC, CA and AB of a triangle (possibly extended). Then AL, BM and CN are concurrent if and only if

$$\frac{BL}{LC} \cdot \frac{CM}{MA} \cdot \frac{AN}{NB} = 1.$$

(Note that BC, for example, is considered to be directed from B to C so that LC, for example, is positive if LC is in the same direction as BC and negative if it is in the opposite direction; see *measure*.) See *Menelaus' Theorem*.

chain rule The following rule that gives the *derivative* of the *composition* of two functions: If $h(x) = (f \circ g)(x) = f(g(x))$ for all x, then $h'(x) = f'(g(x))g'(x)$. For example, if $h(x) = (x^2 + 1)^3$, then $h = f \circ g$, where $f(x) = x^3$ and $g(x) = x^2 + 1$. Then $f'(x) = 3x^2$ and $g'(x) = 2x$. So $h'(x) = 3(x^2 + 1)^2 2x = 6x(x^2 + 1)^2$. Another notation can be used: if $y = f(g(x))$, write $y = f(u)$, where $u = g(x)$. Then the chain rule says that $dy/dx = (dy/du)(du/dx)$. As an example of the use of this notation, suppose that $y = (\sin x)^2$. Then $y = u^2$, where $u = \sin x$. So $dy/du = 2u$ and $du/dx = \cos x$, and hence $dy/dx = 2 \sin x \cos x$.

change of base (of logarithms) See *logarithm*.

change of coordinates (in the plane) The simplest changes from one Cartesian coordinate system to another are *translation of axes* and *rotation of axes*. See also *polar coordinates* for the change from Cartesian coordinates to polar coordinates, and vice versa.

change of coordinates (in 3-dimensional space) The simplest change from one Cartesian coordinate system to another is *translation of axes*. See also *cylindrical polar coordinates* and *spherical polar coordinates* for the change from Cartesian coordinates to those coordinate systems, and vice versa.

change of variable (in integration) See *integration.*

chaos A situation in which a fully deterministic dynamical process can appear to be random and unpredictable due to the sensitive dependence of the process on its starting values and the wide range of qualitatively different behaviours available to the process. This sensitive dependence is often called the **butterfly effect.** A typical example of a chaotic process is that produced by iterations of the function $f(x) = \frac{1}{2}(x - 1/x)$.

characteristic equation See *characteristic polynomial.*

characteristic polynomial Let **A** be a square matrix. Then $\det(\mathbf{A} - \lambda\mathbf{I})$ is a polynomial in λ and is called the **characteristic polynomial** of **A**. The equation $\det(\mathbf{A} - \lambda\mathbf{I}) = 0$ is the **characteristic equation** of **A**, and its roots are the *characteristic values* of **A**. See also *Cayley–Hamilton Theorem.*

characteristic root = *characteristic value.*

characteristic value Let **A** be a square matrix. The roots of the *characteristic equation* $\det(\mathbf{A} - \lambda\mathbf{I}) = 0$ are called the **characteristic values** of **A**. Then λ is a characteristic value of **A** if and only if there is a non-zero vector **x** such that $\mathbf{Ax} = \lambda\mathbf{x}$. Any vector **x** such that $\mathbf{Ax} = \lambda\mathbf{x}$ is called a **characteristic vector** corresponding to the characteristic value λ.

characteristic vector See *characteristic value.*

Chebyshev, Pafnuty Lvovich (1821–1894) Russian mathematician and founder of a notable school of mathematicians in St Petersburg. His name is remembered in results in algebra, analysis and probability theory. In number theory, he proved that, for all $n > 3$, there is at least one prime between n and $2n - 2$.

check digit Suppose that in a *binary code* of length n all the words of length n are possible codewords. Such a code is in no way an *error-detecting code.* If an additional bit is added to each codeword so that the number of 1's in each new codeword is even, the new code is error-detecting. For example, from the code with codewords 00, 01, 10 and 11, the new code with codewords 000, 011, 101 and 110 would be obtained. The additional bit is called a **check digit**, and in this case the construction is a **parity check** (see *parity*). More complicated examples of check digits are often used for the purposes of error-detecting.

chi-squared distribution A type of non-negative continuous probability *distribution*, normally written as the χ^2-distribution, with one parameter ν called the *degrees of freedom.* The distribution (whose precise definition will not be given here) is skewed to the right and has the property that the sum of independent random variables each having a χ^2-distribution also has a χ^2-distribution. It is used in the *chi-squared test* for measuring goodness of fit, in tests on variance and in testing for independence in contingency tables. It has mean ν and variance 2ν. Tables relating to the distribution for different values of ν are available.

chi-squared test A test, normally written as the χ^2-test, to determine how well a set of observations fits a particular discrete distribution or some other given null hypothesis (see *hypothesis testing*). The observed frequencies in different groups are denoted by O_i, and the expected frequencies from the statistical model are denoted by E_i. For each i, the value $(O_i - E_i)^2/E_i$ is calculated, and these are summed. The result is compared with a *chi-squared distribution* with an appropriate number of degrees of freedom. The number of degrees of freedom depends on the number of groups and the number of parameters being estimated. The test requires that the observations are independent and that the sample size and expected frequencies exceed minimum numbers depending on the number of groups.

chord Let A and B be two points on a curve. The straight line through A and B, or the *line segment AB*, is called a **chord**, the word being used when a distinction is to be made between the chord AB and the arc AB.

Chu Shih-chieh (about AD 1300) One of the greatest of Chinese mathematicians, who wrote two major influential texts, the more important being the *Su-yuan yu-chien* ('Precious Mirror of the Four Elements'). Notable are the methods of solving equations by successive approximations and the summation of series using *finite differences*. A diagram of *Pascal's triangle*, as it has come to be called, known in China from before this time, also appears.

circle The **circle** with centre C and radius r is the locus of all points in the plane whose distance from C is equal to r. If C has Cartesian coordinates (a, b), this circle has equation $(x-a)^2 + (y-b)^2 = r^2$. An equation of the form $x^2 + y^2 + 2gx + 2fy + c = 0$ represents a circle if $g^2 + f^2 - c > 0$, and is then an equation of the circle with centre $(-g, -f)$ and radius $\sqrt{g^2 + f^2 - c}$.

The area of a circle of radius r equals πr^2, and the length of the circumference equals $2\pi r$.

circle theorems The following is a summary of some of the theorems that are concerned with properties of a circle:

Let A and B be two points on a circle with centre O. If P is any point on the circumference of the circle and on the same side of the chord AB as O,

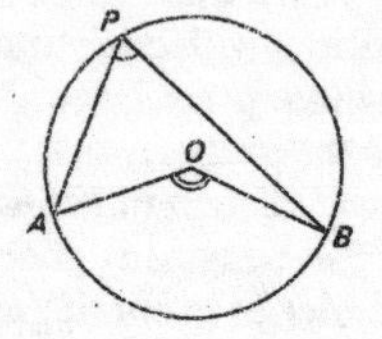

then $\angle AOB = 2\angle APB$. Hence the 'angle at the circumference' $\angle APB$ is independent of the position of *P*.

If *Q* is a point on the circumference and lies on the other side of *AB* from *P*, then $\angle AQB = 180° - \angle APB$. Hence opposite angles of a cyclic quadrilateral add up to 180°. When *AB* is a diameter, the angle at the circumference is the 'angle in a semicircle' and is a right angle. If *T* is any point on the tangent at *A*, then $\angle APB = \angle BAT$.

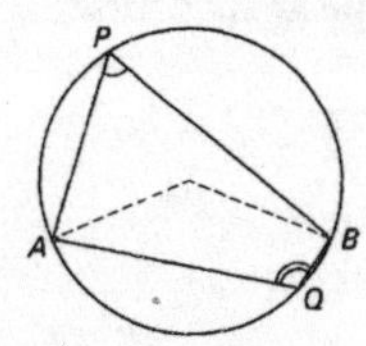

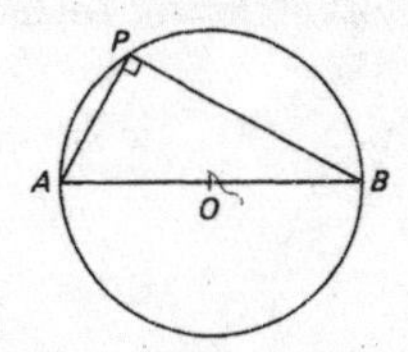

Suppose now that a circle and a point *P* are given. Let any line through *P* meet the circle at points *A* and *B*. Then $PA \,.\, PB$ is constant; that is, the same for all such lines. If *P* lies outside the circle and a line through *P* touches the circle at the point *T*, then $PA \,.\, PB = PT^2$.

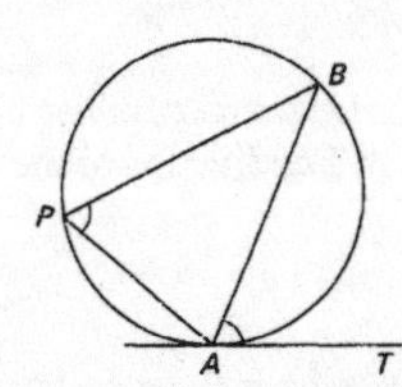

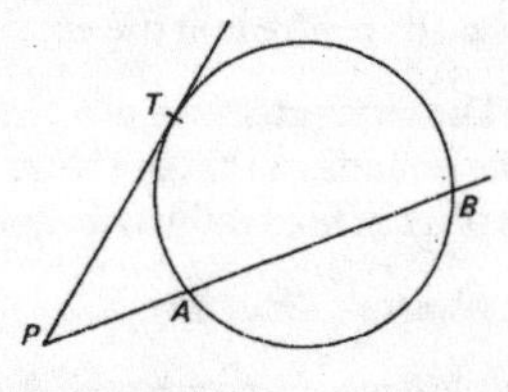

circular function A term used to describe either of the *trigonometric functions* sin and cos. Some authors also apply the term to the trigonometric function tan.

circular motion Motion of a particle in a circular path. Suppose that the path of the particle *P* is a circle in the plane, with centre at the origin *O* and radius r_0. Let $\mathbf{i}$ and $\mathbf{j}$ be unit vectors in the directions of the positive *x*- and *y*-axes. Let $\mathbf{r}$, $\mathbf{v}$ and $\mathbf{a}$ be the position vector, velocity and acceleration of *P*. If *P* has polar coordinates (r_0, θ), then

$$\mathbf{r} = r_0(\mathbf{i}\cos\theta + \mathbf{j}\sin\theta),$$

$$\mathbf{v} = \dot{\mathbf{r}} = r_0(-\dot{\theta}\mathbf{i}\sin\theta + \dot{\theta}\mathbf{j}\cos\theta),$$

$$\mathbf{a} = \ddot{\mathbf{r}} = r_0(-\ddot{\theta}\mathbf{i}\sin\theta - \dot{\theta}^2\mathbf{i}\cos\theta + \ddot{\theta}\mathbf{j}\cos\theta - \dot{\theta}^2\mathbf{j}\sin\theta).$$

Let $\mathbf{e}_r = \mathbf{i}\cos\theta + \mathbf{j}\sin\theta$ and $\mathbf{e}_\theta = -\mathbf{i}\sin\theta + \mathbf{j}\cos\theta$, so that $\mathbf{e}_r$ is a unit vector along *OP* in the direction of increasing *r*, and $\mathbf{e}_\theta$ is a unit vector perpendicular to this in the direction of increasing θ. Then the equations above become

$$\mathbf{r} = r_0\mathbf{e}_r, \qquad \mathbf{v} = \dot{\mathbf{r}} = r_0\dot{\theta}\mathbf{e}_\theta, \qquad \mathbf{a} = \ddot{\mathbf{r}} = -r_0\dot{\theta}^2\mathbf{e}_r + r_0\ddot{\theta}\mathbf{e}_\theta.$$

If the particle, of mass m, is acted on by a force $\mathbf{F}$, where $\mathbf{F} = F_1\mathbf{e}_r + F_2\mathbf{e}_\theta$, then the equation of motion $m\ddot{\mathbf{r}} = \mathbf{F}$ gives $-mr_0\dot{\theta}^2 = F_1$ and $mr_0\ddot{\theta} = F_2$. If the transverse component F_2 of the force is zero, then $\dot{\theta} =$ constant and the particle has constant speed.

See also *angular velocity* and *angular acceleration.*

circumcentre The **circumcentre** of a triangle is the centre of the *circumcircle* of the triangle. It is the point O, shown in the figure, at which the perpendicular bisectors of the sides of the triangle are concurrent.

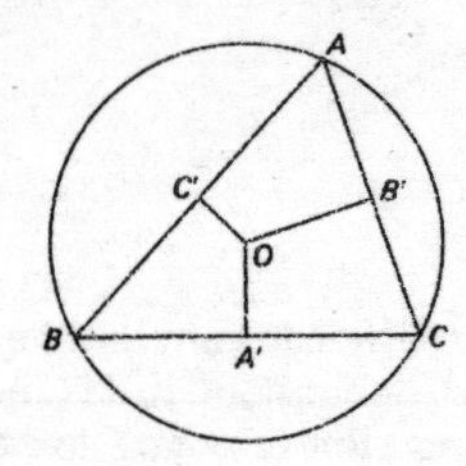

circumcircle The **circumcircle** of a triangle is the circle that passes through the three vertices. Its centre is at the *circumcentre.*

circumference The **circumference** of a circle is the boundary of the circle or the length of the boundary, that is, the perimeter. The (length of the) circumference of a circle of radius r is $2\pi r$.

circumscribing cylinder See *zone.*

cis The notation **cis** θ is sometimes used for $\cos\theta + i\sin\theta$.

class interval Numerical data may be *grouped* by dividing the set of possible values into so-called **class intervals** and counting the number of observations in each interval. For example, if the possible marks obtained in a test lie between 0 and 99, inclusive, groups could be defined by the intervals 0–19, 20–39, 40–59, 60–79 and 80–99. It is often best (but not essential) to take the class intervals to be of equal widths.

closed (under an operation) See *operation.*

closed curve A continuous plane curve that has no ends or, in other words, that begins and ends at the same point.

closed disc See *disc.*

closed half-plane See *half-plane.*

closed half-space See *half-space.*

closed interval The **closed interval** $[a, b]$ is the set

$$\{x \mid x \in \mathbf{R} \text{ and } a \leq x \leq b\}.$$

coaxial Having the same axis.

code See *binary code* and *error-correcting and error-detecting code.*

codeword See *binary code.*

codomain See *function* and *mapping.*

coefficient See *binomial coefficient* and *polynomial.*

coefficient of friction See *friction.*

coefficient of kinetic friction See *friction.*

coefficient of restitution A parameter associated with the behaviour of two bodies during a *collision.* Suppose that two billiard balls are travelling in the same straight line and have velocities u_1 and u_2 before the collision, and velocities v_1 and v_2 after the collision. If the coefficient of restitution is e, then

$$v_2 - v_1 = -e(u_2 - u_1).$$

This formula is **Newton's law of restitution.** The coefficient of restitution always satisfies $0 \leq e \leq 1$. When $e = 0$, the balls remain in contact after the collision. When $e = 1$, the collision is *elastic*: there is no loss of kinetic energy.

It may be convenient to consider a collision as consisting of a deformation phase, during which the shape of each body is deformed, and a restitution phase, during which the shape of each body is completely or partially restored. Newton's law follows from the supposition that, for each body, the *impulse* during restitution is e times the impulse during deformation.

coefficient of skewness See *skewness.*

coefficient of static friction See *friction.*

coefficient of variation A measure of *dispersion* equal to the *standard deviation* of a sample divided by the mean. The value is a dimensionless quantity, not dependent on the units or scale in which the observations are made, and is often expressed as a percentage.

cofactor Let **A** be the square matrix $[a_{ij}]$. The **cofactor** of the entry a_{ij} is equal to $(-1)^{i+j}$ times the *determinant* of the matrix obtained by deleting the i-th row and j-th column of **A**. If **A** is the 3×3 matrix shown, the factor $(-1)^{i+j}$ has the effect of introducing a + or − sign according to the pattern on the right:

$$\mathbf{A} = \begin{bmatrix} a_{11} & a_{12} & a_{13} \\ a_{21} & a_{22} & a_{23} \\ a_{31} & a_{32} & a_{33} \end{bmatrix} \qquad \begin{bmatrix} + & - & + \\ - & + & - \\ + & - & + \end{bmatrix}.$$

So, for example,

$$A_{12} = -\begin{vmatrix} a_{21} & a_{23} \\ a_{31} & a_{33} \end{vmatrix}, \qquad A_{31} = +\begin{vmatrix} a_{12} & a_{13} \\ a_{22} & a_{23} \end{vmatrix}.$$

For a 2 × 2 matrix, the pattern is:

$$\begin{bmatrix} a & b \\ c & d \end{bmatrix} \qquad \begin{bmatrix} + & - \\ - & + \end{bmatrix}.$$

So, the cofactor of a equals d, the cofactor of b equals $-c$, and so on. The following properties hold, for an $n \times n$ matrix A:

(i) The expression $a_{i1}A_{i1} + a_{i2}A_{i2} + \cdots + a_{in}A_{in}$ has the same value for any i, and is the definition of det A, the determinant of A. This particular expression is the evaluation of det A by the i-th row.

(ii) On the other hand, if $i \neq j$, $a_{i1}A_{j1} + a_{i2}A_{j2} + \cdots + a_{in}A_{jn} = 0$.

Results for columns, corresponding to the results (i) and (ii) for rows, also hold.

collinear Any number of points are said to be **collinear** if there is a straight line passing through all of them.

collision A **collision** occurs when two bodies move towards one another and contact takes place between the bodies. The subsequent motion is often difficult to predict. There is normally a loss of kinetic energy. A simple example is a collision between two billiard balls. The behaviour of the two balls after the collision depends upon the *coefficient of restitution*.

column operation See *elementary column operation*.

column matrix A *matrix* with exactly one column; that is, an $m \times 1$ matrix of the form

$$\begin{bmatrix} a_1 \\ a_2 \\ \vdots \\ a_m \end{bmatrix}.$$

Given an $m \times n$ matrix, it may be useful to treat its columns as individual column matrices.

column rank See *rank*.

column stochastic matrix See *stochastic matrix*.

column vector = *column matrix*.

combination = *selection*.

common difference See *arithmetic sequence*.

common fraction A term used to mean *vulgar fraction* or *simple fraction*. Some authors use it to mean *proper fraction*.

common logarithm See *logarithm.*

common perpendicular Let l_1 and l_2 be two straight lines in space that do not intersect and are not parallel. The **common perpendicular** of l_1 and l_2 is the straight line that meets both lines and is perpendicular to both.

common ratio See *geometric sequence.*

commutative The *binary operation* $\circ$ on a set S is **commutative** if, for all a and b in S, $a \circ b = b \circ a$.

commutative ring See *ring.*

commute Let $\circ$ be a *binary operation* on a set S. The elements a and b of S **commute** (under the operation $\circ$) if $a \circ b = b \circ a$. For example, multiplication on the set of all real 2×2 matrices is not commutative, but if A and B are diagonal matrices then A and B commute.

complement, complementation Let A be a subset of some *universal set* E. Then the **complement** of A is the *difference set* $E \setminus A$ (or $E - A$). It may be denoted by A' (or $\bar{A}$) when the universal set is understood or has previously been specified. **Complementation** (the operation of taking the complement) is a unary operation on the set of subsets of a universal set E. The following properties hold:

(i) $E' = \emptyset$ and $\emptyset' = E$.
(ii) For all A, $(A')' = A$.
(iii) For all A, $A \cap A' = \emptyset$ and $A \cup A' = E$.

See also *relative complement.*

complement (for angles) See *complementary angles.*

complementary angles Two angles that add up to a right angle. Each angle is the **complement** of the other.

complementary function See *linear differential equation with constant coefficients.*

complete graph A *simple graph* in which every vertex is joined to every other. The complete graph with n vertices, denoted by K_n, is *regular* of degree $n - 1$ and has $\frac{1}{2}n(n-1)$ edges. See also *bipartite graph.*

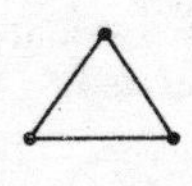

K_3

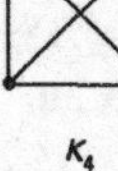

K_4

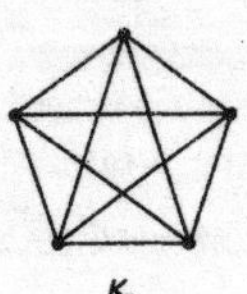

K_5

complete quadrangle The configuration in the plane consisting of four points, no three of which are collinear, together with the six lines joining them in pairs. An example is shown in the figure.

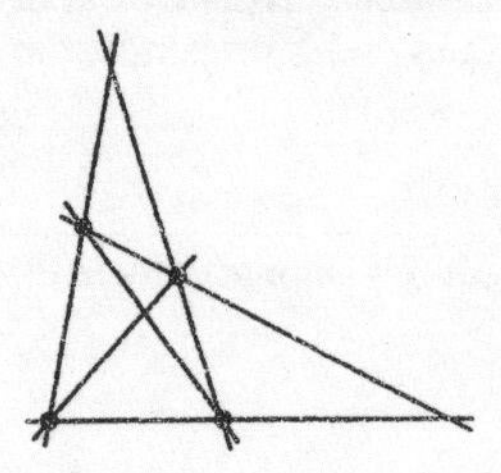

complete quadrilateral The configuration in the plane consisting of four straight lines, no three of which are concurrent, together with the six points in which they intersect in pairs. An example is shown in the figure.

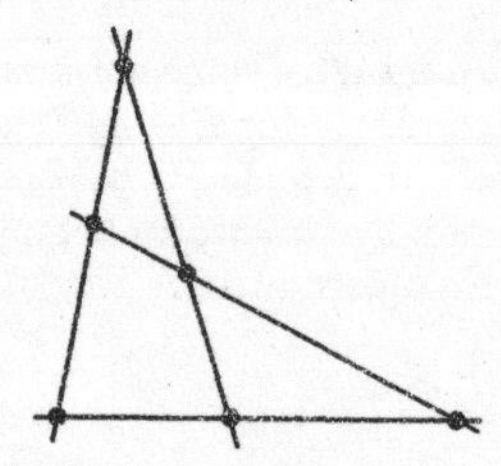

complete set of residues (modulo n) A set of n integers, one from each of the *n residue classes* modulo n. Thus $\{0, 1, 2, 3\}$ is a complete set of residues modulo 4; so too are $\{1, 2, 3, 4\}$ and $\{-1, 0, 1, 2\}$.

completing the square Consider a numerical example: the *quadratic equation* $2x^2 + 5x + 1 = 0$ can be solved by first writing it as

$$x^2 + \frac{5}{2}x = -\frac{1}{2}, \quad \text{and then} \quad \left(x + \frac{5}{4}\right)^2 = -\frac{1}{2} + \frac{25}{16} = \frac{17}{16}.$$

This step is known as **completing the square**: the left-hand side is made into an exact square by adding a suitable constant to both sides. The solution of the quadratic equation can then be accomplished as follows:

$$x + \frac{5}{4} = \pm\frac{\sqrt{17}}{4}, \quad \text{and so} \quad x = \frac{-5 \pm \sqrt{17}}{4}.$$

By proceeding in the same way with $ax^2 + bx + c = 0$, the standard formula for the solution of a quadratic equation can be derived.

complex conjugate = *conjugate* (of a complex number).

complex number There is no real number x such that $x^2 + 1 = 0$. The introduction of a 'new' number i such that $i^2 = -1$ gives rise to further numbers of the form $a + bi$. A number of the form $a + bi$, where a and b are real, is a **complex number**. Since one may take $b = 0$, this includes all the real numbers. The set of all complex numbers is usually denoted by **C**. (The use of j in place of i is quite common.) It is assumed that two such numbers may be added and multiplied using the familiar rules of algebra, with i^2 replaced by -1 whenever it occurs. So,

$$(a + bi) + (c + di) = (a + c) + (b + d)i,$$
$$(a + bi)(c + di) = (ac - bd) + (ad + bc)i.$$

Thus the set **C** of complex numbers is closed under addition and multiplication, and the elements of this enlarged number system satisfy the laws commonly expected of numbers.

The complex number system can be put on a more rigorous basis as follows. Consider the set $\mathbf{R} \times \mathbf{R}$ of all ordered pairs (a, b) of real numbers (see *Cartesian product*). Guided by the discussion above, addition and multiplication are defined on $\mathbf{R} \times \mathbf{R}$ by

$$(a, b) + (c, d) = (a + c, b + d),$$
$$(a, b)(c, d) = (ac - bd, ad + bc).$$

It can be verified that addition and multiplication defined in this way are associative and commutative, that the distributive law holds, that there is a zero element and an identity element, that every element has a negative and every non-zero element has an inverse (see *inverse of a complex number*). This shows that $\mathbf{R} \times \mathbf{R}$ with this addition and multiplication is a *field* whose elements, according to this approach, are called complex numbers. The elements of the form $(a, 0)$ can be seen to behave exactly like the corresponding real numbers a. Moreover, if the element $(0, 1)$ is denoted by i, it is reasonable to write $i^2 = -1$, since $(0, 1)^2 = -(1, 0)$. After providing this rigorous foundation, it is normal to write $a + bi$ instead of (a, b). See also *argument*, *modulus of a complex number* and *polar form of a complex number*.

complex plane Let points in the plane be given coordinates (x, y) with respect to a Cartesian coordinate system. The plane is called the **complex plane** when the point (x, y) is taken to represent the *complex number* $x + yi$.

component (of a compound statement) See *compound statement*.

component (of a graph) A *graph* may be 'in several pieces' and these are called its **components**: two vertices are in the same component if and only if there is a *path* from one to the other. A more precise definition can be given by defining an *equivalence relation* on the set of vertices with u equivalent to v if there is a path from u to v. Then the components are the corresponding *equivalence classes*.

component (of a vector) In a Cartesian coordinate system in 3-dimensional space, let **i**, **j** and **k** be unit vectors along the three coordinate axes. Given a vector **p**, there are unique real numbers x, y and z such that $\mathbf{p} = x\mathbf{i} + y\mathbf{j} + z\mathbf{k}$. Then x, y and z are the **components** of **p** (with respect to the vectors **i**, **j**, **k**). These can be determined by using the *scalar product*: $x = \mathbf{p}.\mathbf{i}$, $y = \mathbf{p}.\mathbf{j}$ and $z = \mathbf{p}.\mathbf{k}$. (See also *direction cosines*.)

More generally, if **u**, **v** and **w** are any 3 non-coplanar vectors, then any vector **p** in 3-dimensional space can be expressed uniquely as $\mathbf{p} = x\mathbf{u} + y\mathbf{v} + z\mathbf{w}$, and x, y and z are called the **components** of **p** with respect to the basis **u**, **v**, **w**. In this case, however, the components cannot be found so simply by using the scalar product.

composite A positive integer is **composite** if it is neither *prime*, nor equal to 1; that is, if it can be written as a product hk, where the integers h and k are both greater than 1.

composition Let $f\colon S \to T$ and $g\colon T \to U$ be *mappings*. With each s in S is associated the element $f(s)$ of T, and hence the element $g(f(s))$ of U. This rule gives a mapping from S to U, which is denoted by $g \circ f$ (read as 'g circle f') and is the **composition** of f and g. Note that f operates first, then g. Thus $g \circ f\colon S \to U$ is defined by $(g \circ f)(s) = g(f(s))$, and exists if and only if the domain of g equals the codomain of f. For example, suppose that $f\colon \mathbf{R} \to \mathbf{R}$ and $g\colon \mathbf{R} \to \mathbf{R}$ are defined by $f(x) = 1 - x$ and $g(x) = x/(x^2 + 1)$. Then $f \circ g\colon \mathbf{R} \to \mathbf{R}$ and $g \circ f\colon \mathbf{R} \to \mathbf{R}$ both exist, and

$$(f \circ g)(x) = 1 - \frac{x}{x^2 + 1}, \qquad (g \circ f)(x) = \frac{1 - x}{(1 - x)^2 + 1}.$$

The term 'composition' may be used for the operation $\circ$ as well as for the resulting function. The composition of mappings is associative: if $f\colon S \to T$, $g\colon T \to U$ and $h\colon U \to V$ are mappings, then $h \circ (g \circ f) = (h \circ g) \circ f$. This means that the mappings $h \circ (g \circ f)$ and $(h \circ g) \circ f$ have the same domain S and the same codomain V and, for all s in S, $(h \circ (g \circ f))(s) = ((h \circ g) \circ f)(s)$.

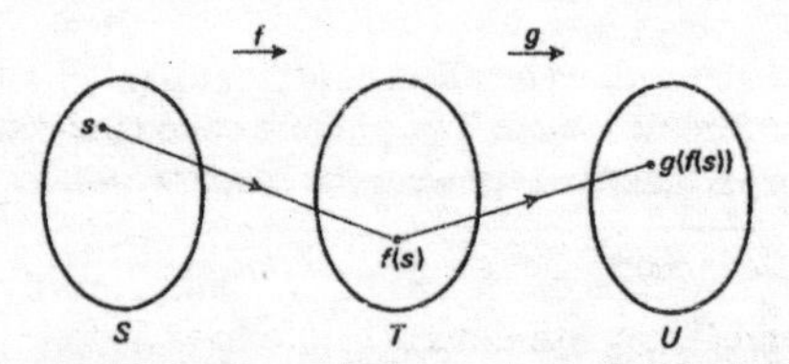

compound fraction A fraction in which the numerator or denominator, or both, contain fractions. For example, $\frac{3}{4}/(1 + \frac{2}{3})$ is a compound fraction. See *simple fraction*.

compound interest Suppose that a sum of money P is invested, attracting interest at i per cent a year. After one year, the amount becomes $P + (i/100)P$. This equals $P(1 + i/100)$, so that adding on i per cent is equivalent to multiplying by $1 + i/100$. When interest is compounded annually, the new amount is used to calculate the interest due in the second year and so, after 2 years, the amount becomes $P(1 + i/100)^2$. After n years, the amount becomes

$$P\left(1+\frac{i}{100}\right)^n.$$

This is the formula for **compound interest**. When points are plotted on graph paper to show how the amount increases, they lie on a curve that illustrates *exponential growth*. This is in contrast to the straight line obtained in the case of *simple interest*.

compound pendulum A pendulum consisting of a **rigid body** that is free to swing about a horizontal axis. Suppose that the centre of gravity of the rigid body is a distance d from the axis, m is the mass of the body, and I is the *moment of inertia* of the body about the axis. The equation of motion can be shown to give $I\ddot{\theta} = -mgd \sin\theta$. As with the *simple pendulum*, this gives approximately simple harmonic motion if θ is small for all time.

compound statement A statement formed from simple statements by the use of words such as 'and', 'or', 'not', 'implies' or their corresponding symbols. The simple statements involved are the **components** of the compound statement. For example, $(p \wedge q) \vee (\neg r)$ is a compound statement built up from the components p, q and r.

concave up and down See *concavity*.

concavity At a point of a graph $y = f(x)$, it may be possible to specify the **concavity** by describing the curve as either concave up or concave down at that point, as follows.

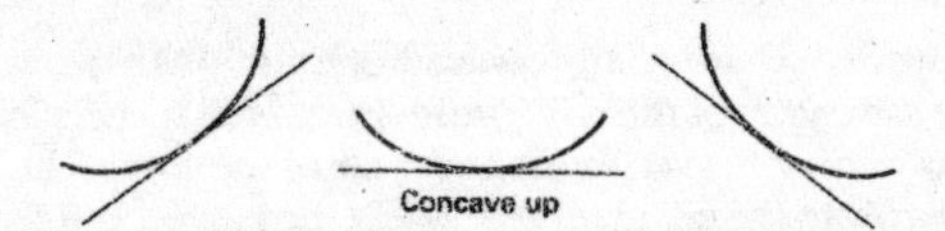
Concave up

If the second derivative $f''(x)$ exists and is positive throughout some neighbourhood of a point a, then $f'(x)$ is strictly increasing in that neighbourhood, and the curve is said to be **concave up** at a. At that point, the graph $y = f(x)$ and its tangent look like one of the cases shown in the first figure. If $f''(a) > 0$ and f'' is continuous at a, it follows that $y = f(x)$ is concave up at a. Consequently, if $f'(a) = 0$ and $f''(a) > 0$, the function f has a *local minimum* at a. Similarly, if $f''(x)$ exists and is negative

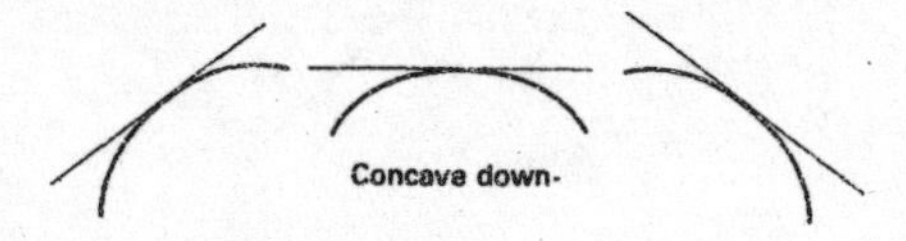
Concave down

throughout some neighbourhood of a, or if $f''(a) < 0$ and f'' is continuous at a, then the graph $y = f(x)$ is **concave down** at a and looks like one of the cases shown in the second figure. If $f'(a) = 0$ and $f''(a) < 0$, the function f has a *local maximum* at a.

concentric Having the same centre.

concurrent Any number of lines are said to be **concurrent** if there is a point through which they all pass.

concyclic A number of points are **concyclic** if there is a circle that passes though all of them.

condition, necessary and **sufficient** The *implication* $q \Rightarrow p$ can be read as 'if q, then p'. When this is true, it may be said that q is a **sufficient condition** for p; that is, the truth of the 'condition' q is sufficient to ensure the truth of p. This means that p is true **if** q is true. On the other hand, when the implication $p \Rightarrow q$ holds, then q is a **necessary condition** for p; that is, the truth of the 'condition' q is a necessary consequence of the truth of p. This means that p is true **only if** q is true. When the implication between p and q holds both ways, p is true **if and only if** q is true, which may be written $p \Leftrightarrow q$. Then q is a **necessary and sufficient condition** for p.

conditional probability For two events A and B, the probability that A occurs, given that B has occurred, is denoted by $\Pr(A \mid B)$, read as 'the probability of A given B'. This is called a **conditional probability**. Provided that $\Pr(B)$ is not zero, $\Pr(A \mid B) = \Pr(A \cap B)/\Pr(B)$. This result is often useful in the following form: $\Pr(A \cap B) = \Pr(B)\Pr(A \mid B)$. If A and B are *independent events*, $\Pr(A \mid B) = \Pr(A)$, and this gives the product law for independent events: $\Pr(A \cap B) = \Pr(A)\Pr(B)$.

cone In elementary work, a **cone** usually consists of a circle as base, a vertex lying directly above the centre of the circle, and the curved surface formed by the line segments joining the vertex to the points of the circle. The distance from the vertex to the centre of the base is the **height**, and the

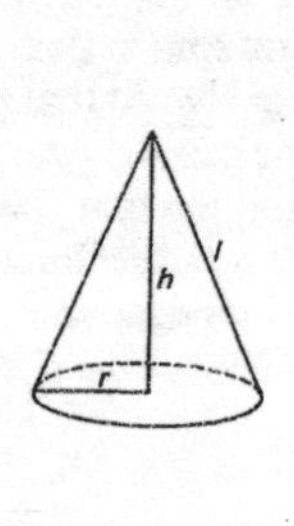

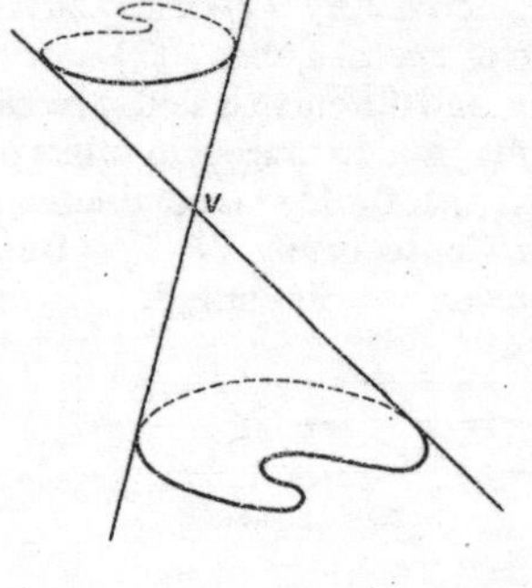

length of any of the line segments is the **slant height**. For a cone with base of radius r, height h and slant height l, the volume equals $\frac{1}{3}\pi r^2 h$ and the area of the curved surface equals $\pi r l$.

In more advanced work, a **cone** is the surface consisting of the points of the lines, called **generators**, drawn through a fixed point V, the **vertex**, and the points of a fixed curve, the generators being extended indefinitely in both directions. Then a **right-circular cone** is a cone in which the fixed curve is a circle and the vertex V lies on the line through the centre of the circle and perpendicular to the plane of the circle. The **axis** of a right-circular cone is the line through V and the centre of the circle, and is perpendicular to the plane of the circle. All the generators make the same angle with the axis; this is the **semi-vertical angle** of the cone. The right-circular cone with vertex at the origin, the z-axis as its axis, and semi-vertical angle α, has equation $x^2 + y^2 = z^2 \tan^2 \alpha$. See also *quadric cone.*

confidence interval An interval, calculated from a sample, which contains the value of a certain population *parameter* with a specified probability. The end-points of the interval are the **confidence limits**. The specified probability is called the **confidence level**. An arbitrary but commonly used confidence level is 95%, which means that there is a one-in-twenty chance that the interval does not contain the true value of the parameter. For example, if $\bar{x}$ is the mean of a sample of n observations taken from a population with a normal distribution with a known standard deviation σ, then

$$\left[\bar{x} - \frac{1.96\sigma}{\sqrt{n}}, \bar{x} + \frac{1.96\sigma}{\sqrt{n}}\right]$$

is a 95% confidence interval for the population mean μ.

confidence level See *confidence interval.*

configuration A particular geometrical arrangement of points, lines, curves and, in 3 dimensions, planes and surfaces.

confocal conics Two *central conics* are **confocal** if they have the same foci. An ellipse and a hyperbola that are confocal intersect at right angles.

conformable Matrices **A** and **B** are **conformable** (for multiplication) if the number of columns of **A** equals the number of rows of **B**. Then **A** has order $m \times n$ and **B** has order $n \times p$, for some m, n and p, and the product **AB**, of order $m \times p$, is defined. See *multiplication* (of matrices).

congruence (modulo n) For each positive integer n, the relation of **congruence** between integers is defined as follows: a **is congruent to** b modulo n if $a - b$ is a multiple of n. This is written $a \equiv b \pmod{n}$. The integer n is the **modulus** of the congruence. Then $a \equiv b \pmod{n}$ if and only if a and b have the same remainder upon division by n. For example, 19 is congruent to 7 modulo 3. The following properties hold, if $a \equiv b \pmod{n}$ and $c \equiv d \pmod{n}$:

(i) $a + c \equiv b + d \pmod{n}$,
(ii) $a - c \equiv b - d \pmod{n}$,
(iii) $ac \equiv bd \pmod{n}$.

It can be shown that congruence modulo n is an *equivalence relation* and so defines a *partition* of the set of integers, where two integers are in the same class if and only if they are congruent modulo n. These classes are the *residue* (or *congruence*) *classes* modulo n.

congruence class = *residue class.*

congruence equation The following are examples of **congruence equations**:

(i) $x + 5 \equiv 3 \pmod{7}$; this has the solution $x \equiv 5 \pmod{7}$.
(ii) $2x \equiv 5 \pmod{4}$; this has no solutions.
(iii) $x^2 \equiv 1 \pmod{8}$; this has solutions $x \equiv 1, 3, 5$ or $7 \pmod{8}$.
(iv) $x^2 + 2x + 3 \equiv 0 \pmod{6}$; this has solutions $x \equiv 1$ or $3 \pmod{6}$.

In seeking solutions to a congruence equation, it is necessary only to consider a *complete set of residues* and find solutions in this set. The examples (i) and (ii) above are **linear congruence equations**. The linear congruence equation $ax \equiv b \pmod{n}$ has a solution if and only if (a, n) divides b, where (a, n) is the *greatest common divisor* of a and n.

congruent (modulo n) See *congruence* (modulo n).

congruent figures Two geometrical figures are **congruent** if they are identical in shape and size. This includes the case when one of them is a mirror image of the other, and so the three triangles shown here are all congruent to each other.

conic, conic section A curve that can be obtained as the plane section of a cone. The figure shows how an ellipse, parabola and hyperbola can be obtained.

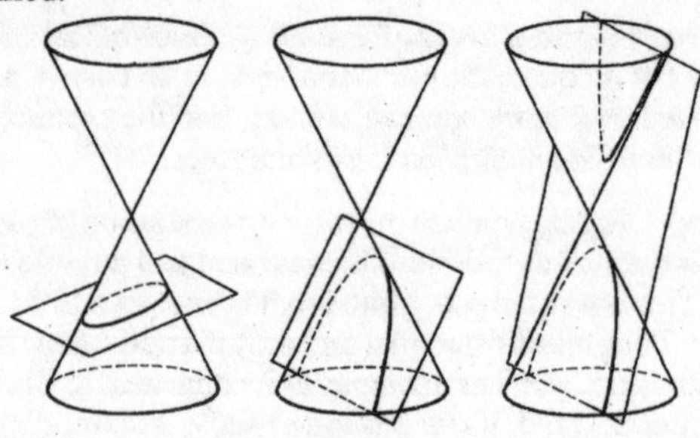

But there are other more convenient characterizations, one of which is by means of the focus and directrix property. Let F be a fixed point (the **focus**) and l a fixed line (the **directrix**), not through F, and let e be a fixed positive number (the **eccentricity**). Then the locus of all points P such that the distance from P to F equals e times the distance from P to l is a curve, and any such curve is a **conic**. The conic is called an *ellipse* if $e < 1$, a *parabola* if $e = 1$ and a *hyperbola* if $e > 1$. Note that a circle is certainly a conic (it is a special case of an ellipse); but it can only be obtained from the focus and directrix property by regarding it as the limiting form of an ellipse as $e \to 0$ and the directrix moves infinitely far away.

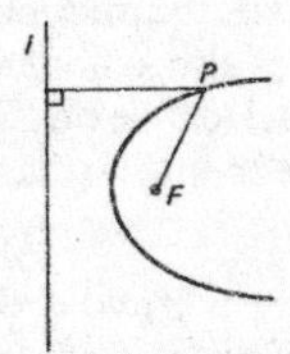

In a Cartesian coordinate system, a conic is a curve that has an equation of the second degree, that is, of the form $ax^2 + 2hxy + by^2 + 2gx + 2gy + c = 0$. This equation represents a parabola if $h^2 = ab$, an ellipse if $h^2 < ab$ and a hyperbola if $h^2 > ab$. It represents a circle if $a = b$ and $h = 0$, and a *rectangular hyperbola* if $a + b = 0$. It represents a pair of straight lines (which may coincide) if $\Delta = 0$, where

$$\Delta = \begin{vmatrix} a & h & g \\ h & b & f \\ g & f & c \end{vmatrix}.$$

The *polar equation* of a conic is normally obtained by taking the origin at a focus of the conic and the direction given by $\theta = 0$ perpendicular to the directrix. Then the equation can be written $l/r = 1 + e\cos\theta$ (all θ such that $\cos\theta \neq -1/e$), where e is the eccentricity and l is another constant.

conical pendulum A pendulum in which a particle, attached by a string of constant length to a fixed point, moves in a circular path in a horizontal plane. Suppose that the string has length l and that the string makes a constant angle α with the vertical. The particle moves in a circular path in a horizontal plane a distance d below the fixed point, where $d = l\cos\alpha$. The period of the conical pendulum is equal to $2\pi\sqrt{d/g}$.

conjugate (of a complex number) For any complex number z, where $z = x + yi$, its **conjugate** $\bar{z}$ (read as 'z bar') is equal to $x - yi$. In the *complex plane*, the points representing a complex number and its conjugate are mirror images with respect to the real axis. The following properties hold:

(i) $\bar{\bar{z}} = z$; so if $z_1 = \overline{z_2}$, then $z_2 = \overline{z_1}$.
(ii) $z + \bar{z}$ is real; if $z = x + yi$, then $z + \bar{z} = 2x$.

(iii) $z\bar{z} = |z|^2$; if $z = x + yi$, then $z\bar{z} = x^2 + y^2$.
(iv) $\overline{z_1 + z_2} = \overline{z_1} + \overline{z_2}$ and $\overline{z_1 - z_2} = \overline{z_1} - \overline{z_2}$.
(v) $\overline{z_1 z_2} = \overline{z_1}\,\overline{z_2}$ and $\overline{(z_1/z_2)} = \overline{z_1}/\overline{z_2}$.

It is an important fact that if the complex number α is a root of a polynomial equation $z^n + a_1 z^{n-1} + \cdots + a_{n-1}z + a_n = 0$, where $a_1, \ldots, a_n$ are real, then $\bar{\alpha}$ is also a root of this equation.

conjugate angles Two angles that add up to four right angles.

conjugate axis See *hyperbola*.

conjugate diameters If l is a diameter of a *central conic*, the midpoints of the chords parallel to l lie on a straight line l', which is also a diameter. It can be shown that the midpoints of the chords parallel to l' lie on l. Then l and l' are **conjugate diameters**. In the case of a circle, conjugate diameters are perpendicular.

conjunction If p and q are statements, then the statement 'p and q', denoted by $p \wedge q$, is the **conjunction** of p and q. For example, if p is 'It is raining' and q is 'It is Monday', then $p \wedge q$ is 'It is raining and it is Monday'. The conjunction of p and q is true only when p and q are both true, and so the *truth table* is as follows:

p	q	$p \wedge q$
T	T	T
T	F	F
F	T	F
F	F	F

connected graph A *graph* in which there is a *path* from any one vertex to any other. So a graph is connected if it is 'all in one piece'; that is, if it has precisely one *component*.

conservation of angular momentum When a particle is acted on by a *central force*, the *angular momentum* of the particle about the centre of the field of force is constant for all time. This fact is called the principle of **conservation of angular momentum**.

Similarly, when the sum of the moments of the forces acting on a rigid body about a fixed point (or the centre of mass) is zero, the angular momentum of the rigid body about the fixed point (or the centre of mass) is conserved.

conservation of energy When all the forces acting on a system are *conservative forces*, $E_k + E_p =$ constant, where E_k is the *kinetic energy* and E_p is the *potential energy*. This fact is known as the **energy equation** or the principle of **conservation of energy**.

From the equation of motion $m\mathbf{a} = \mathbf{F}$ for a particle with mass m moving with acceleration $\mathbf{a}$, it follows that $m\mathbf{a} \cdot \mathbf{v} = \mathbf{F} \cdot \mathbf{v}$ and hence

$(d/dt)(\frac{1}{2}m\mathbf{v}.\mathbf{v}) = \mathbf{F}.\mathbf{v}$. When **F** is a conservative force, the energy equation follows by integration with respect to t.

conservation of linear momentum When the total force acting on a system is zero, the *linear momentum* of the system remains constant for all time. This fact is called the principle of **conservation of linear momentum.**

An application is seen in the recoil of a gun when a shell is fired. Before the firing, the linear momentum of the shell and gun is zero, so the total linear momentum after the firing is also zero. Therefore the mass of the gun must be very much larger than the mass of the shell to ensure that the gun's speed of recoil is very much smaller than the speed of the shell. The principle may also be used, for example, in investigating the collision of two billiard balls.

conservative force A force or field of force is **conservative** if the *work done* by the force as the point of application moves around any closed path is zero. It follows that for a conservative force the work done as the point of application moves from one point to another does not depend on the path taken. For example, the *uniform gravitational force*, the tension in a spring satisfying Hooke's law and the gravitational force given by the *inverse square law of gravitation* are conservative forces. Only for a conservative force can *potential energy* be defined.

conservative strategy In the *matrix game* given by the matrix $[a_{ij}]$, suppose that the players R and C use *pure strategies*. Let m_i be equal to the minimum entry in the i-th row. A **maximin** strategy for R is to choose the r-th row, where $m_r = \max\{m_i\}$. In doing so, R ensures that the smallest pay-off possible is as large as can be. Similarly, let M_j be equal to the maximum entry in the j-th column. A **minimax** strategy for C is to choose the s-th column, where $M_s = \min\{M_j\}$. These are called **conservative** strategies for the two players.

Now let $E(\mathbf{x}, \mathbf{y})$ be the *expectation* when R and C use *mixed strategies* $\mathbf{x}$ and $\mathbf{y}$. Then, for any $\mathbf{x}$, $\min_{\mathbf{y}} E(\mathbf{x}, \mathbf{y})$ is the smallest expectation possible, for all mixed strategies $\mathbf{y}$ that C may use. A **maximin** strategy for R is a strategy $\mathbf{x}$ that maximizes $\min_{\mathbf{y}} E(\mathbf{x}, \mathbf{y})$. Similarly, a **minimax** strategy for C is a strategy $\mathbf{y}$ that minimizes $\max_{\mathbf{x}} E(\mathbf{x}, \mathbf{y})$. By the *Fundamental Theorem of Game Theory*, when R and C use such strategies the expectation takes a certain value, the value of the game.

consistent A set of equations is **consistent** if there is a solution.

consistent estimator See *estimator*.

constant acceleration See *equations of motion with constant acceleration*.

constant function In real analysis, a **constant function** is a *real function* f such that $f(x) = a$ for all x in $\mathbf{R}$, where a, the **value** of f, is a fixed real number.

constant of integration If ϕ is a particular *antiderivative* of a *continuous function* f, then any antiderivative of f differs from ϕ by a constant. It is common practice, therefore, to write

$$\int f(x)\,dx = \phi(x) + c,$$

where c, an arbitrary constant, is the **constant of integration**.

constant of proportionality See *proportion*.

constant speed A particle is moving with **constant speed** if the magnitude of the velocity is independent of time. Thus $\mathbf{v}\,.\,\mathbf{v} =$ constant and this gives $2\mathbf{v}\,.\,(d\mathbf{v}/dt) = 0$. Hence there are three possibilities: $\mathbf{v} = \mathbf{0}$, or $d\mathbf{v}/dt = \mathbf{0}$, or $\mathbf{v}$ is perpendicular to $d\mathbf{v}/dt$. The third possibility shows that the velocity is not necessarily constant. One such example is when a particle is moving in a circle with constant speed, and then the acceleration is perpendicular to the velocity.

constant term See *polynomial*.

construction with ruler and compasses See *duplication of the cube*, *squaring the circle* and *trisection of an angle*.

contact force When two bodies are in contact, each exerts a **contact force** on the other, the two contact forces being equal and opposite. The contact force is the sum of the **frictional force**, which is tangential to the surfaces at the point of contact, and the **normal reaction**, which is normal to the surfaces at the point of contact. If the frictional force is zero, the contact is said to be **smooth** and the surfaces of the two bodies are called smooth. Otherwise, the contact is **rough** and the surfaces are called rough. If the two bodies are moving, or tending to move, relative to each other, the frictional forces will act to oppose the motion. See also *friction*.

contained in, contains It is tempting to say that 'x is contained in S' when $x \in S$, and also to say that 'A is contained in B' if $A \subseteq B$. To distinguish between these two different notions, it is better to say that 'x belongs to S' and to say that 'A is included in B' or 'A is a subset of B'. However, some authors consistently say 'is contained in' for $\subseteq$. Given the same examples, it is similarly tempting to say that 'S contains x' and also that 'B contains A.' It is again desirable to distinguish between the two by saying that 'B includes A' in the second case, though some authors consistently say 'contains' in this situation. The first case is best avoided or else clarified by saying that 'S contains the element x' or 'S contains x as an element.'

contingency table A method of presenting the frequencies of the outcomes from an experiment in which the observations in the sample are categorized according to two criteria. Each cell of the table gives the number of occurrences for a particular combination of categories. A contingency table in which individuals are categorized by sex and hair colour is shown below. A final column giving the row sums and a final row giving the column sums may be added. Then the sum of the final column and the sum of the final row both equal the number in the sample.

	hair colour			
sex	black	blonde	brown	red
male	30	6	22	6
female	24	9	18	5

If the sample is categorized according to three or more criteria, the information can be presented similarly in a number of such tables.

continued fraction An expression of the form $q_1 + 1/b_2$, where $b_2 = q_2 + 1/b_3, b_3 = q_3 + 1/b_4$, and so on, where $q_1, q_2, \ldots$ are integers, usually positive. This can be written

$$q_1 + \cfrac{1}{q_2 + \cfrac{1}{q_3 + \cfrac{1}{q_4 + \cdots}}}$$

or, in a form that is easier to print,

$$q_1 + \frac{1}{q_2+}\,\frac{1}{q_3+}\,\frac{1}{q_4 + \cdots}.$$

If the continued fraction terminates, it gives a rational number. The expression of any given positive rational number as a continued fraction can be found by using the *Euclidean algorithm*. For example, 1274/871 is found, by using the steps which appear in the entry on the Euclidean algorithm, to equal

$$1 + \frac{1}{2+}\,\frac{1}{6+}\,\frac{1}{5}.$$

When the continued fraction continues indefinitely, it represents a real number that is the limit of the sequence

$$q_1,\quad q_1 + \frac{1}{q_2},\quad q_1 + \cfrac{1}{q_2 + \cfrac{1}{q_3}},\quad q_1 + \cfrac{1}{q_2 + \cfrac{1}{q_3 + \cfrac{1}{q_4}}},\ \ldots.$$

For example, it can be shown that

$$1 + \frac{1}{1+}\,\frac{1}{1+}\,\frac{1}{1 + \cdots}$$

is equal to the *golden ratio*, and that the representation of $\sqrt{2}$ as a continued fraction is

$$1 + \frac{1}{2+}\,\frac{1}{2+}\,\frac{1}{2 + \cdots}.$$

continuous data See *data.*

continuous function The *real function* f of one variable is **continuous** at a if $f(x) \to f(a)$ as $x \to a$ (see *limit* (of $f(x)$)). The rough idea is that, close to a, the function has values close to $f(a)$. It means that the function does not suddenly jump at $x = a$ or take widely differing values arbitrarily close to a.

A function f is **continuous in an open interval** if it is continuous at each point of the interval; and f is **continuous on the closed interval** $[a, b]$, where $a < b$, if it is continuous in the open interval (a, b) and if $\lim_{x \to a+} f(x) = f(a)$ and $\lim_{x \to b-} f(x) = f(b)$. The following properties hold:

(i) The sum of two continuous functions is continuous.
(ii) The product of two continuous functions is continuous.
(iii) The quotient of two continuous functions is continuous at any point or in any interval where the denominator is not zero.
(iv) Suppose that f is continuous at a, that $f(a) = b$ and that g is continuous at b. Then h, defined by $h(x) = (g \circ f)(x) = g(f(x))$, is continuous at a.
(v) It can be proved from first principles that the constant functions, and the function f, defined by $f(x) = x$ for all x, are continuous (at any point or in any interval). By using (i), (ii) and (iii), it follows that any *polynomial function* is continuous and that any *rational function* is continuous at any point or in any interval where the denominator is not zero.

The following properties of continuous functions appear to be obvious if a continuous function is thought of as one whose graph is a continuous curve; but rigorous proofs are not elementary, relying as they do on rather deep properties of the real numbers:

(vi) If f is continuous on a closed interval $[a, b]$ and η is any real number between $f(a)$ and $f(b)$, then, for some c in (a, b), $f(c) = \eta$. This is the *Intermediate Value Theorem* or property.
(vii) If f is continuous on a closed interval $[a, b]$, then f is *bounded* on $[a, b]$. Furthermore, if S is the set of values $f(x)$ for x in $[a, b]$ and $M = \sup S$, then there is a ξ in $[a, b]$ such that $f(\xi) = M$ (and similarly for $m = \inf S$). It is said that 'a continuous function on a closed interval attains its bounds'.

continuous random variable See *random variable.*

contrapositive The **contrapositive** of an *implication* $p \Rightarrow q$ is the implication $\neg q \Rightarrow \neg p$. An implication and its contrapositive are *logically equivalent*, so that one is true if and only if the other is. So, in giving a proof of a mathematical result, it may on occasion be more convenient to establish the contrapositive rather than the original form of the theorem. For example, the theorem that if n^2 is odd then n is odd could be proved by showing instead that if n is even then n^2 is even.

converse The **converse** of an *implication* $p \Rightarrow q$ is the implication $q \Rightarrow p$. If an implication is true, then its converse may or may not be true.

convex A plane or solid figure, such as a polygon or polyhedron, is **convex** if the *line segment* joining any two points inside it lies wholly inside it.

convex up and **down** Some authors say that a curve is **convex up** when it is *concave down*, and **convex down** when it is *concave up* (see *concavity*).

coordinates (on a line) One way of assigning coordinates to points on a line is as follows. Make the line into a *directed line* by choosing one direction as the positive direction, running say from x' to x. Take a point O on the line as **origin** and a point A on the line such that OA is equal to the unit length. If P is any point on the line and $OP = x$, then x is the **coordinate** of P in this coordinate system. (Here OP denotes the *measure*.) The coordinate system on the line is determined by the specified direction of the line, the origin and the given unit of length.

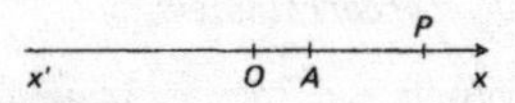

coordinates (in the plane) One way of assigning coordinates to points in the plane is as follows. Take a *directed line* Ox as x-axis and a directed line Oy as y-axis, where the point O is the **origin**, and specify the unit length. For any point P in the plane, let M and N be points on the x-axis and y-axis such that PM is parallel to the y-axis and PN is parallel to the x-axis. If $OM = x$ and $ON = y$, then (x, y) are the **coordinates** of the point P in this coordinate system. The coordinate system is determined by the two directed lines and the given unit length. When the directed lines intersect at a right angle, the system is a **Cartesian**, or **rectangular**, **coordinate system** and (x, y) are **Cartesian coordinates** of P. Normally, Ox and Oy are chosen so that an anticlockwise rotation of one right angle takes the positive x-direction to the positive y-direction.

There are other methods of assigning coordinates to points in the plane. One such is the method of *polar coordinates*.

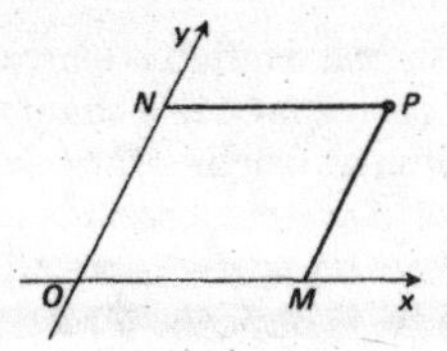

coordinates (in 3-dimensional space) One way of assigning coordinates to points in space is as follows. Take as **axes** three mutually perpendicular *directed lines* Ox, Oy and Oz, intersecting at the point O, the **origin**, and forming a *right-handed system*. Let L be the point where the plane through P, parallel to the plane containing the y-axis and the z-axis, meets

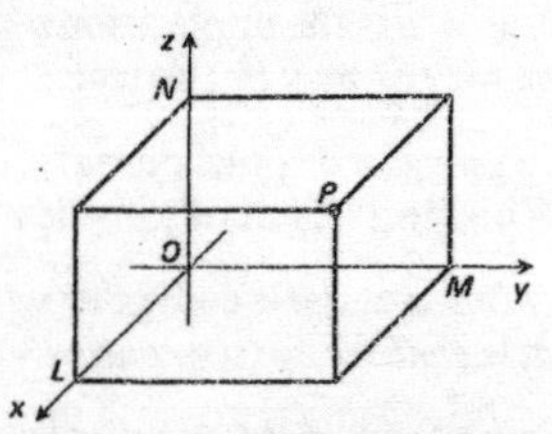

the x-axis. Alternatively, L is the point on the x-axis such that PL is perpendicular to the x-axis. Let M and N be similarly defined points on the y-axis and the z-axis. The points L, M and N are in fact three of the vertices of the *cuboid* with three of its edges along the coordinate axes and with O and P as opposite vertices. If $OL = x$, $OM = y$ and $ON = z$, then (x, y, z) are the **coordinates** of the point P in this **Cartesian coordinate system.**

There are other methods of assigning coordinates to points in space. One is similar to that described above but using oblique axes. Others are by *spherical polar coordinates* and *cylindrical polar coordinates.*

coplanar points and lines A number of points and lines are **coplanar** if there is a plane in which they all lie. Three points are always coplanar: indeed, any three points that are not collinear determine a unique plane that passes through them.

coplanar vectors Let $\overrightarrow{OA}$ and $\overrightarrow{OB}$ be *directed line-segments* representing non-zero, non-parallel *vectors* $\mathbf{a}$ and $\mathbf{b}$. A vector $\mathbf{p}$ is **coplanar** with $\mathbf{a}$ and $\mathbf{b}$ if $\mathbf{p}$ can be represented by a directed line-segment $\overrightarrow{OP}$, where P lies in the plane determined by O, A and B. The vector $\mathbf{p}$ is coplanar with $\mathbf{a}$ and $\mathbf{b}$ if and only if there exist scalars λ and μ such that $\mathbf{p} = \lambda\mathbf{a} + \mu\mathbf{b}$.

coprime = *relatively prime.*

Coriolis force See *fictitious force.*

corollary A result that follows from a theorem almost immediately, often without further proof.

correlation Between two random variables, the **correlation** is a measure of the extent to which a change in one tends to correspond to a change in the other. The correlation is high or low depending on whether the relationship between the two is close or not. If the change in one corresponds to a change in the other in the same direction, there is **positive correlation**, and there is a **negative correlation** if the changes are in opposite directions. *Independent random variables* have zero correlation. One measure of correlation between the random variables X and Y is the **correlation coefficient** ρ defined by

$$\rho = \frac{\mathrm{Cov}(X, Y)}{\sqrt{\mathrm{Var}(X)\,\mathrm{Var}(Y)}}$$

(see *covariance* and *variance*). This satisfies $-1 \leq \rho \leq 1$. If X and Y are linearly related, then $\rho = -1$ or $+1$.

For a sample of n paired observations $(x_1, y_1), (x_2, y_2), \ldots, (x_n, y_n)$, the **(sample) correlation coefficient** is equal to

$$\frac{\sum(x_i - \bar{x})(y_i - \bar{y})}{\sqrt{(\sum(x_i - \bar{x})^2)(\sum(y_i - \bar{y})^2)}}.$$

Note that the existence of some correlation between two variables need not imply that the link between the two is one of cause and effect.

correspondence See *one-to-one correspondence.*

corresponding angles See *transversal.*

cosecant See *trigonometric function.*

cosech, cosh See *hyperbolic function.*

cosine See *trigonometric function.*

cosine rule See *triangle.*

cotangent See *trigonometric function.*

coth See *hyperbolic function.*

countable A set X is **countable** if there is a *one-to-one correspondence* between X and a subset of the set of natural numbers. Thus a countable set is either finite or *denumerable.* Some authors use 'countable' to mean denumerable.

countably infinite = *denumerable.*

counterexample Let $p(x)$ be a mathematical sentence involving a symbol x, so that, when x is a particular element of some universal set, $p(x)$ is a statement that is either true or false. What may be of concern is the proving or disproving of the supposed theorem that $p(x)$ is true for all x in the universal set. The supposed theorem can be shown to be false by producing just one particular element of the universal set to serve as x that makes $p(x)$ false. The particular element produced is a **counterexample**. For example, let $p(x)$ say that $\cos x + \sin x = 1$, and consider the supposed theorem that $\cos x + \sin x = 1$ for all real numbers x. This is demonstrably false (though $p(x)$ may be true for some values of x) because $x = \pi/4$ is a counterexample: $\cos(\pi/4) + \sin(\pi/4) \neq 1$.

couple A system of forces whose sum is zero. The simplest example is a pair of equal and opposite forces acting at two different points, B and C. It can be shown that the *moment* of a couple about any point A is independent of the position of A.

covariance The **covariance** of two random variables X and Y, denoted by $\mathrm{Cov}(X, Y)$, is equal to $\mathrm{E}((X - \mu_X)(Y - \mu_Y))$, where μ_X and μ_Y are the population means of X and Y respectively (see *expected value*). If X and Y are *independent random variables*, then $\mathrm{Cov}(X, Y) = 0$. For computational purposes, note that $\mathrm{E}((X-\mu_X)(Y-\mu_Y)) = \mathrm{E}(XY) - \mu_X\mu_Y$. For a sample of n paired observations $(x_1, y_1), (x_2, y_2), \ldots, (x_n, y_n)$, the **sample covariance** is equal to

$$\frac{\sum(x_i - \bar{x})(y_i - \bar{y})}{n}.$$

Cramer, Gabriel (1704–1752) Swiss mathematician whose introduction to algebraic curves, published in 1750, contains the so-called *Cramer's rule*. The rule was known earlier by Maclaurin.

Cramer's rule Consider a set of n linear equations in n unknowns $x_1, x_2, \ldots, x_n$, written in matrix form as $\mathbf{Ax} = \mathbf{b}$. When $\mathbf{A}$ is *invertible*, the set of equations has a unique solution $\mathbf{x} = \mathbf{A}^{-1}\mathbf{b}$. Since $\mathbf{A}^{-1} = (1/\det \mathbf{A}) \operatorname{adj} \mathbf{A}$, this gives the solution

$$\mathbf{x} = \frac{(\operatorname{adj} \mathbf{A})\mathbf{b}}{\det \mathbf{A}},$$

which may be written

$$x_j = \frac{b_1 A_{1j} + b_2 A_{2j} + \cdots + b_n A_{nj}}{\det \mathbf{A}} \qquad (j = 1, \ldots, n);$$

using the entries of $\mathbf{b}$ and the *cofactors* of $\mathbf{A}$. This is **Cramer's rule**. Note that here the numerator is equal to the determinant of the matrix obtained by replacing the j-th column of $\mathbf{A}$ by the column $\mathbf{b}$. For example, this gives the solution of

$$ax + by = h,$$
$$cx + dy = k,$$

when $ad - bc \neq 0$, as

$$x = \begin{vmatrix} h & b \\ k & d \end{vmatrix} \bigg/ \begin{vmatrix} a & b \\ c & d \end{vmatrix} = \frac{hd - bk}{ad - bc},$$

$$y = \begin{vmatrix} a & h \\ c & k \end{vmatrix} \bigg/ \begin{vmatrix} a & b \\ c & d \end{vmatrix} = \frac{ak - hc}{ad - bc}.$$

critical damping See *damped oscillations*.

critical path analysis Suppose that the vertices of a *network* represent steps in a process, and the weights on the arcs represent the times that must elapse between steps. **Critical path analysis** is a method of determining the longest path in the network, and hence of finding the least time in which the whole process can be completed.

critical point = *stationary point*.

critical region See *hypothesis testing*.

critical value = *stationary value.*

cross-product = *vector product.*

cube A solid figure bounded by six square faces. It has eight vertices and twelve edges.

cube root See *n-th root.*

cube root of unity A complex number z such that $z^3 = 1$. The three cube roots of unity are 1, ω and ω^2, where

$$\omega = e^{2\pi i/3} = \cos\frac{2\pi}{3} + i\sin\frac{2\pi}{3} = -\frac{1}{2} + \frac{\sqrt{3}}{2}i,$$

$$\omega^2 = e^{4\pi i/3} = \cos\frac{4\pi}{3} + i\sin\frac{4\pi}{3} = -\frac{1}{2} - \frac{\sqrt{3}}{2}i.$$

Properties: (i) $\omega^2 = \bar{\omega}$ (see *conjugate*), (ii) $1 + \omega + \omega^2 = 0$.

cubic equation A polynomial equation of degree three.

cubic polynomial A polynomial of degree three.

cuboctahedron One of the *Archimedean solids,* with 6 square faces and 8 triangular faces. It can be formed by cutting off the corners of a cube to obtain a polyhedron whose vertices lie at the midpoints of the edges of the original cube. It can also be formed by cutting off the corners of an *octahedron* to obtain a polyhedron whose vertices lie at the midpoints of the edges of the original octahedron.

cuboid A *parallelepiped* all of whose faces are rectangles.

cumulative distribution function For a random variable X, the **cumulative distribution function** (or **c.d.f.**) is the function F defined by $F(x) = \Pr(X \leq x)$. Thus, for a discrete random variable,

$$F(x) = \sum_{x_i \leq x} p(x_i),$$

where p is the *probability mass function,* and, for a continuous random variable,

$$F(x) = \int_{-\infty}^{x} f(t)\,dt,$$

where f is the *probability density function.*

cumulative frequency The sum of the frequencies of all the values up to a given value. If the values $x_1, x_2, \ldots, x_n$, in ascending order, occur with frequencies $f_1, f_2, \ldots, f_n$, respectively, then the cumulative frequency at x_i is equal to $f_1 + f_2 + \cdots + f_i$. Cumulative frequencies may be similarly obtained for *grouped data.*

cumulative frequency distribution For discrete data, the information consisting of the possible values and the corresponding *cumulative frequencies* is called the **cumulative frequency distribution**. For *grouped data*, it gives the information consisting of the groups and the corresponding cumulative frequencies. It may be presented in a table or in a diagram similar to a histogram.

cup The operation $\cup$ (see *union*) is read by some as '**cup**', this being derived from the shape of the symbol, and in contrast to *cap* ($\cap$).

curve sketching When a graph $y = f(x)$ is to be sketched, what is generally required is a sketch showing the general shape of the curve and the behaviour at points of special interest. The different parts of the graph do not have to be to scale. It is normal to investigate the following: *symmetry*, *stationary points*, intervals in which the function is always increasing or always decreasing, *asymptotes* (vertical, horizontal and slant), *concavity*, *points of inflexion*, points of intersection with the axes, and the gradient at points of interest.

cycle (in graph theory) In a *graph*, a **cycle** is a sequence $v_0, e_1, v_1, \ldots, e_k, v_k$ $(k \geq 1)$ of alternately vertices and edges (where e_i is an edge joining v_{i-1} and v_i), with all the edges different and all the vertices different, except that $v_0 = v_k$. See *Hamiltonian graph* and *tree*.

cycle (in mechanics) See *period*, *periodic*.

cycle, cyclic arrangement The arrangement of a number of objects that may be considered to be positioned in a circle may be called a **cycle** or **cyclic arrangement**. The arrangement of *a*, *b*, *c* and *d* in that order round a circle is considered the same as *b*, *c*, *d* and *a* in that order round a circle. The cycle or cyclic arrangement of *a*, *b*, *c*, *d* may be written $[a, b, c, d]$, where it is understood that $[a, b, c, d] = [b, c, d, a]$, for example, but that $[a, d, c, b]$ is different.

cyclic group Let *a* be an element of a *multiplicative group* *G*. The elements a^r, where *r* is an integer (positive, zero or negative), form a *subgroup* of *G*, called the subgroup generated by *a*. A group *G* is **cyclic** if there is an element *a* in *G* such that the subgroup generated by *a* is the whole of *G*. If *G* is a finite cyclic group with identity element *e*, the set of elements of *G* may be written $\{e, a, a^2, \ldots, a^{n-1}\}$, where $a^n = e$ and *n* is the smallest such positive integer. If *G* is an infinite cyclic group, the set of elements may be written $\{\ldots, a^{-2}, a^{-1}, e, a, a^2, \ldots\}$.

By making appropriate changes, a cyclic *additive group* (or group with any other operation) can be defined. For example, the set $\{0, 1, 2, \ldots, n-1\}$ with addition modulo *n* is a cyclic group, and the set of all integers with addition is an infinite cyclic group. Any two cyclic groups of the same order are *isomorphic*.

cyclic polygon A polygon whose vertices lie on a circle. From one of the *circle theorems*, it follows that opposite angles of a cyclic quadrilateral add up to 180°.

cycloid The curve traced out by a point on the circumference of a circle that rolls without slipping along a straight line. With suitable axes, the cycloid has *parametric equations* $x = a(t - \sin t), y = a(1 - \cos t)$ $(t \in \mathbf{R})$, where a is a constant (equal to the radius of the rolling circle). In the figure, $OA = 2\pi a$.

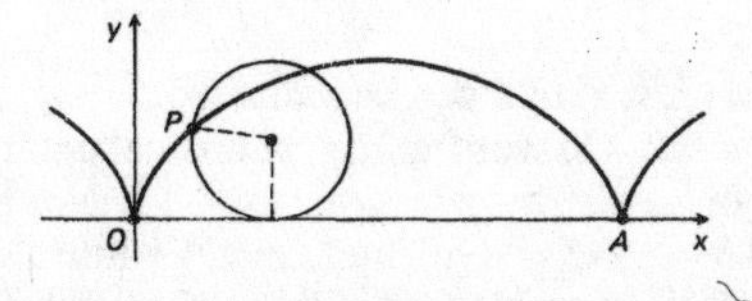

cylinder In elementary work, a **cylinder**, if taken, say, with its axis vertical, would be reckoned to consist of a circular base, a circular top of the same size, and the curved surface formed by the vertical line segments joining them. For a cylinder with base of radius r and height h, the volume equals $\pi r^2 h$, and the area of the curved surface equals $2\pi rh$.

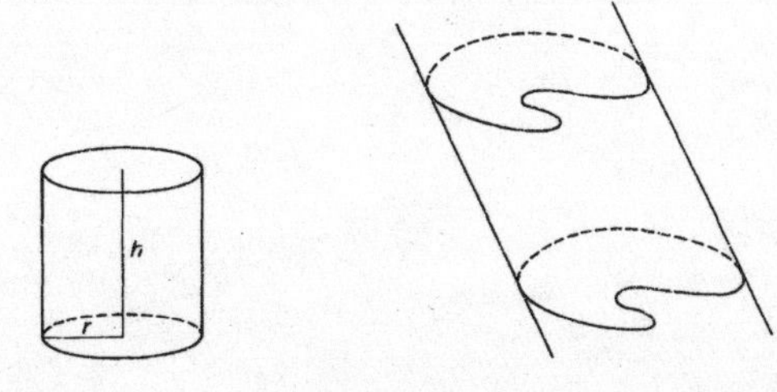

In more advanced work, a **cylinder** is a surface, consisting of the points of the lines, called **generators**, drawn through the points of a fixed curve and parallel to a fixed line, the generators being extended indefinitely in both directions. Then a **right-circular cylinder** is one in which the fixed curve is a circle and the fixed line is perpendicular to the plane of the circle. The **axis** of a right-circular cylinder is the line through the centre of the circle and perpendicular to the plane of the circle, that is, parallel to the generators.

cylindrical polar coordinates Suppose that three mutually perpendicular directed lines Ox, Oy and Oz, intersecting at the point O and forming a right-handed system, are taken as coordinate axes. For any point P, let M and N be the projections of P onto the xy-plane and the z-axis respectively. Then $ON = PM = z$, the z-coordinate of P. Let $\rho = |PN|$, the distance of P from the z-axis, and let ϕ be the angle $\angle xOM$ in radians $(0 \leq \phi < 2\pi)$. Then (ρ, ϕ, z) are the **cylindrical polar coordinates** of P. (It should be noted that the points of the z-axis give no value for ϕ.) The two coordinates (ρ, ϕ) can be seen as polar coordinates of the point M and, as with polar coordinates, $\phi + 2k\pi$, where k is an integer, may be allowed in place of ϕ.

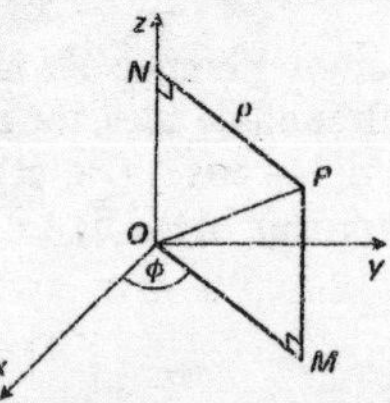

The Cartesian coordinates (x, y, z) of P can be found from (ρ, ϕ, z) by: $x = \rho \cos \phi$, $y = \rho \sin \phi$, and $z = z$. Conversely, the cylindrical polar coordinates can be found from (x, y, z) by: $\rho = \sqrt{x^2 + y^2}$, ϕ is such that $\cos \phi = x/\sqrt{x^2 + y^2}$ and $\sin \phi = y/\sqrt{x^2 + y^2}$, and $z = z$. Cylindrical polar coordinates can be useful in treating problems involving right-circular cylinders. Such a cylinder with its axis along the z-axis then has equation $\rho =$ constant.

D

damped oscillations *Oscillations* in which the amplitude decreases with time. Consider the equation of motion $m\ddot{x} = -kx - c\dot{x}$, where the first term on the right-hand side arises from an elastic restoring force satisfying *Hooke's law*, and the second term arises from a *resistive force*. The constants k and c are positive. The form of the general solution of this linear differential equation depends on the *auxiliary equation* $m\alpha^2 + c\alpha + k = 0$. When $c^2 < 4mk$, the auxiliary equation has non-real roots and damped oscillations occur. This is a case of **weak damping**. When $c^2 = 4mk$, the auxiliary equation has equal roots and **critical damping** occurs: oscillation just fails to take place. When $c^2 > 4mk$, there is **strong damping**: the resistive force is so strong that no oscillations take place.

dashpot A device consisting of a cylinder containing a liquid through which a piston moves, used for damping vibrations.

data The observations gathered from an experiment, survey or observational study. Often the data are a randomly selected sample from an underlying *population*. Numerical data are **discrete** if the underlying population is finite or countably infinite and are **continuous** if the underlying population forms an interval, finite or infinite. Data are **nominal** if the observations are not numerical or quantitative, but are descriptive and have no natural order. Data specifying country of origin, type of vehicle or subject studied, for example, are nominal.

Note that the word 'data' is plural. The singular 'datum' may be used for a single observation.

deca- = *deka-*.

decagon A ten-sided polygon.

deceleration A slowing down. Suppose that a particle is moving in a straight line. If the particle is moving in the positive direction, then it experiences a deceleration when the acceleration is negative. Notice, however, that this is not true if the particle is moving in the negative direction (see *acceleration*).

deci- Prefix used with *SI units* to denote multiplication by 10^{-1}.

decile See *quantile*.

decimal fraction A *fraction* expressed by using *decimal representation*, as opposed to a *vulgar fraction*. For example, $\frac{3}{4}$ is a vulgar fraction; 0.75 is a decimal fraction.

decimal places In *rounding* or *truncation* of a number to n **decimal places**, the original is replaced by a number with just n digits after the decimal point. When the rounding or truncation takes place to the left of the decimal point, a phrase such as 'to the nearest 10' or 'to the nearest 1000' has to be used. To say that $a = 1.9$ to 1 decimal place means that the exact value of a becomes 1.9 after rounding to 1 decimal place, and so $1.85 \leq a \leq 1.95$.

decimal representation Any real number a between 0 and 1 has a **decimal representation**, written $.d_1d_2d_3\ldots$, where each d_i is one of the digits $0, 1, 2, \ldots, 9$; this means that

$$a = d_1 \times 10^{-1} + d_2 \times 10^{-2} + d_3 \times 10^{-3} + \cdots.$$

This notation can be extended to enable any positive real number to be written as

$$c_nc_{n-1}\ldots c_1c_0.d_1d_2d_3\ldots$$

using, for the integer part, the normal representation $c_nc_{n-1}\ldots c_1c_0$ to base 10 (see *base*). If, from some stage on, the representation consists of the repetition of a string of one or more digits, it is called a **recurring** or **repeating decimal**. For example, the recurring decimal .12748748748... can be written $.12\dot{7}4\dot{8}$, where the dots above indicate the beginning and end of the repeating string. The repeating string may consist of just one digit, and then, for example, .16666... is written $.1\dot{6}$. If the repeating string consists of a single zero, this is generally omitted and the representation may be called a **terminating decimal**.

The decimal representation of any real number is unique except that, if a number can be expressed as a terminating decimal, it can also be expressed as a decimal with a recurring 9. Thus .25 and $.24\dot{9}$ are representations of the same number. The numbers that can be expressed as recurring (including terminating) decimals are precisely the *rational numbers*.

decreasing function A *real function* f is **decreasing** in or on an interval I if $f(x_1) \geq f(x_2)$ whenever x_1 and x_2 are in I with $x_1 < x_2$. Also, f is **strictly decreasing** if $f(x_1) > f(x_2)$ whenever $x_1 < x_2$.

decreasing sequence A sequence $a_1, a_2, a_3, \ldots$ is said to be **decreasing** if $a_i \geq a_{i+1}$ for all i, and **strictly decreasing** if $a_i > a_{i+1}$ for all i.

Dedekind, (Julius Wilhelm) Richard (1831–1916) German mathematician who developed a formal construction of the real numbers from the rational numbers by means of the so-called Dedekind cut. This new approach to *irrational numbers*, contained in the very readable paper *Continuity and Irrational Numbers*, was an important step towards the formalization of mathematics. He also proposed a definition of infinite sets that was taken up by Cantor, with whom he developed a lasting friendship.

definite integral See *integral*.

degenerate conic A *conic* that consists of a pair of (possibly coincident) straight lines. The equation $ax^2 + 2hxy + by^2 + 2gx + 2fy + c = 0$ represents a degenerate conic if $\Delta = 0$, where

$$\Delta = \begin{vmatrix} a & h & g \\ h & b & f \\ g & f & c \end{vmatrix}.$$

degenerate quadric The *quadric* with equation $ax^2 + by^2 + cz^2 + 2fyz + 2gzx + 2hxy + 2ux + 2vy + 2wz + d = 0$ is **degenerate** if $\Delta = 0$, where

$$\Delta = \begin{vmatrix} a & h & g & u \\ h & b & f & v \\ g & f & c & w \\ u & v & w & d \end{vmatrix}.$$

The non-degenerate quadrics are the *ellipsoid*, the *hyperboloid of one sheet*, the *hyperboloid of two sheets*, the *elliptic paraboloid* and the *hyperbolic paraboloid*.

degree (angular measure) The method of measuring angles in **degrees** dates back to Babylonian mathematics around 2000 BC. A complete revolution is divided into 360 degrees (°); a right angle measures 90°. Each degree is divided into 60 **minutes** (′) and each minute into 60 **seconds** (″). In more advanced work, angles should be measured in *radians*.

degree (of a polynomial) See *polynomial*.

degree (of a vertex of a graph) The **degree** of a vertex v of a *graph* is the number of edges ending at v. (If *loops* are allowed, each loop joining v to itself contributes two to the degree of v.)

In the graph on the left, the vertices u, v, w and x have degrees 2, 2, 3 and 1. The graph on the right has vertices v_1, v_2, v_3 and v_4 with degrees 5, 4, 6 and 5.

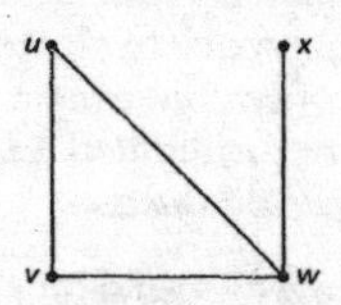

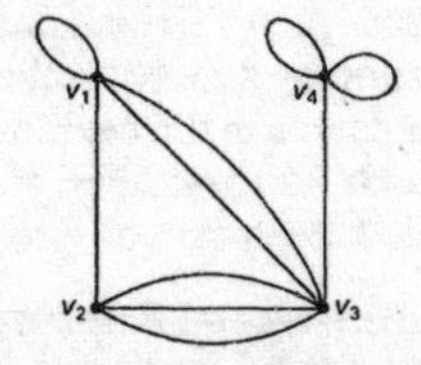

degrees of freedom (in mechanics) The number of **degrees of freedom** of a body is the minimum number of independent coordinates required to describe the position of the body at any instant, relative to a frame of reference. A particle in straight-line motion or circular motion has one degree of freedom. So too does a rigid body rotating about a fixed axis. A particle moving in a plane, such as a projectile, or a particle moving on a cylindrical or spherical surface has two degrees of freedom. A rigid body in general motion has six degrees of freedom.

degrees of freedom (in statistics) A positive integer normally equal to the number of independent observations in a sample minus the number of population parameters to be estimated from the sample. When the *chi-squared test* is applied to a *contingency table* with h rows and k columns, the number of degrees of freedom equals $(h-1)(k-1)$.

deka- Prefix used with *SI units* to denote multiplication by 10.

Delian (altar) problem Another name for the problem of the *duplication of the cube*. In 428 BC, an oracle at Delos ordered that the altar of Apollo should be doubled in volume as a means of bringing a certain plague to an end.

De Moivre, Abraham (1667–1754) Prolific mathematician, born in France, who later settled in England. In *De Moivre's Theorem*, he is remembered for his use of complex numbers in trigonometry. But he was also the author of two notable early works on probability. His *Doctrine of Chances* of 1718, examines numerous problems and develops a number of principles, such as the notion of independent events and the product law. Later work contains the result known as *Stirling's formula* and probably the first use of the normal frequency curve.

De Moivre's Theorem From the definition of *multiplication* (of a complex number), it follows that $(\cos\theta_1 + i\sin\theta_1)(\cos\theta_2 + i\sin\theta_2) = \cos(\theta_1+\theta_2) + i\sin(\theta_1+\theta_2)$. This leads to the following result known as **De Moivre's Theorem**, which is crucial to any consideration of the powers z^n of a complex number z:

THEOREM: For all positive integers n, $(\cos\theta + i\sin\theta)^n = \cos n\theta + i\sin n\theta$.

The result is also true for negative (and zero) integer values of n, and this may be considered as either included in or forming an extension of De Moivre's Theorem.

De Morgan, Augustus (1806–1871) British mathematician and logician who was responsible for developing a more symbolic approach to algebra, and who played a considerable role in the beginnings of symbolic logic. His name is remembered in *De Morgan's laws*, which he formulated. In an article of 1838, he clarified the notion of *mathematical induction*.

De Morgan's laws For all sets A and B (subsets of a *universal set*), $(A\cup B)' = A'\cap B'$ and $(A\cap B)' = A'\cup B'$. These are **De Morgan's laws**.

denominator See *fraction*.

density The average **density** of a body is the ratio of its mass to its volume. In general, the density of a body may not be constant throughout the body. The density at a point P, denoted by $\rho(P)$, is equal to the limit as $\Delta V \to 0$ of $\Delta m/\Delta V$, where ΔV is the volume of a small region containing P and Δm is the mass of the part of the body occupying that small region.

Consider a rod of length l, with density $\rho(x)$ at the point a distance x from one end of the rod. Then the mass m of the rod is given by

$$m = \int_0^l \rho(x)\, dx.$$

In the same way, the mass of a lamina or a 3-dimensional rigid body, with density $\rho(\mathbf{r})$ at the point with position vector $\mathbf{r}$, is given by

$$m = \int_V \rho(\mathbf{r})\, dV,$$

with, as appropriate, a double or triple integral over the region V occupied by the body.

denumerable A set X is **denumerable** if there is a *one-to-one correspondence* between X and the set of natural numbers. It can be shown that the set of rational numbers is denumerable but that the set of real numbers is not. Some authors use 'denumerable' to mean *countable.*

dependent events See *independent events.*

dependent variable In statistics, the variable which is thought might be influenced by certain other *explanatory variables.* In *regression,* a relationship is sought between the dependent variable and the explanatory variables. The purpose is normally to enable the value of the dependent variable to be predicted from given values of the explanatory variables.

derivative For the *real function* f, if $(f(a+h)-f(a))/h$ has a *limit* as $h \to 0$, this limit is the *derivative* of f at a and is denoted by $f'(a)$. (The term 'derivative' may also be used loosely for the *derived function.*)

Consider the graph $y = f(x)$. If (x, y) are the coordinates of a general point P on the graph, and $(x+\Delta x, y+\Delta y)$ are those of a nearby point Q on the graph, it can be said that a change Δx in x produces a change Δy in y. The quotient $\Delta y/\Delta x$ is the gradient of the chord PQ. Also, $\Delta y = f(x+\Delta x) - f(x)$. So the derivative of f at x is the limit of the quotient $\Delta y/\Delta x$ as $\Delta x \to 0$. This limit can be denoted by dy/dx, which is thus an alternative notation for $f'(x)$. The notation y' is also used.

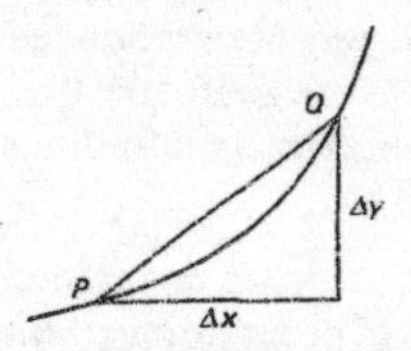

The derivative $f'(x)$ may be denoted by $(d/dx)(\quad)$, where the brackets contain a formula for $f(x)$. Some authors use the notation df/dx. The derivative $f'(a)$ gives the gradient of the curve $y = f(x)$, and hence the gradient of the tangent to the curve, at the point given by $x = a$. Suppose now, with a different notation, that x is a function of t, where t is some measurement of time. Then the derivative dx/dt, which is the *rate of change*

of x with respect to t, may be denoted by $\dot{x}$. The derivatives of certain common functions are given in the Table of Derivatives (Appendix 2). See also *differentiation*, *left and right derivative*, *higher derivative* and *partial derivative*.

derived function The function f', where $f'(x)$ is the *derivative* of f at x, is the **derived function** of f. See also *differentiation*.

derived unit See *SI units*.

Desargues, Girard (1591–1661) French mathematician and engineer whose work on conics and the result known as *Desargues's Theorem* were to become the basis of the subject known as projective geometry. His 1639 book was largely ignored, partly because of the obscurity of the language. It was nearly 200 years later that projective geometry developed and the beauty and importance of his ideas were recognized.

Desargues's Theorem The following theorem of projective geometry:

THEOREM: Consider two triangles lying in the plane or positioned in 3-dimensional space. If the lines joining corresponding vertices are concurrent, then corresponding sides intersect in points that are collinear, and conversely.

In detail, suppose that one triangle has vertices A, B and C, and the other has vertices A', B' and C'. The theorem states that if AA', BB' and CC' are concurrent, then BC and $B'C'$ intersect at a point L, CA and $C'A'$ intersect at M, and AB and $A'B'$ intersect at N, where L, M and N are collinear.

Perhaps surprisingly, the case of three dimensions is the easier to prove. The theorem can be taken as a theorem of Euclidean geometry if suitable amendments are made to cover the possibility that points of intersection do not exist because lines are parallel. It played an important role in the emergence of projective geometry.

Descartes, René (1596–1650) French philosopher and mathematician who in mathematics is known mainly for his methods of applying algebra to geometry, from which analytic geometry developed. He expounded these in *La Géométrie*, in which he was also concerned to use geometry to solve algebraic problems. Though named after him, Cartesian coordinates did not in fact feature in his work.

descriptive statistics The part of the subject of statistics concerned with describing the basic statistical features of a set of observations. Simple numerical summaries, using notions such as *mean*, *range* and *standard deviation*, together with appropriate diagrams such as *histograms*, are used to present an overall impression of the data.

determinant For the square matrix **A**, the **determinant** of **A**, denoted by det **A** or $|\mathbf{A}|$, can be defined as follows. Consider, in turn, 1×1, 2×2, 3×3, and $n \times n$ matrices.

The determinant of the 1×1 matrix $[a]$ is simply equal to a. If $\mathbf{A}$ is the 2×2 matrix below, then $\det \mathbf{A} = ad - bc$, and the determinant can also be written as shown:

$$\mathbf{A} = \begin{bmatrix} a & b \\ c & d \end{bmatrix}, \qquad \det \mathbf{A} = \begin{vmatrix} a & b \\ c & d \end{vmatrix}.$$

If $\mathbf{A}$ is a 3×3 matrix $[a_{ij}]$, then $\det \mathbf{A}$, which may be denoted by

$$\begin{vmatrix} a_{11} & a_{12} & a_{13} \\ a_{21} & a_{22} & a_{23} \\ a_{31} & a_{32} & a_{33} \end{vmatrix},$$

is given by

$$\det \mathbf{A} = a_{11} \begin{vmatrix} a_{22} & a_{23} \\ a_{32} & a_{33} \end{vmatrix} - a_{12} \begin{vmatrix} a_{21} & a_{23} \\ a_{31} & a_{33} \end{vmatrix} + a_{13} \begin{vmatrix} a_{21} & a_{22} \\ a_{31} & a_{32} \end{vmatrix}.$$

Notice how each 2×2 determinant occurring here is obtained by deleting the row and column containing the entry by which the 2×2 determinant is multiplied. This expression for the determinant of a 3×3 matrix can be written $a_{11}A_{11} + a_{12}A_{12} + a_{13}A_{13}$, where A_{ij} is the *cofactor* of a_{ij}. This is the evaluation of $\det \mathbf{A}$, 'by the first row'. In fact, $\det \mathbf{A}$ may be found by using evaluation by any row or column: for example, $a_{31}A_{31} + a_{32}A_{32} + a_{33}A_{33}$ is the evaluation by the third row, and $a_{12}A_{12} + a_{22}A_{22} + a_{32}A_{32}$ is the evaluation by the second column. The determinant of an $n \times n$ matrix $\mathbf{A}$ may be defined similarly, as $a_{11}A_{11} + a_{12}A_{12} + \cdots + a_{1n}A_{1n}$, and the same value is obtained using a similar evaluation by any row or column. The following properties hold:

(i) If two rows (or two columns) of a square matrix $\mathbf{A}$ are identical, then $\det \mathbf{A} = 0$.

(ii) If two rows (or two columns) of a square matrix $\mathbf{A}$ are interchanged, then only the sign of $\det \mathbf{A}$ is changed.

(iii) The value of $\det \mathbf{A}$ is unchanged if a multiple of one row is added to another row, or if a multiple of one column is added to another column.

(iv) If $\mathbf{A}$ and $\mathbf{B}$ are square matrices of the same order, then $\det(\mathbf{AB}) = (\det \mathbf{A})(\det \mathbf{B})$.

(v) If $\mathbf{A}$ is *invertible*, then $\det(\mathbf{A}^{-1}) = (\det \mathbf{A})^{-1}$.

(vi) If $\mathbf{A}$ is an $n \times n$ matrix, then $\det k\mathbf{A} = k^n \det \mathbf{A}$.

In particular cases, the determinant of a given matrix may be evaluated by using operations of the kind described in (iii) to produce a matrix whose determinant is easier to evaluate.

developable surface A surface that can be rolled out flat onto a plane without any distortion. Examples are the cone and cylinder. On any map of the world, distances are inevitably inaccurate because a sphere is not a developable surface.

diagonal entry For a square matrix $[a_{ij}]$, the **diagonal entries** are the entries $a_{11}, a_{22}, \ldots, a_{nn}$, which form the main diagonal.

diagonal matrix A square matrix in which all the entries not in the main diagonal are zero.

diameter A **diameter** of a circle or *central conic* is a line through the centre. If the line meets the circle or central conic at P and Q, then the line segment PQ may also be called a diameter. The term also applies in both senses to a sphere or *central quadric*.

In the case of a circle or sphere, all such line segments have the same length. This length is also called the diameter of the circle or sphere, and is equal to twice the radius.

die (plural: dice) A small cube with its faces numbered from 1 to 6. When the die is thrown, the probability that any particular number from 1 to 6 is obtained on the face landing uppermost is $\frac{1}{6}$.

difference equation Let $u_0, u_1, u_2, \ldots, u_n, \ldots$ be a sequence (where it is convenient to start with a term u_0). If the terms satisfy the **first-order difference equation** $u_{n+1} + au_n = 0$, it is easy to see that $u_n = A(-a)^n$, where $A\ (= u_0)$ is arbitrary.

Suppose that the terms satisfy the **second-order difference equation** $u_{n+2} + au_{n+1} + bu_n = 0$. Let α and β be the roots of the quadratic equation $x^2 + ax + b = 0$. It can be shown that (i) if $\alpha \neq \beta$, then $u_n = A\alpha^n + B\beta^n$, and (ii) if $\alpha = \beta$, then $u_n = (A + Bn)\alpha^n$, where A and B are arbitrary constants. The *Fibonacci sequence* is given by the difference equation $u_{n+2} = u_{n+1} + u_n$, with $u_0 = 1$ and $u_1 = 1$, and the above method gives

$$u_n = \frac{1}{2}\left(\frac{1+\sqrt{5}}{2}\right)^n + \frac{1}{2}\left(\frac{1-\sqrt{5}}{2}\right)^n.$$

Difference equations, also called **recurrence relations**, do not necessarily have constant coefficients like those considered above; they have similarities with differential equations. By using the notation $\Delta u_n = u_{n+1} - u_n$, difference equations can be written in terms of *finite differences*. Indeed, difference equations may arise from the consideration of finite differences.

difference of two squares Since $a^2 - b^2 = (a - b)(a + b)$, any expression with the form of the left-hand side, known as the **difference of two squares**, can be factorized into the form of the right-hand side.

difference quotient = *Newton quotient*.

difference set The **difference** $A \setminus B$ of sets A and B (subsets of a *universal set*) is the set consisting of all elements of A that are not elements of B. The notation $A - B$ is also used. The set is represented by the shaded region of the *Venn diagram* shown in the figure.

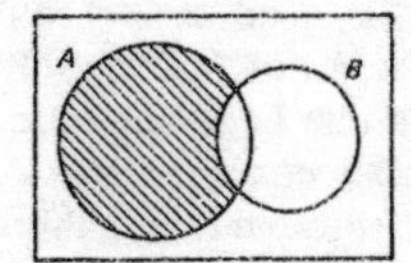

differentiable function The *real function* f of one variable is **differentiable** at a if $(f(a+h)-f(a))/h$ has a limit as $h \to 0$; that is, if the *derivative* of f at a exists. The rough idea is that a function is differentiable if it is possible to define the gradient of the graph $y=f(x)$ and hence define a tangent at the point. The function f is differentiable in an open interval if it is differentiable at every point in the interval; and f is differentiable on the closed interval $[a, b]$, where $a < b$, if it is differentiable in (a, b) and if the *right derivative* of f at a and the *left derivative* of f at b exist.

differential calculus The part of mathematics that develops from the definition of the *derivative* of a function or the gradient of a graph. The derivative is obtained as the limit of the *Newton quotient*, and this is equivalent to the notion of the gradient of a graph as the limit of the gradient of a chord of the graph. From another point of view, the subject is concerned essentially with the *rate of change* of one quantity with respect to another.

differential coefficient = *derivative.*

differential equation Suppose that y is a function of x and that y', $y'', \dots, y^{(n)}$ denote the *derivatives* $dy/dx, d^2y/dx^2, \dots, d^ny/dx^n$. An **ordinary differential equation** is an equation involving $x, y, y', y'', \dots$. (The term 'ordinary' is used here to make the distinction from **partial differential equations**, which involve *partial derivatives* and which will not be discussed here.) The **order** of the differential equation is the order n of the highest derivative $y^{(n)}$ that appears.

The problem of solving a differential equation is to find functions y whose derivatives satisfy the equation. In certain circumstances, it can be shown that a differential equation of order n has a **general solution** (that is, a function y, involving n arbitrary constants) that gives all the solutions. A solution given by some set of values of the arbitrary constants is a **particular solution**. Here are some examples of differential equations and their general solutions, where A, B and C are arbitrary constants:

(i) $y' - y = 3$ has general solution $y = Ae^x - 3$.
(ii) $y' = (2x+3y+2)/(4x+6y-3)$ has general solution $\ln|2x+3y| = 2y - x + C$.
(iii) $y'' + y = 0$ has general solution $y = A\cos x + B\sin x$.
(iv) $y'' - 2y' - 3y = e^{-x}$ has general solution $y = Ae^{-x} + Be^{3x} - \frac{1}{4}xe^{-x}$.

Example (ii) shows that it is not necessarily possible to express y explicitly as a function of x.

A differential equation of the first order (that is, of order one) can be expressed in the form $dy/dx = f(x, y)$. Whether or not it can be solved depends upon the function f. Among those that may be solvable are *separable*, *homogeneous* and *linear first-order differential equations*. Among higher-order differential equations that may be solvable reasonably easily are *linear differential equations with constant coefficients*.

differentiation The process of obtaining the *derived function* f' from the function f, where $f'(x)$ is the *derivative* of f at x. The derivatives of certain common functions are given in the Table of Derivatives (Appendix 2), and from these many other functions can be differentiated using the following rules of differentiation:

(i) If $h(x) = kf(x)$ for all x, where k is a constant, then $h'(x) = kf'(x)$.

(ii) If $h(x) = f(x) + g(x)$ for all x, then $h'(x) = f'(x) + g'(x)$.

(iii) The **product rule**: If $h(x) = f(x)g(x)$ for all x, then

$$h'(x) = f(x)g'(x) + f'(x)g(x).$$

(iv) The **reciprocal rule**: If $h(x) = 1/f(x)$ and $f(x) \neq 0$ for all x, then

$$h'(x) = -\frac{f'(x)}{(f(x))^2}.$$

(v) The **quotient rule**: If $h(x) = f(x)/g(x)$ and $g(x) \neq 0$ for all x, then

$$h'(x) = \frac{g(x)f'(x) - f(x)g'(x)}{(g(x))^2}.$$

(vi) The *chain rule*: If $h(x) = (f \circ g)(x) = f(g(x))$ for all x, then $h'(x) = f'(g(x))g'(x)$.

digit A symbol used in writing numbers in their *decimal representation* or to some other *base*. In decimal notation, the digits used are 0, 1, 2, 3, 4, 5, 6, 7, 8 and 9. In *hexadecimal* notation, the digits are 0, 1, 2, 3, 4, 5, 6, 7, 8, 9, A, B, C, D, E and F. In *binary* notation, there are just the two digits 0 and 1.

digraph A **digraph** (or **directed graph**) consists of a number of **vertices**, some of which are joined by **arcs**, where an arc, or **directed edge**, joins one vertex to another and has an arrow on it to indicate its direction. The arc from the vertex u to the vertex v may be denoted by the ordered pair (u, v). The

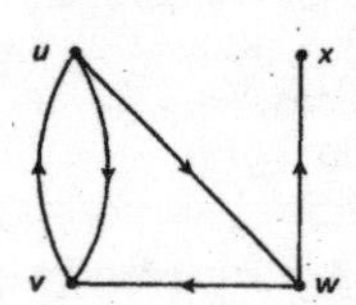

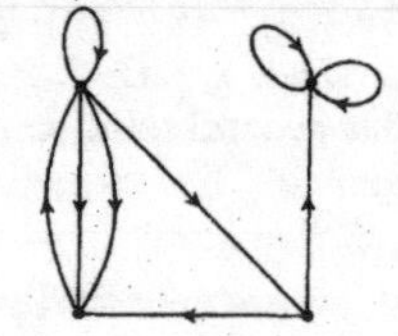

digraph with vertices u, v, w, x and arcs $(u, v), (u, w), (v, u), (w, v), (w, x)$ is shown in the figure, on the left.

As for *graphs*, and with a similar terminology, there may be multiple arcs and loops. A digraph with multiple arcs and loops is shown in the figure, on the right.

dilatation A **dilatation** of the plane from O with scale factor c ($\neq 0$) is the *transformation* of the plane in which the origin O is mapped to itself and a point P is mapped to the point P', where O, P and P' are collinear and $OP' = cOP$. This is given in terms of Cartesian coordinates by $x' = cx$, $y' = cy$.

dimensions In mechanics, physical quantities can be described in terms of the basic **dimensions** of mass M, length L and time T, using positive and negative indices. For example, the following have the dimensions given: area, L^2; velocity, LT^{-1}; force, MLT^{-2}; linear momentum, MLT^{-1}; energy, ML^2T^{-2}; and power, ML^2T^{-3}. The notation has similarities with that of SI units.

Diophantine equation An algebraic equation in one or more unknowns, with integer coefficients, for which integer solutions are required. A great variety of Diophantine equations have been studied. Some have infinitely many solutions, some have finitely many and some have no solutions. For example:

(i) $14x + 9y = 1$ has solutions $x = 2 + 9t$, $y = -3 - 14t$ (where t is any integer).

(ii) $x^2 + 1 = 2y^4$ has two solutions $x = 1, y = 1$ and $x = 239$, $y = 13$.

(iii) $x^3 + y^3 = z^3$ has no solutions.

See also *Hilbert's tenth problem* and *Pell's equation*.

Diophantus of Alexandria (about AD 250) Greek mathematician whose work displayed an algebraic approach to the solution of equations in one or more unknowns, unlike earlier Greek methods that were more geometrical. In the books of *Arithmetica* that survive, particular numerical examples of more than 100 problems are solved, probably to indicate the general methods of solution. These are mostly of the kind now referred to as *Diophantine equations*.

directed graph = *digraph*.

directed line A straight line with a specified direction along the line. The specified direction may be called the **positive direction** and the opposite the **negative direction**. It may be convenient to distinguish the ends of the line by labelling them x' and x, where the positive direction runs from x' to x. Alternatively, a directed line may be denoted by Ox, where O is a point on the line and the positive direction runs towards the end x.

directed line-segment If A and B are two points on a straight line, the part of the line between and including A and B, together with a specified direction along the line, is a **directed line-segment**. Thus $\overrightarrow{AB}$ is the directed line-segment from A to B, and $\overrightarrow{BA}$ is the directed line-segment from B to A. See also *vector*.

direction cosines In a Cartesian coordinate system in 3-dimensional space, a certain direction can be specified as follows. Take a point P such that $\overrightarrow{OP}$ has the given direction and $|OP| = 1$. Let α, β and γ be the three angles $\angle xOP$, $\angle yOP$ and $\angle zOP$, measured in radians ($0 \leq \alpha \leq \pi, 0 \leq \beta \leq \pi$, $0 \leq \gamma \leq \pi$). Then $\cos\alpha$, $\cos\beta$ and $\cos\gamma$ are the **direction cosines** of the given direction or of $\overrightarrow{OP}$. They are not independent, however, since $\cos^2\alpha + \cos^2\beta + \cos^2\gamma = 1$. Point P has coordinates $(\cos\alpha, \cos\beta, \cos\gamma)$ and, using the standard unit vectors **i**, **j** and **k** along the coordinate axes, the position vector **p** of P is given by $\mathbf{p} = (\cos\alpha)\mathbf{i} + (\cos\beta)\mathbf{j} + (\cos\gamma)\mathbf{k}$. So the direction cosines are the components of **p**. The direction cosines of the x-axis are 1, 0, 0; of the y-axis, 0, 1, 0; and of the z-axis, 0, 0, 1.

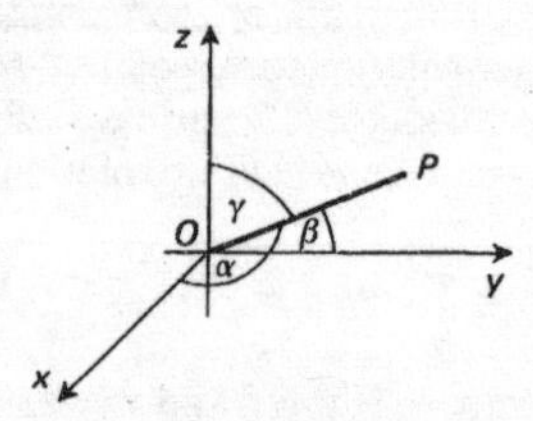

direction ratios Suppose that a direction has *direction cosines* $\cos\alpha$, $\cos\beta$, $\cos\gamma$. Any triple of numbers l, m, n, not all zero, such that $l = k\cos\alpha$, $m = k\cos\beta$, $n = k\cos\gamma$, are called **direction ratios** of the given direction. Since $\cos^2\alpha + \cos^2\beta + \cos^2\gamma = 1$, it follows that

$$\cos\alpha = \frac{\pm l}{\sqrt{l^2 + m^2 + n^2}}, \quad \cos\beta = \frac{\pm m}{\sqrt{l^2 + m^2 + n^2}},$$

$$\cos\gamma = \frac{\pm n}{\sqrt{l^2 + m^2 + n^2}},$$

where either the + sign or the − sign is taken throughout. So any triple of numbers, not all zero, determine two possible sets of direction cosines, corresponding to opposite directions. The triple l, m, n are said to be direction ratios of a straight line when they are direction ratios of either direction of the line.

directly proportional See *proportion*.

direct proof For a theorem that has the form $p \Rightarrow q$, a **direct proof** is one that supposes p and shows that q follows. Compare this with an *indirect proof*.

directrix (plural: directrices) See *conic, ellipse, hyperbola* and *parabola.*

Dirichlet, Peter Gustav Lejeune (1805–1859) German mathematician who was professor at the University of Berlin before succeeding Gauss at the University of Göttingen. He proved that in any arithmetic series $a, a+d, a+2d, \ldots$, where a and d are relatively prime, there are infinitely many primes. He gave the modern definition of a *function.* In more advanced work, he was concerned to see analysis applied to number theory and mathematical physics.

disc The circle, centre C, with coordinates (a, b) and radius r, has equation $(x-a)^2+(y-b)^2=r^2$. The set of points (x, y) in the plane such that $(x-a)^2+(y-b)^2<r^2$ forms the interior of the circle, and may be called the **open disc**, centre C and radius r. The **closed disc**, centre C and radius r, is the set of points (x, y) such that $(x-a)^2+(y-b)^2 \leq r^2$.

discrete data See *data.*

discrete random variable See *random variable.*

discriminant For the *quadratic equation* $ax^2+bx+c=0$, the quantity b^2-4ac is the **discriminant.** The equation has two distinct real roots, equal roots (that is, one root) or no real roots according to whether the discriminant is positive, zero or negative.

disjoint Sets A and B are **disjoint** if they have no elements in common; that is, if $A \cap B = \emptyset$.

disjunction If p and q are statements, then the statement 'p or q', denoted by $p \vee q$, is the **disjunction** of p and q. For example, if p is 'It is raining' and q is 'It is Monday' then $p \vee q$ is 'It is raining or it is Monday'. To be quite clear, notice that $p \vee q$ means 'p or q or both': the disjunction of p and q is true when *at least* one of the statements p and q is true. The *truth table* is therefore as follows:

p	q	$p \vee q$
T	T	T
T	F	T
F	T	T
F	F	F

dispersion A measure of **dispersion** is a way of describing how scattered or spread out the observations in a sample are. The term is also applied similarly to a random variable. Common measures of dispersion are the *range, interquartile range, mean absolute deviation, variance* and *standard deviation.* The range may be unduly affected by odd high and low values. The mean absolute deviation is difficult to work with because of the absolute value signs. The standard deviation is in the same units as the data, and it is this that is most often used. The interquartile range may be appropriate when the median is used as the measure of *location*.

displacement Suppose that a particle is moving in a straight line, with a point O on the line taken as the origin and one direction along the line taken as positive. Let $|OP|$ be the distance between O and P, where P is the position of the particle at time t. Then the **displacement** x is equal to $|OP|$ if $\overrightarrow{OP}$ is in the positive direction and equal to $-|OP|$ if $\overrightarrow{OP}$ is in the negative direction. Indeed, the displacement is equal to the *measure OP*.

In the preceding paragraph, a common convention has been followed, in which the unit vector $\mathbf{i}$ in the positive direction along the line has been suppressed. Displacement is in fact a vector quantity, and in the 1-dimensional case above is equal to $x\mathbf{i}$.

When the motion is in 2 or 3 dimensions, vectors are used explicitly. The displacement is a vector giving the change in the position of a particle. If the particle P moves from the point A to the point B, not necessarily along a straight line, the displacement is equal to the position vector of B relative to A, that is, the vector represented by the directed line-segment $\overrightarrow{AB}$.

dissection (of an interval) = *partition* (of an interval).

dissipative force A force that causes a loss of *energy* (considered as consisting of kinetic energy and potential energy). A *resistive force* is dissipative because the work done by it is negative.

distance (in the complex plane) If P_1 and P_2 represent the complex numbers z_1 and z_2, the **distance** $|P_1P_2|$ is equal to $|z_1 - z_2|$, the *modulus* of $z_1 - z_2$.

distance between two codewords The **distance** between two codewords in a *binary code* is the number of bits in which the two codewords differ. For example, the distance between 010110 and 001100 is 3 because they differ in the second, third and fifth bits. If the distance between any two different codewords in a binary code is at least 3, the code is an *error-correcting code* capable of correcting any one error.

distance between two lines (in 3-dimensional space) Let l_1 and l_2 be lines in space that do not intersect. There are two cases. If l_1 and l_2 are parallel, the **distance** between the two lines is the length of any line segment N_1N_2, with N_1 on l_1 and N_2 on l_2, perpendicular to both lines. If l_1 and l_2 are not parallel, there are unique points N_1 on l_1 and N_2 on l_2 such that the length of the line segment N_1N_2 is the shortest possible. The length $|N_1N_2|$ is the distance between the two lines. In fact, the line N_1N_2 is the *common perpendicular* of l_1 and l_2.

distance between two points (in the plane) Let A and B have coordinates (x_1, y_1) and (x_2, y_2). It follows from Pythagoras' Theorem that the **distance** $|AB|$ is equal to $\sqrt{(x_2 - x_1)^2 + (y_2 - y_1)^2}$.

distance between two points (in 3-dimensional space) Let A and B have coordinates (x_1, y_1, z_1) and (x_2, y_2, z_2). Then the **distance** $|AB|$ is equal to

$$\sqrt{(x_2 - x_1)^2 + (y_2 - y_1)^2 + (z_2 - z_1)^2}.$$

distance between two points (in n-dimensional space) See *n-dimensional space*.

distance from a point to a line (in the plane) The **distance** from the point P to the line l is the shortest distance between P and a point on l. It is equal to $|PN|$, where N is the point on l such that the line PN is perpendicular to l. If P has coordinates (x_1, y_1) and l has equation $ax + by + c = 0$, then the distance from P to l is equal to

$$\frac{|ax_1 + by_1 + c|}{\sqrt{a^2 + b^2}},$$

where $|ax_1 + by_1 + c|$ is the *absolute value* of $ax_1 + by_1 + c$.

distance from a point to a plane (in 3-dimensional space) The **distance** from the point P to the plane p is the shortest distance between P and a point in p, and is equal to $|PN|$, where N is the point in p such that the line PN is normal to p. If P has coordinates (x_1, y_1, z_1) and p has equation $ax + by + cz + d = 0$, the distance from P to p is equal to

$$\frac{|ax_1 + by_1 + cz_1 + d|}{\sqrt{a^2 + b^2 + c^2}},$$

where $|ax_1 + by_1 + cz_1 + d|$ is the *absolute value* of $ax_1 + by_1 + cz_1 + d$.

distance–time graph A graph that shows *displacement* plotted against time for a particle moving in a straight line. Let $x(t)$ be the displacement of the particle at time t. The distance–time graph is the graph $y = x(t)$, where the t-axis is horizontal and the y-axis is vertical with the positive direction upwards. The gradient at any point is equal to the velocity of the particle at that time. (Here a common convention has been followed, in which the unit vector $\mathbf{i}$ in the positive direction along the line has been suppressed. The displacement of the particle is in fact a vector quantity equal to $x(t)\mathbf{i}$, and the velocity of the particle is a vector quantity equal to $\dot{x}(t)\mathbf{i}$.)

distribution The **distribution** of a random variable is concerned with the way in which the probability of its taking a certain value, or a value within a certain interval, varies. It may be given by the *cumulative distribution function*. More commonly, the distribution of a discrete random variable is given by its *probability mass function* and that of a continuous random variable by its *probability density function*.

distribution-free methods = *non-parametric methods*.

distribution function = *cumulative distribution function*.

distributive Suppose that $\circ$ and $*$ are *binary operations* on a set S. Then $\circ$ is **distributive** over $*$ if, for all a, b and c in S,

$$a \circ (b * c) = (a \circ b) * (a \circ c) \quad \text{and} \quad (a * b) \circ c = (a \circ c) * (b \circ c).$$

If the two operations are multiplication and addition, 'the distributive laws' normally means those that say that multiplication is distributive over addition.

divides Let a and b be integers. Then a **divides** b (which may be written as $a \mid b$) if there is an integer c such that $ac = b$. It is said that a is a **divisor** or **factor** of b, that b is **divisible** by a, and that b is a **multiple** of a.

divisible See *divides*.

Division Algorithm The following theorem of elementary number theory:

THEOREM: For integers a and b, with $b > 0$, there exist unique integers q and r such that $a = bq + r$, where $0 \leq r < b$.

In the division of a by b, the number q is the **quotient** and r is the **remainder**.

divisor See *divides*.

divisor of zero If in a *ring* there are non-zero elements a and b such that $ab = 0$, then a and b are **divisors of zero**. For example, in the ring of 2×2 real matrices,

$$\begin{bmatrix} 0 & 1 \\ 0 & 0 \end{bmatrix}\begin{bmatrix} 1 & 0 \\ 0 & 0 \end{bmatrix} = \begin{bmatrix} 0 & 0 \\ 0 & 0 \end{bmatrix},$$

and so each of the matrices on the left-hand side is a divisor of zero. In the ring $\mathbf{Z}_6$, consisting of the set $\{0, 1, 2, 3, 4, 5\}$ with addition and multiplication modulo 6, the element 4 is a divisor of zero since $4 \,.\, 3 = 0$.

dodecagon A twelve-sided polygon.

dodecahedron (plural: dodecahedra) A *polyhedron* with twelve faces, often assumed to be regular. The regular dodecahedron is one of the *Platonic solids*, and its faces are regular pentagons. It has 20 vertices and 30 edges.

Dodgson, Charles Lutwidge (1832–1898) British mathematician and logician, better known under the pseudonym Lewis Carroll as the author of *Alice's Adventures in Wonderland*. He was lecturer in mathematics at Christ Church, Oxford, and wrote a number of minor books on mathematics.

domain See *function* and *mapping*.

dot product = *scalar product*.

double root See *root*.

doubly stochastic matrix See *stochastic matrix*.

drag See *aerodynamic drag*.

dummy variable A variable appearing in an expression is a **dummy variable** if the letter being used could equally well be replaced by another letter. For example, the two expressions

$$\int_0^1 x^2\,dx \quad \text{and} \quad \int_0^1 t^2\,dt$$

represent the same definite integral, and so x and t are dummy variables. Similarly, the summation

$$\sum_{r=1}^{5} r^2$$

denotes the sum $1^2 + 2^2 + 3^2 + 4^2 + 5^2$, and would still do so if the letter r were replaced by the letter s, say; so r here is a dummy variable.

duplication of the cube One of the problems that Greek geometers attempted (like *squaring the circle* and *trisection of an angle*) was to find a construction, with ruler and pair of compasses, to obtain the side of a cube whose volume was twice the volume of a given cube. This is equivalent to finding a geometrical construction to obtain a length of $\sqrt[3]{2}$ from a given unit length. Now constructions of the kind envisaged can only give lengths belonging to a class of numbers obtained, essentially, by addition, subtraction, multiplication, division and the taking of square roots. Since $\sqrt[3]{2}$ does not belong to this class of numbers, the duplication of the cube is impossible.

e The number that is the base of natural logarithms. There are several ways of defining it. Probably the most satisfactory is this. First, define ln as in approach **2** to the *logarithmic function*. Then define exp as the inverse function of ln (see approach **2** to the *exponential function*). Then define e as equal to exp 1. This amounts to saying that e is the number that makes

$$\int_1^e \frac{1}{t}\,dt = 1.$$

It is necessary to go on to show that e^x and $\exp x$ are equal and so are identical as functions, and also that ln and $\log_e$ are identical functions.

The number e has important properties derived from some of the properties of ln and exp. For example,

$$e = \lim_{h\to 0}(1+h)^{1/h} = \lim_{n\to\infty}\left(1+\frac{1}{n}\right)^n.$$

Also, e is the sum of the series

$$1+\frac{1}{1!}+\frac{1}{2!}+\cdots+\frac{1}{n!}+\cdots.$$

Another approach, but not a recommended one, is to make one of these properties the definition of e. Then $\exp x$ would be defined as e^x, $\ln x$ would be defined as its inverse function, and the properties of these functions would have to be proved.

The value of e is 2.718 281 83 (to 8 decimal places). The proof that e is *irrational* is comparatively easy. In 1873, Hermite proved that e is *transcendental*, and his proof was subsequently simplified by Hilbert.

Earth Our particular planet in the solar system. The Earth is often assumed to be a sphere with a radius of approximately 6400 kilometres. A more accurate accepted value is 6371 km. A better approximation is to say that the Earth is in the shape of an *oblate spheroid* with an equatorial radius of 6378 km and a polar radius of 6357 km. The mass of the Earth is 5.976×10^{24} kg.

eccentricity See *conic*, *ellipse* and *hyperbola*.

echelon form Suppose that a row of a matrix is called zero if all its entries are zero. Then a matrix is in **echelon form** if (i) all the zero rows come below the non-zero rows, and (ii) the first non-zero entry in each non-zero row is 1 and occurs in a column to the right of the leading 1 in the row above. For example, these two matrices are in echelon form:

$$\begin{bmatrix} 1 & 6 & -1 & 4 & 2 \\ 0 & 0 & 1 & 2 & -3 \\ 0 & 0 & 0 & 1 & 5 \end{bmatrix}, \qquad \begin{bmatrix} 1 & 6 & -1 & 4 & 2 \\ 0 & 1 & 2 & -3 & 5 \\ 0 & 0 & 0 & 0 & 0 \end{bmatrix}.$$

Any matrix can be transformed to a matrix in echelon form using *elementary row operations*, by a method known as *Gaussian elimination*. The solutions of a set of linear equations may be investigated by transforming the *augmented matrix* to echelon form. Further elementary row operations may be used to transform a matrix to *reduced echelon form*. A set of linear equations is said to be in echelon form if its augmented matrix is.

edge (of a graph), **edge-set** See *graph*.

efficiency See *machine*.

effort See *machine*.

eigenvalue, eigenvector = *characteristic value, characteristic vector*.

Einstein, Albert (1879–1955) Outstanding mathematical physicist whose work was the single most important influence on the development of physics since Newton. Born in Ulm in Germany, he lived in Switerland and Germany before moving to the United States in 1933. He was responsible in 1905 for the Special Theory of Relativity and in 1916 for the General Theory. He made a fundamental contribution to the birth of quantum theory and had an important influence on thermodynamics. He is perhaps most widely known for his equation $E = mc^2$, quantifying the equivalence of matter and energy. He regarded himself as a physicist rather than as a mathematician, but his work has triggered off many developments in modern mathematics. Contrary to the popular image of a white-haired professor scribbling incomprehensible symbols on a blackboard, Einstein's great strength was his ability to ask simple questions and give simple answers. In that way, he changed our view of the universe and our concepts of space and time.

elastic A body is said to be **elastic** if, after being deformed by forces applied to it, it is able to regain its original shape as soon as the deforming forces cease to act.

elastic collision A *collision* in which there is no loss of kinetic energy.

elastic string A string that can be extended, but not compressed, and that resumes its natural length as soon as the forces applied to extend it are removed. How the *tension* in the string varies with the *extension* may be complicated. In the simplest mathematical model, it is assumed that the tension is proportional to the extension; that is, that *Hooke's law* holds. A particle suspended from a fixed support by an elastic string performs *simple harmonic motion* in the same way as a particle suspended by a spring does, provided that the amplitude of the oscillations is sufficiently small for the string never to go slack.

element An object in a set is an **element** of that set.

elementary column operation One of the following operations on the columns of a matrix:

(i) interchange two columns,
(ii) multiply a column by a non-zero scalar,
(iii) add a multiple of one column to another column.

An elementary column operation can be produced by post-multiplication by the appropriate *elementary matrix*.

elementary function Any of the following real functions: the *rational functions*, the *trigonometric functions*, the *logarithmic* and *exponential functions*, the functions f defined by $f(x) = x^{m/n}$ (where m and n are non-zero integers), and all those functions that can be obtained from these by using addition, subtraction, multiplication, division, *composition* and the taking of *inverse functions*.

elementary matrix A square matrix obtained from the identity matrix **I** by an *elementary row operation*. Thus there are three types of elementary matrix. Examples of each type are:

$$\text{(i)}\begin{bmatrix}1&0&0&0&0&0&0\\0&0&0&0&1&0&0\\0&0&1&0&0&0&0\\0&0&0&1&0&0&0\\0&1&0&0&0&0&0\\0&0&0&0&0&1&0\\0&0&0&0&0&0&1\end{bmatrix},\quad \text{(ii)}\begin{bmatrix}1&0&0&0&0&0&0\\0&1&0&0&0&0&0\\0&0&-3&0&0&0&0\\0&0&0&1&0&0&0\\0&0&0&0&1&0&0\\0&0&0&0&0&1&0\\0&0&0&0&0&0&1\end{bmatrix},$$

$$\text{(iii)}\begin{bmatrix}1&0&0&0&0&0&0\\0&1&0&0&4&0&0\\0&0&1&0&0&0&0\\0&0&0&1&0&0&0\\0&0&0&0&1&0&0\\0&0&0&0&0&1&0\\0&0&0&0&0&0&1\end{bmatrix}.$$

The matrix (i) is obtained from **I** by interchanging the second and fifth rows, matrix (ii) by multiplying the third row by -3, and matrix (iii) by adding 4 times the fifth row to the second row. Pre-multiplication of an $m \times n$ matrix **A** by an $m \times m$ elementary matrix produces the result of the corresponding row operation on **A**.

Alternatively, an elementary matrix can be seen as one obtained from the identity matrix by an *elementary column operation*; and post-multiplication of an $m \times n$ matrix **A** by an $n \times n$ elementary matrix produces the result of the corresponding column operation on **A**.

elementary row operation One of the following operations on the rows of a matrix:

(i) interchange two rows,
(ii) multiply a row by a non-zero scalar,
(iii) add a multiple of one row to another row.

An elementary row operation can be produced by a pre-multiplication by the appropriate *elementary matrix*. Elementary row operations are applied to the *augmented matrix* of a set of linear equations to transform it into *echelon form* or *reduced echelon form*. Each elementary row operation corresponds to an operation on the set of linear equations that does not alter the solution set of the equations.

ellipse A particular 'oval' shape, obtained, it could be said, by stretching or squashing a circle. If it has length $2a$ and width $2b$, its area equals πab.

In more advanced work, a more precise definition of an ellipse is required. One approach is to define it as a *conic* with eccentricity less than 1. Thus it is the locus of all points P such that the distance from P to a fixed point F_1 (the **focus**) equals e (< 1) times the distance from P to a fixed line l_1 (the **directrix**). It turns out that there is another point F_2 and another line l_2 such that the same locus would be obtained with these as focus and directrix. An ellipse is also the conic section that results when a plane cuts a cone in such a way that a finite section is obtained (see *conic*).

The line through F_1 and F_2 is the **major axis**, and the points V_1 and V_2 where it cuts the ellipse are the **vertices**. The length $|V_1V_2|$ is the **length of the major axis** and is usually taken to be $2a$. The midpoint of V_1V_2 is the **centre** of the ellipse. The line through the centre perpendicular to the major axis is the **minor axis**, and the distance, usually taken to be $2b$, between the points where it cuts the ellipse is the **length of the minor axis**. The three constants a, b and e are related by $b^2 = a^2(1 - e^2)$ or, in another form, $e^2 = 1 - b^2/a^2$. The eccentricity e determines the shape of the ellipse. The value $e = 0$ is permitted and gives rise to a circle, though this requires the directrices to be infinitely far away and invalidates the focus and directrix approach.

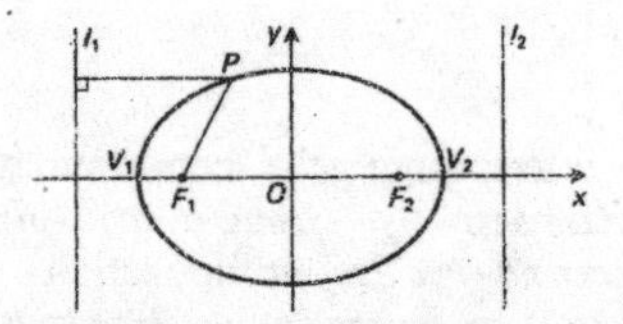

By taking a coordinate system with origin at the centre of the ellipse and x-axis along the major axis, the foci have coordinates $(ae, 0)$ and $(-ae, 0)$, the directrices have equations $x = a/e$ and $x = -a/e$, and the ellipse has equation

$$\frac{x^2}{a^2} + \frac{y^2}{b^2} = 1,$$

where $a > b > 0$. When investigating the properties of an ellipse, it is a common practice to choose this convenient coordinate system. It may be useful to take $x = a\cos\theta$, $y = b\sin\theta$ $(0 \le \theta < 2\pi)$ as *parametric equations*.

An ellipse with its centre at the origin and its major axis, of length $2a$, along the y-axis instead has equation $y^2/a^2 + x^2/b^2 = 1$, where $a > b > 0$, and its foci are at $(0, ae)$ and $(0, -ae)$.

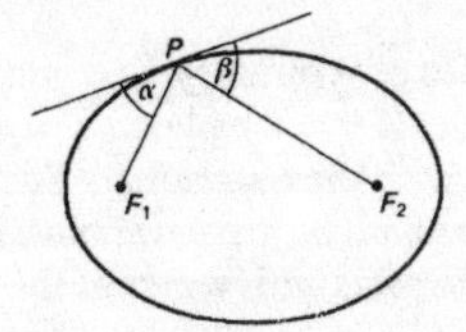

The ellipse has two important properties:

(i) If P is any point of the ellipse with foci F_1 and F_2 and length of major axis $2a$, then $|PF_1| + |PF_2| = 2a$. The fact that an ellipse can be seen as the locus of all such points is the basis of a practical method of drawing an ellipse using a string between two points.

(ii) For any point P on the ellipse, let α be the angle between the tangent at P and the line PF_1, and β the angle between the tangent at P and the line PF_2, as shown in the figure; then $\alpha = \beta$. This property is analogous to that of the parabolic reflector (see *parabola*).

ellipsoid A *quadric* whose equation in a suitable coordinate system is

$$\frac{x^2}{a^2} + \frac{y^2}{b^2} + \frac{z^2}{c^2} = 1.$$

The three axial planes are planes of symmetry. All non-empty plane sections are ellipses.

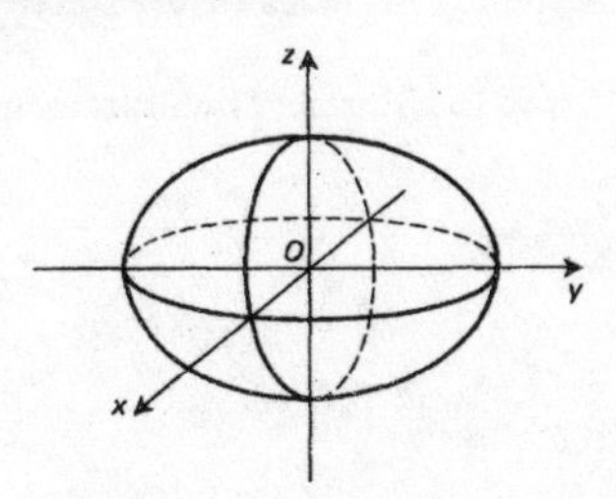

elliptic cylinder A *cylinder* in which the fixed curve is an *ellipse* and the fixed line to which the generators are parallel is perpendicular to the plane of the ellipse. It is a *quadric*, and in a suitable coordinate system has equation

$$\frac{x^2}{a^2} + \frac{y^2}{b^2} = 1.$$

elliptic geometry See *non-Euclidean geometry*.

elliptic paraboloid A *quadric* whose equation in a suitable coordinate system is

$$\frac{x^2}{a^2}+\frac{y^2}{b^2}=\frac{2z}{c}.$$

Here the yz-plane and the zx-plane are planes of symmetry. Sections by planes $z = k$, where $k \geq 0$, are ellipses (circles if $a = b$); planes $z = k$, where $k < 0$, have no points of intersection with the paraboloid. Sections by planes parallel to the yz-plane and to the zx-plane are parabolas. Planes through the z-axis cut the paraboloid in parabolas with vertex at the origin.

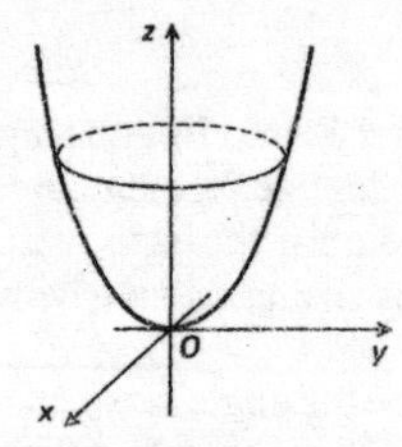

empty set The set, denoted by Ø, with no elements in it. Consequently, its *cardinality*, $n(\varnothing)$, is zero.

end-point A number defining one end of an *interval* on the real line. Each of the finite intervals $[a, b]$, (a, b), $[a, b)$ and $(a, b]$ has two end-points, a and b. Each of the infinite intervals $[a, \infty)$, (a, ∞), $(-\infty, a]$ and $(-\infty, a)$ has one end-point, a.

energy Mechanics is, in general, concerned with two forms of energy: *kinetic energy* and *potential energy*. When there is a transference of energy of one of these forms into energy of another form such as heat or noise, there may be said to be a loss of energy.

Energy has dimensions ML^2T^{-2}, and the SI unit of measurement is the *joule*.

energy equation See *conservation of energy*.

enneagon A nine-sided polygon.

entry See *matrix*.

enumerable = *denumerable*.

epicycloid The curve traced out by a point on the circumference of a circle rolling round the outside of a fixed circle. When the two circles have the same radius, the curve is a *cardioid*.

equality (of complex numbers) See *equating real and imaginary parts*.

equality (of matrices) Matrices **A** and **B**, where $\mathbf{A} = [a_{ij}]$ and $\mathbf{B} = [b_{ij}]$, are **equal** if and only if they have the same order and $a_{ij} = b_{ij}$ for all i and j.

equality (of sets) Sets A and B are **equal** if they consist of the same elements. In order to establish that $A = B$, a technique that can be useful is to show, instead, that both $A \subseteq B$ and $B \subseteq A$.

equality (of vectors) See *vector.*

equating coefficients Let $f(x)$ and $g(x)$ be polynomials, and let

$$f(x) = a_n x^n + a_{n-1}x^{n-1} + \cdots + a_1 x + a_0,$$
$$g(x) = b_n x^n + b_{n-1}x^{n-1} + \cdots + b_1 x + b_0,$$

where it is not necessarily assumed that $a_n \neq 0$ and $b_n \neq 0$. If $f(x) = g(x)$ for all values of x, then $a_n = b_n, a_{n-1} = b_{n-1}, \ldots, a_1 = b_1, a_0 = b_0$. Using this fact is known as **equating coefficients.** The result is obtained by applying the *Fundamental Theorem of Algebra* to the polynomial $h(x)$, where $h(x) = f(x) - g(x)$. If $h(x) = 0$ for all values of x (or, indeed, for more than n values of x), the only possibility is that $h(x)$ is the zero polynomial with all its coefficients zero. The method can be used, for example, to find numbers A, B, C and D such that

$$x^3 = A(x-1)(x-2)(x-3) + B(x-1)(x-2) + C(x-1) + D$$

for all values of x. It is often used to find the unknowns in *partial fractions.*

equating real and imaginary parts Complex numbers $a + bi$ and $c + di$ are **equal** if and only if $a = c$ and $b = d$. Using this fact is called **equating real and imaginary parts.** For example, if $(a + bi)^2 = 5 + 12i$, then $a^2 - b^2 = 5$ and $2ab = 12$.

equation of motion An equation, based on the second of *Newton's laws of motion*, that governs the motion of a particle. In vector form, the equation is $m\ddot{\mathbf{r}} = \mathbf{F}$, where $\mathbf{F}$ is the total force acting on the particle of mass m. In Cartesian coordinates, this is equivalent to the three differential equations $m\ddot{x} = F_1, m\ddot{y} = F_2$ and $m\ddot{z} = F_3$, where $\mathbf{F} = F_1\mathbf{i} + F_2\mathbf{j} + F_3\mathbf{k}$. In cylindrical polar coordinates, taken here to be (r, θ, z), it is equivalent to $m(\ddot{r} - r\dot{\theta}^2) = F_1, m(r\ddot{\theta} + 2\dot{r}\dot{\theta}) = F_2$ and $m\ddot{z} = F_3$, where now $\mathbf{F} = F_1\mathbf{e}_r + F_2\mathbf{e}_\theta + F_3\mathbf{k}$, and $\mathbf{e}_r = \mathbf{i}\cos\theta + \mathbf{j}\sin\theta$ and $\mathbf{e}_\theta = -\mathbf{i}\sin\theta + \mathbf{j}\cos\theta$ (see *radial and transverse components*).

equations of motion with constant acceleration Equations relating to an object moving in a straight line with constant acceleration a. Let u be the initial velocity, v the final velocity, t the time of travel and s the displacement from the starting point. Then

$$v = u + at, \quad s = ut + \tfrac{1}{2}at^2, \quad s = \tfrac{1}{2}(u+v)t, \quad v^2 = u^2 + 2as.$$

These equations may be applied to a particle falling under *gravity* near the Earth's surface. If the positive direction is taken to be downwards, then $a = g$. When a particle is projected vertically upwards from the ground, it may be appropriate to take the positive direction upwards, and then $a = -g$.

equiangular spiral A curve whose equation in polar coordinates is $r = ae^{k\theta}$, where a (> 0) and k are constants. Let O be the origin and P be any point on the curve. The curve derives its name from the property that the angle α between OP and the tangent at P is constant. In fact, $k = \cot\alpha$. The equation can be written $\ln r = k\theta + b$, and the curve is also called the **logarithmic spiral.**

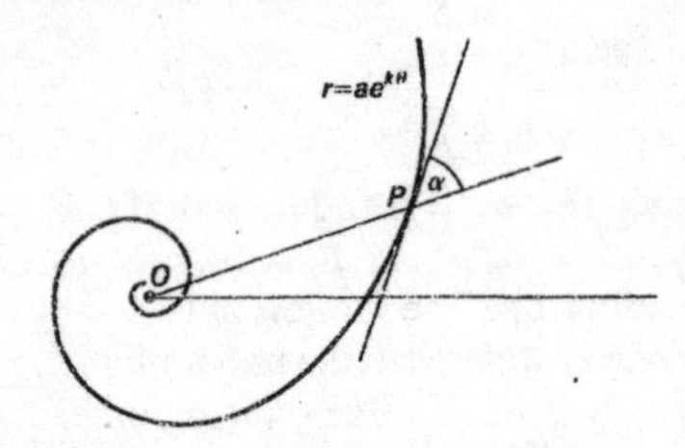

equilateral A polygon is **equilateral** if all its sides have the same length. In an equilateral triangle, the three angles are all equal and so each equals 60°.

equilibrium A particle is in **equilibrium** when it is at rest and the total force acting on it is zero for all time. These conditions are equivalent to saying that $\dot{\mathbf{r}} = \mathbf{0}$ and $\ddot{\mathbf{r}} = \mathbf{0}$ for all time, where $\mathbf{r}$ is the position vector of the particle.

Consider a rigid body experiencing a system of forces. Let $\mathbf{F}$ be the total force, and $\mathbf{M}$ the moment of the forces about the origin. The rigid body is in **equilibrium** when it is at rest and $\mathbf{F} = \mathbf{0}$ and $\mathbf{M} = \mathbf{0}$ for all time.

An **equilibrium position** for a body is a position in which it can be in *equilibrium*. A body is in **stable** equilibrium if, following a small change in its position, it returns to the equilibrium position. It is in **unstable** equilibrium if, following a small change in its position, it continues to move further from the equilibrium position. It is in **neutral** equilibrium if, following a small change in its position, it neither returns to nor moves further from the equilibrium position.

equivalence class For an *equivalence relation* $\sim$ on a set S, an **equivalence class** $[a]$ is the set of elements of S equivalent to a; that is to say, $[a] = \{x \mid x \in S \text{ and } a \sim x\}$. It can be shown that if two equivalence classes have an element in common, then the two classes are, as sets, equal. The collection of distinct equivalence classes having the property that every element of S belongs to exactly one of them is a *partition* of S.

equivalence relation A *binary relation* $\sim$ on a set S that is *reflexive, symmetric* and *transitive*. For an equivalence relation $\sim$, a is said to be **equivalent** to b when $a \sim b$. It is an important fact that, from an equivalence relation on S, *equivalence classes* can be defined to obtain a *partition* of S.

equivalent See *equivalence relation*.

Eratosthenes of Cyrene (about 275–195 BC) Greek astronomer and mathematician who was the first to calculate the size of the Earth by making measurements of the angle of the Sun at two different places a known distance apart. His other achievements include measuring the tilt of the Earth's axis. He is also credited with the method known as the *sieve of Eratosthenes.*

error Let x be an approximation to a value X. According to some authors, the **error** is $X - x$; for example, when 1.9 is used as an approximation for 1.875, the error equals −0.025. Others define the error to be $x - X$. Whichever of these definitions is used, the error can be positive or negative. Yet other authors define the error to be $|X - x|$, the difference between the true value and the approximation, in which case the error is always greater than or equal to zero. When contrasted with *relative error*, the error may be called the **absolute error**.

error-correcting and **error-detecting code** A *code* is said to be **error-detecting** if any one error in a codeword results in a word that is not a codeword, so that the receiver knows that an error has occurred. A code is **error-correcting** if, when any one error occurs in a codeword, it is possible to decide which codeword was intended. Certain error-correcting codes may not be able to detect errors if more than one error occurs in a codeword; other error-correcting codes can be constructed that can detect and correct more than one error in a codeword. See also *distance between two codewords.*

escape speed For a celestial body, the minimum speed with which an object must be projected away from its surface so that the object does not return again because of gravity. The escape speed for the Earth is $\sqrt{2gR}$, where R is the radius of the Earth, and is approximately 11.2 kilometres per second. Similarly, the escape speed associated with a spherical celestial body of mass M and radius R is $\sqrt{2GM/R}$, where G is the *gravitational constant.*

escribed circle = *excircle.*

estimate The value of an *estimator* calculated from a particular sample. If the estimate is a single figure, it is a **point estimate**; if it is an interval, such as a *confidence interval*, it is an **interval estimate**.

estimation The process of determining as nearly as possible the value of a population *parameter* by using an *estimator*. The value of an estimator found from a particular sample is called an *estimate.*

estimator A *statistic* used to estimate the value of a population *parameter.* An estimator X of a parameter θ is **consistent** if the probability of the difference between the two exceeding an arbitrarily small fixed number tends to zero as the sample size increases indefinitely. An estimator X is an **unbiased** estimator of the parameter θ if $\mathrm{E}(X) = \theta$, and it is **biased** if not (see *expected value*). The **best** unbiased estimator is the unbiased estimator with the minimum *variance*. The **relative efficiency** of two unbiased estimators X and Y is the ratio $\text{Var}(Y)/\text{Var}(X)$ of their variances.

Estimators may be found in different ways, including the method of maximum likelihood (see *likelihood*), and the method of moments (see *moment*).

Euclid (about 300 BC) Outstanding mathematician of Alexandria, author of what may well be the second most influential book in Western Culture: the *Elements*. Little is known about Euclid himself, and it is not clear to what extent the book describes original work and to what extent it is a textbook. The *Elements* develops a large section of elementary geometry by rigorous logic starting from 'undeniable' axioms. It includes his proof that there are infinitely many primes, the *Euclidean algorithm*, the derivation of the five *Platonic solids*, and much more. It has served for two millenia as a model of what pure mathematics is about.

Euclidean Algorithm A process, based on the *Division Algorithm*, for finding the *greatest common divisor* (a, b) of two positive integers a and b. Assuming that $a > b$, write $a = bq_1 + r_1$, where $0 \leq r_1 < b$. If $r_1 = 0$, the g.c.d. (a, b) is equal to b; if $r_1 \neq 0$, then $(a, b) = (b, r_1)$, so the step is repeated with b and r_1 in place of a and b. After further repetitions, the last non-zero remainder obtained is the required g.c.d. For example, for $a = 1274$ and $b = 871$, write

$$\begin{aligned} 1274 &= 1 \times 871 + 403, \\ 871 &= 2 \times 403 + 65, \\ 403 &= 6 \times 65 + 13, \\ 65 &= 5 \times 13, \end{aligned}$$

and then $(1274, 871) = (871, 403) = (403, 65) = (65, 13) = 13$.

The algorithm also enables s and t to be found such that the g.c.d. can be expressed as $sa + tb$: use the equations in turn to express each remainder in this form. Thus,

$$\begin{aligned} 403 &= 1274 - 1 \times 871 = a - b, \\ 65 &= 871 - 2 \times 403 = b - 2(a - b) = 3b - 2a, \\ 13 &= 403 - 6 \times 65 = (a - b) - 6(3b - 2a) = 13a - 19b. \end{aligned}$$

Eudoxus of Cnidus (about 380 BC) Greek mathematician and astronomer, one of the greatest of antiquity. All his original works are lost, but it is known from later writers that he was responsible for the work in Book 5 of Euclid's *Elements*. This work of his was a precise and rigorous development of the real number system in the language of his day. The significance of his sophisticated ideas was not really appreciated until the nineteenth century. He also developed methods of determining areas with curved boundaries.

Euler, Leonhard (1707–1783) Beyond comparison, the most prolific of famous mathematicians. He was born in Switzerland but is most closely associated with the Berlin of Frederick the Great and the St Petersburg of Catherine the Great. He worked in a highly productive period when the newly developed calculus was being extended in all directions at once, and he made contributions to most areas of mathematics, pure and applied. Euler, more than any other individual, was responsible for notation that is standard today. Among his contributions to the language are the basic symbols, π, e and i, the summation notation $\sum$ and the standard function notation $f(x)$. His *Introductio in analysin infinitorum* was the most important mathematics text of the late eighteenth century. From the vast bulk of his work, one famous result of which he was justifiably proud is this:

$$1+\frac{1}{2^2}+\frac{1}{3^2}+\cdots+\frac{1}{n^2}+\cdots=\frac{\pi^2}{6}.$$

Eulerian graph One area of graph theory is concerned with the possibility of travelling around a *graph*, going along edges in such a way as to use every edge exactly once. A *connected graph* is called **Eulerian** if there is a sequence $v_0, e_1, v_1, \ldots, e_k, v_k$ of alternately vertices and edges (where e_i is an edge joining v_{i-1} and v_i), with $v_0 = v_k$ and with every edge of the graph occurring exactly once. Simply put, it means that 'you can draw the graph without taking your pencil off the paper or retracing any lines, ending at your starting-point'. The name arises from Euler's consideration of the problem of whether the *bridges of Königsberg* could be crossed in this way. It can be shown that a connected graph is Eulerian if and only if every vertex has even degree.

Euler line In a triangle, the *circumcentre* O, the *centroid* G and the *orthocentre* H lie on a straight line called the **Euler line**. On this line, $OG : GH = 1 : 2$. The centre of the *nine-point circle* also lies on the Euler line.

Euler's constant Let $a_n = 1+\frac{1}{2}+\frac{1}{3}+\cdots+(1/n) - \ln n$. This sequence has a limit whose value is known as **Euler's constant**, γ; that is, $a_n \to \gamma$. The value equals 0.577 215 66, to 8 decimal places. It is not known whether γ is *rational* or *irrational*.

Euler's formula The name given to the equation $\cos\theta + i\sin\theta = e^{i\theta}$, a special case of which gives $e^{i\pi}+1=0$.

Euler's function For a positive integer n, let $\phi(n)$ be the number of positive integers less than n that are *relatively prime* to n. For example, $\phi(12) = 4$, since four numbers, 1, 5, 7 and 11, are relatively prime to 12. This function ϕ, defined on the set of positive integers, is **Euler's function**. It can be shown that, if the prime decomposition of n is $n = p_1^{\alpha_1} p_2^{\alpha_2} \ldots p_r^{\alpha_r}$, then

$$\begin{aligned}\phi(n) &= p_1^{\alpha_1-1} p_2^{\alpha_2-1} \cdots p_r^{\alpha_r-1}(p_1-1)(p_2-1)\cdots(p_r-1),\\ &= n\left(1-\frac{1}{p_1}\right)\left(1-\frac{1}{p_2}\right)\cdots\left(1-\frac{1}{p_r}\right).\end{aligned}$$

Euler proved the following extension of *Fermat's Little Theorem*: If n is a positive integer and a is any integer such that $(a, n) = 1$, then $a^{\phi(n)} \equiv 1 \pmod{n}$.

Euler's Theorem If a *planar graph* G is drawn in the plane, so that no two edges cross, the plane is divided into a number of regions which may be called 'faces'. **Euler's Theorem** (for planar graphs) is the following:

THEOREM: Let G be a connected planar graph drawn in the plane. If there are v vertices, e edges and f faces, then $v - e + f = 2$.

An application of this gives **Euler's Theorem** (for polyhedra):

THEOREM: If a convex polyhedron has v vertices, e edges and f faces, then $v - e + f = 2$.

For particular polyhedra, it is easy to confirm the result stated in the theorem. For example, a cube has $v = 8, e = 12, f = 6$, and a tetrahedron has $v = 4, e = 6, f = 4$.

even function The *real function* f is an **even function** if $f(-x) = f(x)$ for all x (in the domain of f). Thus the graph $y = f(x)$ of an even function has the y-axis as a line of symmetry. For example, f is an even function when $f(x)$ is defined as any of the following: $5, x^2, x^6 - 4x^4 + 1, 1/(x^2 - 3), \cos x$.

event A subset of the *sample space* relating to an experiment. For example, suppose that the sample space for an experiment in which a coin is tossed three times is { HHH, HHT, HTH, HTT, THH, THT, TTH, TTT }, and let A = { HHH, HHT, HTH, THH }. Then A is the event in which at least two 'heads' are obtained. If, when the experiment is performed, the outcome is one that belongs to A, then A is said to have occurred. The *intersection* $A \cap B$ of two events is the event that can be described by saying that 'both A and B occur'. The *union* $A \cup B$ of two events is the event that 'either A or B occurs'. Taking the sample space as the universal set, the *complement* A' of A is the event that 'A does not occur'.

The probability $\Pr(A)$ of an event A is often of interest. The following laws hold:

(i) $\Pr(A \cup B) = \Pr(A) + \Pr(B) - \Pr(A \cap B)$.

(ii) When A and B are *mutually exclusive* events, $\Pr(A \cup B) = \Pr(A) + \Pr(B)$.

(iii) When A and B are *independent* events, $\Pr(A \cap B) = \Pr(A)\Pr(B)$.

(iv) $\Pr(A') = 1 - \Pr(A)$.

exa- Prefix used with *SI units* to denote multiplication by 10^{18}.

excentre See *excircle*.

excircle An **excircle** of a triangle is a circle that lies outside the triangle and touches the three sides, two of them extended. There are three excircles. The centre of an excircle is an **excentre** of the triangle. Each excentre is the point of intersection of the bisector of the interior angle at one vertex and the bisectors of the exterior angles at the other two vertices.

excluded middle See *principle of the excluded middle.*

existential quantifier See *quantifier.*

expectation (of a matrix game) Suppose that, in the *matrix game* given by the $m \times n$ matrix $[a_{ij}]$, the *mixed strategies* **x** and **y** for the two players are given by $\mathbf{x} = (x_1, x_2, \ldots, x_m)$, and $\mathbf{y} = (y_1, y_2, \ldots, y_n)$. The **expectation** $E(\mathbf{x}, \mathbf{y})$ is given by

$$E(\mathbf{x}, \mathbf{y}) = \sum_{i=1}^{m} \sum_{j=1}^{n} x_i a_{ij} y_j.$$

If $A = [a_{ij}]$, and **x** and **y** are written as column matrices, $E(\mathbf{x}, \mathbf{y}) = \mathbf{x}^T A\mathbf{y}$. The expectation can be said, loosely, to give the average pay-off each time when the game is played many times with the two players using these mixed strategies.

expectation (of a random variable) = *expected value.*

expected value The **expected value** $E(X)$ of a random variable X is a value that gives the mean value of the distribution, and is defined as follows. For a discrete random variable X, $E(X) = \sum p_i x_i$, where $p_i = \Pr(X = x_i)$. For a continuous random variable X,

$$E(X) = \int_{-\infty}^{\infty} x f(x)\, dx,$$

where f is the *probability density function* of X. The following laws hold:

(i) $E(aX + bY) = aE(X) + bE(Y)$.
(ii) When X and Y are *independent*, $E(XY) = E(X)E(Y)$.

The following is a simple example. Let X be the number obtained when a die is thrown. Let $x_i = i$, for $i = 1, 2, \ldots, 6$. Then $p_i = \frac{1}{6}$, for all i. So $E(X) = \frac{1}{6} \times 1 + \frac{1}{6} \times 2 + \cdots + \frac{1}{6} \times 6 = 3.5$.

explanatory variable One of the variables that it is thought might influence the value of the *dependent* variable in a statistical model.

exponent = *index*. See also *floating-point notation.*

exponential decay Suppose that $y = Ae^{kt}$, where A (> 0) and k are constants, and t represents some measurement of time (see *exponential growth*). When $k < 0$, y can be said to be exhibiting **exponential decay**. In such circumstances, the length of time it takes for y to be reduced to half its value is the same, whatever the value. This length of time, called the **half-life**, is a useful measure of the rate of decay. It is applicable, for example, to the decay of radioactive isotopes.

exponential distribution The continuous probability *distribution* with *probability density function* f given by $f(x) = \lambda \exp(-\lambda x)$, where λ is a positive parameter, and $x \geq 0$. It has mean $1/\lambda$ and variance $1/\lambda^2$. The

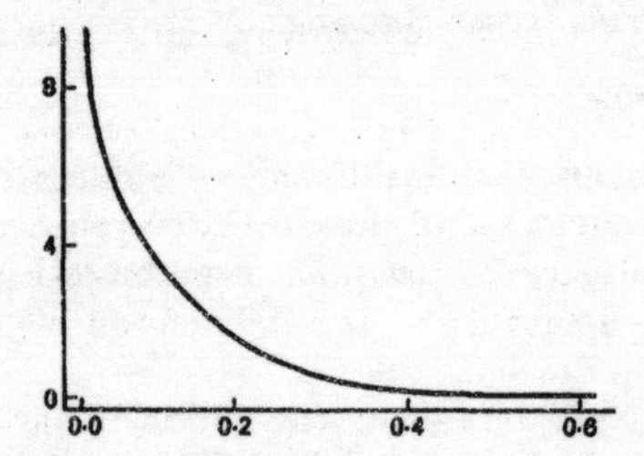

time between events that occur randomly but at a constant rate has an exponential distribution. The distribution is skewed to the right. The figure shows the probability density function of the exponential distribution with $\lambda = 10$.

exponential function The function f such that $f(x) = e^x$, or $\exp x$, for all x in **R**. The two notations arise from different approaches described below, but are used interchangeably. Among the important properties that the exponential function has are the following:

(i) $\exp(x + y) = (\exp x)(\exp y)$, $\exp(-x) = 1/\exp x$ and $(\exp x)^r = \exp rx$. (These hold by the usual rules for indices once the equivalence of $\exp x$ and e^x has been established.)

(ii) The exponential function is the *inverse function* of the *logarithmic function*: $y = \exp x$ if and only if $x = \ln y$.

(iii) $\dfrac{d}{dx}(\exp x) = \exp x$.

(iv) $\exp x$ is the sum of the series $1 + \dfrac{x}{1!} + \dfrac{x^2}{2!} + \cdots + \dfrac{x^n}{n!} + \cdots$.

(v) As $n \to \infty$, $\left(1 + \dfrac{x}{n}\right)^n \to \exp x$.

Three approaches can be used:

1. Suppose that the value of e has already been obtained independently. Then it is possible to define e^x, the exponential function to base e, by using approach 1 to the *exponential function to base a*. Then $\exp x$ can be taken to mean just e^x. The problem with this approach is its reliance on a prior

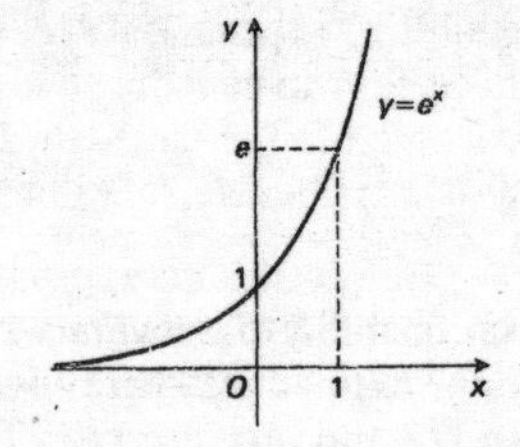

definition of e and the difficulty of subsequently proving some of the other properties of exp.

2. Define ln as in approach **2** to the *logarithmic function*, and take exp as its inverse function. It is then possible to define the value of e as exp 1, establish the equivalence of exp x and e^x, and prove the other properties. This is widely held to be the most satisfactory approach mathematically, but it has to be admitted that it is artificial and does not match up with any of the ways in which exp is usually first encountered.

3. Some other property of exp may be used as a definition. It may be defined as the unique function that satisfies the differential equation $dy/dx = y$ (that is, as a function that is equal to its own derivative), with $y = 1$ when $x = 0$. Alternatively, property (iv) or (v) above could be taken as the definition of exp x. In each case, it has to be shown that the other properties follow.

exponential function to base a Let a be a positive number not equal to 1. The **exponential function to base *a*** is the function f such that $f(x) = a^x$ for all x in **R**. This must be clearly distinguished from what is commonly called 'the' *exponential function*. The graphs $y = 2^x$ and $y = (\frac{1}{2})^x$ illustrate the essential difference between the cases when $a > 1$ and $a < 1$. See also *exponential growth* and *exponential decay*.

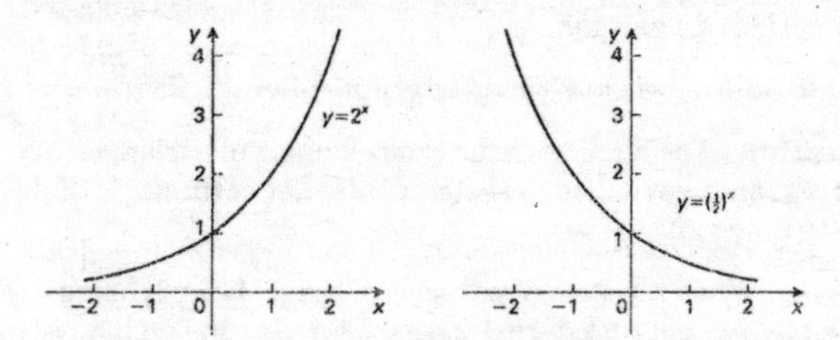

Clarifying just what is meant by a^x can be done in two ways:

1. The familiar rules for indices (see *index*) give a meaning to a^x for rational values of x. For x not rational, take a sequence of rationals that approximate more and more closely to x. For example, when $x = \sqrt{2}$, such a sequence could be 1.4, 1.41, 1.414, ... Now each of the values $a^{1.4}$, $a^{1.41}$, $a^{1.414}$, . . . has a meaning, since in each case the index is rational. It can be proved that this sequence of values has a limit, and this limit is then taken as the definition of $a^{\sqrt{2}}$. The method is applicable for any real value of x.

2. Alternatively, suppose that exp has been defined (say by approach **2** to the *exponential function*) and that ln is its *inverse function*. Then the following can be taken as a definition: $a^x = \exp(x \ln a)$. This approach is less elementary, but really more satisfactory than **1**. It follows that $\ln(a^x) = x \ln a$, as would be expected, and the following can be proved:

(i) $a^{x+y} = a^x a^y, a^{-x} = 1/a^x$ and $(a^x)^y = a^{xy}$.

(ii) When n is a positive integer, a^n, defined in this way, is indeed equal to the product $a \times a \times \cdots \times a$ with n occurrences of a, and $a^{1/n}$ is equal to $\sqrt[n]{a}$.

(iii) $\frac{d}{dx}(a^x) = a^x \ln a$.

exponential growth When $y = Ae^{kt}$, where A (> 0) and k are constants, and t represents some measurement of time, y can be said to be exhibiting **exponential growth**. This occurs when $dy/dt = ky$; that is, when the *rate of change* of the quantity y at any time is proportional to the value of y at that time. When $k > 0$, then y is growing larger with x, and moreover the rate at which y is increasing increases with x. In fact, any quantity with exponential growth (with $k > 0$) ultimately outgrows any quantity growing linearly or in proportion to a fixed power of t. When $k < 0$, the term *exponential decay* may be used.

extension The difference $x - l$ between the actual length x of a string or spring and its natural length l. The extension of a spring is negative when the spring is compressed.

exterior See *Jordan Curve Theorem*.

exterior angle (of a polygon) The angle between one side and the extension of an adjacent side of a polygon.

exterior angle (with respect to a transversal of a pair of lines) See *transversal*.

external bisector The bisector of the exterior angle of a triangle (or polygon) is sometimes called the **external bisector** of the angle of the triangle (or polygon).

external force When a system of particles or a rigid body is being considered as a whole, an **external force** is a force acting on the system from outside. Compare *internal force*.

extrapolation Suppose that certain values $f(x_0), f(x_1), \ldots, f(x_n)$ of a function f are known, where $x_0 < x_1 < \cdots < x_n$. A method of finding from these an approximation for $f(x)$, for a given value of x that lies outside the interval $[x_0, x_n]$, is called **extrapolation**. Such methods are normally far less reliable than *interpolation*, in which x lies between x_0 and x_n.

factor See *divides*.

factorial For a positive integer n, the notation $n!$ (read as 'n **factorial**') is used for the product $n(n-1)(n-2)\cdots\times 2\times 1$. Thus $4! = 4\times 3\times 2\times 1 = 24$ and $10! = 10\times 9\times 8\times 7\times 6\times 5\times 4\times 3\times 2\times 1 = 3\,628\,800$. Also, by definition, $0! = 1$.

Factor Theorem The following result, which is an immediate consequence of the *Remainder Theorem*:

THEOREM: Let $f(x)$ be a polynomial. Then $x-h$ is a factor of $f(x)$ if and only if $f(h) = 0$.

The theorem is valuable for finding factors of polynomials. For example, to factorize $2x^3+3x^2-12x-20$, look first for possible factors $x-h$, where h is an integer. Here h must divide 20. Try possible values for h, and calculate $f(h)$. It is found that $f(-2) = -16+12+24-20 = 0$, and so $x+2$ is a factor. Now divide the polynomial by this factor to obtain a quadratic which it may be possible to factorize further.

family A set whose elements are themselves sets may be called a **family**. In certain other circumstances, for example where less formal language is appropriate, the word 'family' may be used as an alternative to 'set'.

family of distributions A set of distributions which have the same general mathematical formula. A member of the family is obtained by choosing specific values for the *parameters* in the formula.

***F*-distribution** A non-negative continuous *distribution* formed from the ratio of the distributions of two independent random variables with *chi-squared distributions*, each divided by its degrees of freedom. The mean is $\dfrac{\nu_2}{\nu_2-2}$ and the variance is $\dfrac{2\nu_2^2(\nu_2+\nu_1-2)}{\nu_1(\nu_2-2)^2(\nu_2-4)}$, where ν_1 and ν_2 are the degrees of freedom of the numerator and denominator respectively. It is used to test the hypothesis that two normally distributed random variables have the same variance, and in *regression* to test the relationship between an explanatory variable and the dependent variable. The distribution is skewed to the right. Tables relating to the distribution are available.

feasible region See *linear programming*.

femto- Prefix used with *SI units* to denote multiplication by 10^{-15}.

Fermat, Pierre de (1601–1665) Leading mathematician of the first half of the seventeenth century, remembered chiefly for his work in the theory of numbers, including *Fermat's Little Theorem* and what is known as *Fermat's*

Last Theorem. His work on tangents was an acknowledged inspiration to Newton in the latter's development of the calculus. Fermat introduced coordinates as a means of studying curves. Professionally he was a judge in Toulouse, and to mathematicians he is the 'Prince of Amateurs'.

Fermat prime A *prime* of the form $2^{2^r}+1$. At present, the only known primes of this form are those given by $r = 0, 1, 2, 3$ and 4.

Fermat's Last Theorem The statement that, for all integers $n > 2$, the equation $x^n + y^n = z^n$ has no solution in positive integers. Fermat wrote in the margin of a book that he had a proof of this, but as he never repeated the claim it is likely that he realized the incompleteness of his supposed proof. Since Fermat's time, the 'theorem' has been treated as a conjecture and much research has been done on the problem. The discovery of a proof was announced in 1994. It is, not surprisingly, long and highly advanced but, if correct, is likely to be simplified in due course.

Fermat's Little Theorem The name sometimes given to the following result:

THEOREM: Let p be a prime, and let a be an integer not divisible by p. Then $a^{p-1} \equiv 1 \pmod{p}$.

Sometimes the name is given instead to the following, which is a corollary of the preceding result:

THEOREM: If p is a prime and a is any integer, then $a^p \equiv a \pmod{p}$.

Feuerbach's Theorem See *nine-point circle*.

Fibonacci (about 1170–1250) Pseudonym of one of the first European mathematicians to emerge after the Dark Ages. An Italian merchant by the name of Leonardo of Pisa, he was one of those who introduced the Hindu–Arabic number system to Europe. He strongly advocated this system in *Liber abaci*, published in 1202, which also contained problems including one that gives rise to the *Fibonacci numbers*. Other writings of his deal with Euclidean geometry and *Diophantine equations*.

Fibonacci number One of the numbers in the **Fibonacci sequence** 1, 1, 2, 3, 5, 8, 13, . . . , where each number after the second is the sum of the two preceding numbers in the sequence. This sequence has many interesting properties. For instance, the sequence consisting of the ratios of one Fibonacci number to the previous one, $\frac{1}{1}, \frac{2}{1}, \frac{3}{2}, \frac{5}{3}, \frac{8}{5}, \frac{13}{8}, \ldots$, has the limit τ, the *golden ratio*. See also *difference equation* and *generating function*.

Fibonacci sequence See *Fibonacci number*.

fictitious force A force that may be assumed to exist by an observer whose *frame of reference* is accelerating relative to an inertial frame of reference. Suppose, for example, that a frame of reference with origin O is rotating relative to an inertial frame of reference with the same origin. A particle P subject to a certain total force satisfies Newton's second law of motion, relative to the inertial frame of reference. To the observer in the rotating frame, the particle appears to satisfy an equation of motion that is Newton's second law of motion with additional terms. The observer may suppose that these terms are explained by certain fictitious forces. When these forces are assumed to exist, Newton's laws appear to hold in the non-inertial frame of reference.

Consider the special case in which the rotating frame of reference has a constant angular velocity and the particle is moving in a plane perpendicular to the angular velocity of the rotating frame of reference. One fictitious force is in the direction along *OP* and is called the **centrifugal force**. This is the force outwards that is believed to exist by a rider on a roundabout. The second fictitious force is perpendicular to the path of *P* as seen by the observer in the rotating frame of reference and is called the **Coriolis force**.

To an observer standing on the Earth, which is rotating about its axis, an object such as an intercontinental missile appears to deviate from its path due to the Coriolis force. The deviation is to the right in the northern hemisphere and to the left in the southern hemisphere. This force, first described by the French mathematician and engineer, Gustave-Gaspard de Coriolis (1792–1843), also has important applications to the movement of air masses in meteorology.

Similar fictitious forces may arise whenever the observer's frame of reference is accelerating relative to an inertial frame of reference, as when a passenger in an accelerating lift witnesses a ceiling tile fall from the roof of the cabin.

field A commutative ring with identity (see *ring*) with the following additional property:

10. For each a ($\neq 0$), there is an element a^{-1} such that $a^{-1}a = 1$.

(The axiom numbering here follows on from that used for ring and *integral domain*.) From the defining properties of a field, Axioms **1** to **8** and Axiom **10**, it can be shown that $ab = 0$ only if $a = 0$ or $b = 0$. Thus Axiom **9** holds, and so any field is an integral domain. Familiar examples of fields are the set **Q** of rational numbers, the set **R** of real numbers and the set **C** of complex numbers, each with the usual addition and multiplication. Another example is $\mathbf{Z}_p$, consisting of the set $\{ 0, 1, 2, \ldots, p-1 \}$ with addition and multiplication modulo p, where p is a prime.

field of force A **field of force** is said to exist when a force acts at any point of a region of space. A particle placed at any point of the region then experiences the force, which may depend on the position and on time. Examples are gravitational, electric and magnetic fields of force.

Fields Medal A prize awarded for outstanding achievements in mathematics, considered by mathematicians to be equivalent to a Nobel Prize. Medals are awarded to individuals at successive International Congresses of Mathematicians, normally held at four-year intervals. The proposal was made by J. C. Fields to found two gold medals, using funds remaining after the financing of the Congress in Toronto in 1924. The first two medals were presented at the Congress in Oslo in 1936. In some instances, three or four medals have been awarded. It has been the practice to make the awards to mathematicians under the age of 40.

figurate numbers Numerous sequences of numbers associated with different geometrical figures were considered special by the Pythagoreans and other early mathematicians, and these are known loosely as **figurate numbers**. They include the *perfect squares*, the *triangular numbers* and the *tetrahedral numbers*, and others known as the pentagonal and the hexagonal numbers.

finite differences Let $x_0, x_1, x_2, \ldots, x_n$ be equally spaced values, so that $x_i = x_0 + ih$, for $i = 1, 2, \ldots, n$. Suppose that the values $f_0, f_1, \ldots, f_n$ are known, where $f_i = f(x_i)$, for some function f. The first differences are defined, for $i = 0, 1, 2, \ldots, n-1$, by $\Delta f_i = f_{i+1} - f_i$. The second differences are defined by $\Delta^2 f_i = \Delta(\Delta f_i) = \Delta f_{i+1} - \Delta f_i$ and, in general, the k-th differences are defined by $\Delta^k f_i = \Delta(\Delta^{k-1} f_i) = \Delta^{k-1} f_{i+1} - \Delta^{k-1} f_i$. For a polynomial of degree n, the $(n+1)$-th differences are zero.

These **finite differences** may be displayed in a table, as in the following example. Alongside it is a numerical example.

x_0	f_0				1.0	1.000			
		Δf_0					0.331		
x_1	f_1		$\Delta^2 f_0$		1.1	1.331		0.066	
		Δf_1		$\Delta^3 f_0$			0.397		0.006
x_2	f_2		$\Delta^2 f_1$		1.2	1.728		0.072	
		Δf_2					0.469		
x_3	f_3				1.3	2.197			

With such tables it should be appreciated that if the values $f_0, f_1, f_2, \ldots, f_n$ are *rounded* values then increasingly serious errors result in the succeeding columns.

Numerical methods using finite differences have been extensively developed. They may be used for *interpolation*, as in the *Gregory–Newton forward difference formula*, for finding a polynomial that approximates to a given function, or for estimating derivatives from a table of values.

finite sequence See *sequence*.

finite series See *series*.

first derivative A term used for the *derivative* when it is being contrasted with *higher derivatives*.

Fisher, Ronald Aylmer (1890–1962) British geneticist and statistician who established methods of designing experiments and analysing results that have been extensively used ever since. His influential book on statistical methods appeared in 1925. He developed the *t-test* and the use of *contingency tables*, and is responsible for the method known as *analysis of variance*.

fixed point See *transformation* (of the plane).

fixed-point iteration To find a root of an equation $f(x) = 0$ by the method of **fixed-point iteration**, the equation is first rewritten in the form $x = g(x)$. Starting with an initial approximation x_0 to the root, the values $x_1, x_2, x_3, \ldots$ are calculated using $x_{n+1} = g(x_n)$. The method is said to converge if these values tend to a limit α. If they do, then $\alpha = g(\alpha)$ and so α is a root of the original equation.

A root of $x = g(x)$ occurs where the graph $y = g(x)$ meets the line $y = x$. It can be shown that, if $|g'(x)| < 1$ in an interval containing both the root and the value x_0, the method will converge, but not if $|g'(x)| > 1$. This can be illustrated in figures such as those shown below, which are for cases in which $g'(x)$ is positive. For example, the equation $x^3 - x - 1 = 0$ has a root between 1 and 2, so take $x_0 = 1.5$. The equation can be written in the form $x = g(x)$ in several ways, such as (i) $x = x^3 - 1$ or (ii) $x = (x+1)^{1/3}$. In case (i), $g'(x) = 3x^2$, which does not satisfy $|g'(x)| < 1$ near x_0. In case (ii), $g'(x) = \frac{1}{3}(x+1)^{-2/3}$ and $g'(1.5) \approx 0.2$, so it is likely that with this formulation the method converges.

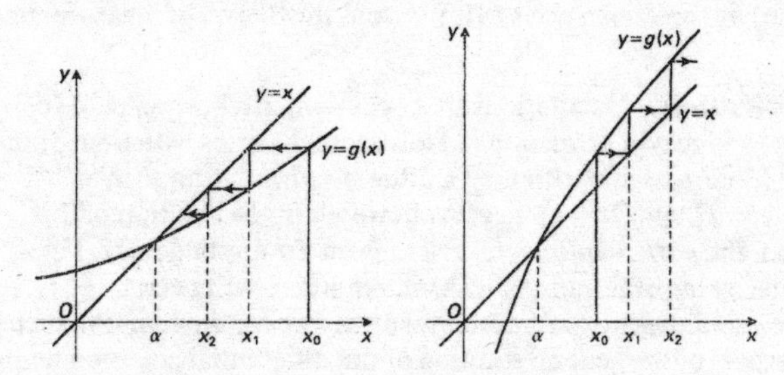

fixed-point notation See *floating-point notation*.

floating-point notation A method of writing real numbers, used in computing, in which a number is written as $a \times 10^n$, where $0.1 \leq a < 1$ and n is an integer. The number a is called the **mantissa** and n is the **exponent**. Thus 634.8 and 0.002 34 are written as 0.6348×10^3 and 0.234×10^{-2}. (There is also a base 2 version similar to the base 10 version just described.)

This is in contrast to **fixed-point notation**, in which all numbers are given by means of a fixed number of digits with a fixed number of digits after the decimal point. For example, if numbers are given by means of 8 digits with four of them after the decimal point, the two numbers above would be written (with an approximation) as 0634.8000 and 0000.0023. Integers are likely to be written in fixed-point notation; consequently, in the context of computers, some authors use 'fixed-point' to mean 'integer'.

fluent See *fluxion*.

fluxion In Newton's work on calculus, he thought of the variable x as a 'flowing quantity' or **fluent**. The rate of change of x was called the **fluxion** of x, denoted by $\dot{x}$.

focus (plural: foci) See *conic*, *ellipse*, *hyperbola* and *parabola*.

foot of the perpendicular See *projection* (of a point on a line) and *projection* (of a point on a plane).

force In the real world, many different kinds of force are part of everyday life. A human being or an animal may use muscles to apply a force to move or try to move an object. An engine may produce a force that can be applied to turn a wheel. Commonly experienced forces are the force due to gravity, which acts throughout the region occupied by an object, forces that act between two bodies that are in contact, forces within a body that deform or restore its shape, and electrical and magnetic forces.

In a mathematical model, a force has a magnitude, a direction and a *point of application*. It acts at a point and may be represented by a vector whose length is the magnitude of the force and whose direction is the direction of the force.

Force has the dimensions MLT^{-2}, and the SI unit of measurement is the *newton*.

forced oscillations Oscillations that occur when a body capable of oscillating is subject to an applied force which varies with time. If the applied force is itself oscillatory, a differential equation such as $m\ddot{x} + kx = F_0 \sin(\Omega t + \epsilon)$ may be obtained. In the solution of this equation, the *particular integral* arises from the applied force. For a particular value of Ω, namely $\sqrt{k/m}$, *resonance* will occur.

If the oscillations are *damped* as well as forced, the *complementary function* part of the general solution of the differential equation tends to zero as t tends to infinity, and the particular integral arising from the applied force describes the eventual motion.

forward difference formula See *Gregory–Newton forward difference formula*.

Foucault pendulum A pendulum consisting of a heavy bob suspended by a long inextensible string from a fixed point, free to swing in any direction, designed to demonstrate the rotation of the Earth. In the original experiment in 1851, the French physicist Jean-Bernard-Léon Foucault (1819–1868) suspended a bob of 28 kilograms by a wire 67 metres long from the dome of the Panthéon in Paris. If the experiment were set up at the north or south pole, the vertical plane of the swinging bob would appear to an observer fixed on the Earth to precess or rotate once a day. At a location of latitude λ in the northern hemisphere, the vertical plane of the swinging bob would precess in a clockwise direction with an angular speed of $\omega \sin \lambda$, where $\omega = 7.29 \times 10^{-5}$ rad s^{-1}, the angular speed of rotation of the Earth. The period of precession in Oxford, for example, would be about $30\frac{1}{2}$ hours.

Four Colour Theorem It has been observed by map-makers through the centuries that any geographical map (that is, a division of the plane into regions) can be coloured with just four colours in such a way that no two neighbouring regions have the same colour. A proof of this, the **Four Colour Theorem**, was sought by mathematicians from about the 1850s. In 1890 Heawood proved that five colours would suffice, but it was not until 1976 that Appel and Haken proved the Four Colour Theorem itself. Initially, some mathematicians were sceptical of the proof because it relied, in an essential way, on a massive amount of checking of configurations by computer that could not easily be verified independently. However, the proof is now generally accepted and considered a magnificent achievement.

four-group See *Klein four-group*.

Fourier, (Jean Baptiste) Joseph, Baron (1768–1830) French engineer and mathematician, best known in mathematics for his fundamental contributions to the theory of heat conduction and his study of trigonometric series. These so-called Fourier series are of immense importance in physics, engineering and other disciplines, as well as being of great mathematical interest.

fourth root of unity A complex number z such that $z^4 = 1$. There are 4 fourth roots of unity and they are 1, i, -1 and $-i$. (See *n-th root of unity*.)

fractal A set of points whose *fractal dimension* is not an integer or, loosely, any set of similar complexity. Fractals are typically sets with infinitely complex structure and usually possess some measure of self-similarity, whereby any part of the set contains within it a scaled-down version of the whole set. Examples are the *Cantor set* and the *Koch curve*.

fractal dimension One of the many extensions of the notion of dimension to objects for which the traditional concept of dimension is not appropriate. The fractal dimension may have a non-integer value. The *Koch curve* has dimension $\ln 4/\ln 3 \approx 1.26$. Being between 1 and 2, this reflects the fact that the set is, as it were, too 'thick' to count as a curve and too 'thin' to count as an area. The *Cantor set* has dimension $\ln 2/\ln 3$. Fractal dimension has found many practical applications in the analysis of chaotic or noisy processes (see *chaos*).

fraction The **fraction** a/b, where a and b are positive integers, was historically obtained by dividing a unit length into b parts and taking a of these parts. The number a is the **numerator** and the number b is the **denominator**. It is a **proper fraction** if $a < b$ and an **improper fraction** if $a > b$. Any fraction can be expressed as $c + d/e$, where c is an integer and

d/e is a proper fraction, and in this form it is called a **mixed fraction**. For example, $3\frac{1}{2}$ is a mixed fraction (equal to 7/2).

fractional part For any real number x, its **fractional part** is equal to $x - [x]$, where $[x]$ is the *integer part* of x. It may be denoted by $\{x\}$. The fractional part r of any real number always satisfies $0 \leq r < 1$.

frame of reference In mechanics, a means by which an observer specifies positions and describes the motion of bodies. For example, an observer may use a Cartesian coordinate system or a polar coordinate system. In some circumstances, it may be useful to consider two or more different frames of reference, each with its own observer. One frame of reference, its origin and axes, may be moving relative to another. The motion of a particle, for example, as it appears to one observer will be different from the motion as seen by the other observer.

A frame of reference in which *Newton's laws of motion* hold is called an **inertial frame (of reference)**. Any frame of reference that is at rest or moving with constant velocity relative to an inertial frame is an inertial frame. A frame of reference that is accelerating or rotating with respect to an inertial frame is not an inertial frame.

A frame of reference fixed on the Earth is not an inertial frame because of the rotation of the Earth. However, such a frame of reference may be assumed to be an inertial frame in problems where the rotation of the Earth has little effect.

Frege, (Friedrich Ludwig) Gottlob (1848–1925) German mathematician and philosopher, founder of the subject of mathematical logic. In his works of 1879 and 1884, he developed the fundamental ideas, invented the standard notation of *quantifiers* and variables, and studied the foundations of arithmetic. Not widely recognized at the time, his work was disseminated primarily through others such as Peano and Russell.

frequency (in mechanics) When *oscillations*, or cycles, occur with period T, the **frequency** is equal to $1/T$. The frequency is equal to the number of oscillations or cycles that take place per unit time.

Frequency has the dimensions T^{-1}, and the SI unit of measurement is the *hertz*.

frequency (in statistics) The number of times that a particular value occurs as an observation. In *grouped data*, the frequency corresponding to a group is the number of observations that lie in that group. If numerical data are grouped by means of class intervals, the frequency corresponding to a class interval is the number of observations in that interval.

frequency distribution For nominal or discrete data, the information consisting of the possible values and the corresponding *frequencies* is called the **frequency distribution**. For *grouped data*, it gives the information consisting of the groups and the corresponding frequencies. It may be presented in a table or in a diagram such as a *bar chart*, *histogram* or *stem-and-leaf plot*.

friction Suppose that two bodies are in contact, and that the frictional force and the normal reaction have magnitudes F and N respectively (see *contact force*). The **coefficient of static friction** μ_s is the ratio F/N in the limiting case when the bodies are just about to move relative to each other. Thus if the bodies are at rest relative to each other, $F \leq \mu_s N$. The **coefficient of kinetic friction** μ_k is the ratio F/N when the bodies are sliding; that is, in contact and moving relative to each other. These coefficients of friction depend on the materials of which the bodies are made. Normally, μ_k is somewhat less than μ_s. See also *angle of friction*.

frictional couple A *couple* created by a pair of equal and opposite *frictional forces*. A frictional couple may occur, for example, when a rigid body rotates about an axis.

frictional force See *contact force*.

frustum (plural: frusta) A **frustum** of a *right-circular cone* is the part between two parallel planes perpendicular to the axis. Suppose that the planes are a distance h apart, and that the circles that form the top and bottom of the frustum have radii a and b. Then the volume of the frustum equals $\frac{1}{3}\pi h(a^2 + ab + b^2)$. Let l be the **slant height** of the frustum; that is, the length of the part of a generator between the top and bottom of the frustum. Then the area of the curved surface of the frustum equals $\pi(a + b)l$.

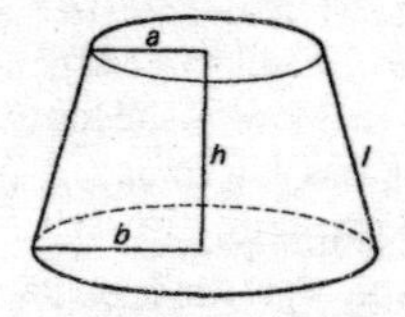

F-test A test that uses the *F-distribution*.

fulcrum See *lever*.

function A **function** f from S to T, where S and T are non-empty sets, is a rule that associates with each element of S (the **domain**) a unique element of T (the **codomain**). Thus it is the same thing as a *mapping*. The word 'function' tends to be used when the domain S is the set $\mathbf{R}$ of real numbers, or some subset of $\mathbf{R}$, and the codomain T is $\mathbf{R}$ (see *real function*). The notation $f: S \to T$, read as 'f from S to T', is used. If $x \in S$, then $f(x)$ is the **image** of x under f. The subset of T consisting of those elements that are images of elements of S under f, that is, the set $\{ y \mid y = f(x), \text{for some } x \text{ in } S \}$, is the **range** of f. If $f(x) = y$, it is said that f **maps** x to y, written $f: x \mapsto y$. If the graph of f is then taken to be $y = f(x)$, it may be said that y is a function of x. When $x = a$, $f(a)$ is the corresponding **value** of the function.

Fundamental Theorem of Algebra The following important theorem in mathematics, concerned with the roots of polynomial equations:

THEOREM: Every polynomial equation

$$a_n z^n + a_{n-1} z^{n-1} + \cdots + a_1 z + a_0 = 0,$$

where the a_i are real or complex numbers and $a_n \neq 0$, has a root in the set of complex numbers.

It follows that, if $f(z) = a_n z^n + a_{n-1} z^{n-1} + \cdots + a_1 z + a_0$, there exist complex numbers $\alpha_1, \alpha_2, \ldots, \alpha_n$ (not necessarily distinct) such that

$$f(z) = a_n (z - \alpha_1)(z - \alpha_2) \ldots (z - \alpha_n).$$

Hence the equation $f(z) = 0$ cannot have more than n distinct roots.

Fundamental Theorem of Arithmetic = *Unique Factorization Theorem.*

Fundamental Theorem of Calculus A sound approach to integration defines the *integral*

$$\int_a^b f(x)\,dx$$

as the limit, in a certain sense, of a sum. That this can be evaluated, when f is continuous, by finding an *antiderivative* of f, is the result embodied in the so-called **Fundamental Theorem of Calculus**. It establishes that integration is the reverse process to differentiation:

THEOREM: If f is continuous on $[a, b]$ and ϕ is a function such that $\phi'(x) = f(x)$ for all x in $[a, b]$, then

$$\int_a^b f(x)\,dx = \phi(b) - \phi(a).$$

Fundamental Theorem of Game Theory The following theorem, also known as the 'Minimax Theorem', due to von Neumann:

THEOREM: Suppose that, in a matrix game, $E(\mathbf{x}, \mathbf{y})$ is the expectation, where $\mathbf{x}$ and $\mathbf{y}$ are mixed strategies for the two players. Then

$$\max_{\mathbf{x}} \min_{\mathbf{y}} E(\mathbf{x}, \mathbf{y}) = \min_{\mathbf{y}} \max_{\mathbf{x}} E(\mathbf{x}, \mathbf{y}).$$

By using a *maximin strategy*, one player, R, ensures that the expectation is at least as large as the left-hand side of the equation appearing in the theorem. Similarly, by using a *minimax strategy*, the other player, C, ensures that the expectation is less than or equal to the right-hand side of the equation. Such strategies may be called **optimal strategies** for R and C. Since, by the theorem, the two sides of the equation are equal, then if R and C use optimal strategies the expectation is equal to the common value, which is called the **value** of the game.

For example, consider the game given by the matrix

$$\begin{bmatrix} 4 & 2 \\ 3 & 4 \end{bmatrix}.$$

If $\mathbf{x}^* = (\frac{1}{3}, \frac{2}{3})$, it can be shown that $E(\mathbf{x}^*, \mathbf{y}) \geq 10/3$ for all $\mathbf{y}$. Also, if $\mathbf{y}^* = (\frac{2}{3}, \frac{1}{3})$, then $E(\mathbf{x}, \mathbf{y}^*) \leq 10/3$ for all $\mathbf{x}$. It follows that the value of the game is 10/3, and $\mathbf{x}^*$ and $\mathbf{y}^*$ are optimal strategies for the two players.

G

g See *gravity.*

Galileo Galilei (1564–1642) Italian mathematician, astronomer and physicist who established the method of studying dynamics by a combination of theory and experiment. He formulated and verified by experiment the law $s = \frac{1}{2}at^2$ of uniform acceleration for falling bodies, and derived the parabolic path of a projectile. He developed the telescope and was the first to use it to make significant and outstanding astronomical observations. In later life, his support for the Copernican theory that the planets travel round the Sun resulted in conflict with the Church and consequent trial and house arrest.

Galois, Évariste (1811–1832) French mathematician who had made major contributions to the theory of equations before he died at the age of 20, shot in a duel. His work developed the necessary group theory to deal with the question of whether an equation can be solved algebraically. He spent the night before the duel writing a letter containing notes of his discoveries.

Galton, Francis (1822–1911) British explorer and anthropologist, a cousin of Charles Darwin. His primary interest was eugenics. He made contributions to statistics in the areas of *regression* and *correlation.*

game An attempt to represent and analyse mathematically some conflict situation in which the outcome depends on the choices made by the opponents. The applications of **game theory** are not primarily concerned with recreational activities. Games may be used to investigate problems in business, personal relationships, military manœuvres and other areas involving decision-making. One particular kind of game for which the theory has been well developed is the *matrix game.*

Gauss, Carl Friedrich (1777–1855) German mathematician and astronomer, perhaps the greatest pure mathematician of all time. He also made enormous contributions to other parts of mathematics, physics and astronomy. He was highly talented as a child. At the age of 18, he invented the method of least squares and made the new discovery that a 17-sided regular polygon could be constructed with ruler and compasses. By the age of 24, he was ready to publish his *Disquisitiones arithmeticae*, a book that was to have a profound influence on the theory of numbers. In this, he proved the *Fundamental Theorem of Arithmetic* and the *Fundamental Theorem of Algebra.* In later work, he developed the theory of curved surfaces using methods now known as differential geometry. His work on complex functions was fundamental but, like his discovery of *non-Euclidean geometry,* it was not published at the time. He introduced what is now known to statisticians as the *Gaussian distribution*. His memoir on potential theory was just one of his contributions to applied mathematics. In astronomy, his great powers of mental calculation allowed him to calculate the orbits of comets and asteroids from limited observational data.

Gaussian distribution = *normal distribution*.

Gaussian elimination A particular systematic procedure for solving a set of linear equations in several unknowns. This is normally carried out by applying *elementary row operations* to the augmented matrix

$$\begin{bmatrix} a_{11} & a_{12} & \dots & a_{1n} & b_1 \\ a_{21} & a_{22} & \dots & a_{2n} & b_2 \\ \vdots & \vdots & \ddots & \vdots & \vdots \\ a_{m1} & a_{m2} & \dots & a_{mn} & b_m \end{bmatrix}$$

to transform it to *echelon form*. The method is to divide the first row by a_{11} and then subtract suitable multiples of the first row from the subsequent rows, to obtain a matrix of the form

$$\begin{bmatrix} 1 & a'_{12} & \dots & a'_{1n} & b'_1 \\ 0 & a'_{22} & \dots & a'_{2n} & b'_2 \\ \vdots & \vdots & \ddots & \vdots & \vdots \\ 0 & a'_{m2} & \dots & a'_{mn} & b'_m \end{bmatrix}.$$

(If $a_{11} = 0$, it is necessary to interchange two rows first.) The first row now remains untouched and the process is repeated with the remaining rows, dividing the second row by a'_{22} to produce a 1, and subtracting suitable multiples of the new second row from the subsequent rows to produce zeros below that 1. The method continues in the same way. The essential point is that the corresponding set of equations at any stage has the same solution set as the original. (See also *simultaneous linear equations*.)

Gauss–Jordan elimination An extension of the method of *Gaussian elimination*. At the stage when the i-th row has been divided by a suitable value to obtain a 1, suitable multiples of this row are subtracted, not only from subsequent rows, but also from preceding rows to produce zeros both below and above the 1. The result of this systematic method is that the augmented matrix is transformed into *reduced echelon form*. As a method for solving *simultaneous linear equations*, Gauss–Jordan elimination in fact requires more work than Gaussian elimination followed by *back-substitution*, and so it is not in general recommended.

g.c.d. = *greatest common divisor*.

general solution See *differential equation*.

generating function The power series $G(x)$, where

$$G(x) = g_0 + g_1x + g_2x^2 + g_3x^3 + \cdots,$$

is the **generating function** for the infinite sequence $g_0, g_1, g_2, g_3, \ldots$. (Notice that it is convenient here to start the sequence with a term with subscript 0). Such power series can be manipulated algebraically, and it can be shown, for example, that

$$\frac{1}{1-x} = 1 + x + x^2 + x^3 + \cdots,$$
$$\frac{1}{(1-x)^2} = 1 + 2x + 3x^2 + 4x^3 + \cdots.$$

Hence, $1/(1-x)$ and $1/(1-x)^2$ are the generating functions for the sequences $1, 1, 1, 1, \ldots$ and $1, 2, 3, 4, \ldots$, respectively.

The *Fibonacci sequence* $f_0, f_1, f_2, \ldots$ is given by $f_0 = 1, f_1 = 1$, and $f_{n+2} = f_{n+1} + f_n$. It can be shown that the generating function for this sequence is $1/(1 - x - x^2)$.

The use of generating functions enables sequences to be handled concisely and algebraically. A *difference equation* for a sequence can lead to an equation for the corresponding generating function, and the use of *partial fractions*, for example, may then lead to a formula for the n-th term of the sequence.

generator See *cone* and *cylinder*.

geodesic A curve on a surface, joining two given points, that is the shortest curve between the two points. On a sphere, a geodesic is an arc of a great circle through the two given points. This arc is unique unless the two points are *antipodal*.

geometrical representation (of a vector) = *representation* (of a vector).

geometric distribution The discrete probability *distribution* for the number of experiments required to achieve the first success in a sequence of independent experiments, all with the same probability p of success. The *probability mass function* is given by $\Pr(X = r) = p(1-p)^{r-1}$, for $r = 1, 2, \ldots$. It has mean $1/p$ and variance $(1-p)/p^2$.

geometric mean See *mean*.

geometric sequence A finite or infinite sequence $a_1, a_2, a_3, \ldots$ with a **common ratio** r, so that $a_2/a_1 = r, a_3/a_2 = r, \ldots$. The first term is usually denoted by a. For example, $3, 6, 12, 24, 48, \ldots$ is the geometric sequence with $a = 3, r = 2$. In such a geometric sequence, the n-th term a_n is given by $a_n = ar^{n-1}$.

geometric series A series $a_1 + a_2 + a_3 + \cdots$ (which may be finite or infinite) in which the terms form a *geometric sequence*. Thus the terms have a **common ratio** r with $a_k/a_{k-1} = r$ for all k. If the first term a_1 equals a, then $a_k = ar^{k-1}$. Let s_n be the sum of the first n terms, so that $s_n = a + ar + ar^2 + \cdots + ar^{n-1}$. Then s_n is given (when $r \neq 1$) by the formulae

$$s_n = \frac{a(1 - r^n)}{1 - r} = \frac{a(r^n - 1)}{r - 1}.$$

If the common ratio r satisfies $-1 < r < 1$, then $r^n \to 0$ and it can be seen that $s_n \to a/(1 - r)$. The value $a/(1 - r)$ is called the **sum to infinity** of the series $a + ar + ar^2 + \cdots$. In particular, for $-1 < x < 1$, the geometric series $1 + x + x^2 + \cdots$ has sum to infinity equal to $1/(1 - x)$. For example, putting $x = \frac{1}{2}$, the series $1 + \frac{1}{2} + \frac{1}{4} + \frac{1}{8} + \cdots$ has sum 2. If $x \leq -1$ or $x \geq 1$, then s_n does not tend to a limit and the series has no sum to infinity.

giga- Prefix used with *SI units* to denote multiplication by 10^9.

Gödel, Kurt (1906–1978) Logician and mathematician who showed that the consistency of elementary arithmetic could not be proved from within the system itself. This result followed from his proof that any formal axiomatic system contains undecidable propositions. It undermined the hopes of those who had been attempting to determine axioms from which all mathematics could be deduced. Born in Brno, he was at the University of Vienna from 1930 until he emigrated to the United States in 1940.

Goldbach, Christian (1690–1764) Mathematician born in Prussia, who later became professor in St Petersburg and tutor to the Tsar in Moscow. *Goldbach's conjecture*, for which he is remembered, was proposed in 1742 in a letter to Euler.

Goldbach's conjecture The conjecture that every even integer greater than 2 is the sum of two *primes*. Neither proved nor disproved, Goldbach's conjecture remains one of the most famous unsolved problems in number theory.

golden ratio, golden rectangle See *golden section*.

golden section A line segment is divided in **golden section** if the ratio of the whole length to the larger part is equal to the ratio of the larger part to the smaller part. This definition implies that, if the smaller part has unit length and the larger part has length τ, then $(\tau + 1)/\tau = \tau/1$. It follows that $\tau^2 - \tau - 1 = 0$, which gives $\tau = \frac{1}{2}(1 + \sqrt{5}) = 1.6180$, to 4 decimal places. This number τ is the **golden ratio**. A **golden rectangle**, whose sides are in this ratio, has throughout history been considered to have a particularly pleasing shape. It has the property that the removal of a square from one end of it leaves a rectangle that has the same shape.

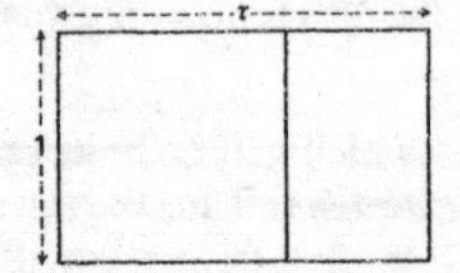

goodness-of-fit test = *chi-squared test*.

googol A fanciful name for the number 10^{100}, written in decimal notation as a 1 followed by 100 zeros.

googolplex The number equal to the googolth power of 10, written in decimal notation as a 1 followed by a *googol* of zeros.

Gosset, William Sealy (1876–1937) British industrial scientist and statistician best known for his discovery of the *t-distribution*. His statistical work was motivated by his research for the brewery firm that he was with all his life. The most important of his papers, which were published under the pseudonym '**Student**', appeared in 1908.

gradient (of a curve) The **gradient** of a curve at a point P may be defined as equal to the gradient of the *tangent* to the curve at P. This definition presupposes an intuitive idea of what it means for a line to touch a curve. At a more advanced level, it is preferable to define the gradient of a curve by the methods of *differential calculus*. In the case of a graph $y = f(x)$, the gradient is equal to $f'(x)$, the value of the derivative. The tangent at P can then be defined as the line through P whose gradient equals the gradient of the curve.

gradient (of a straight line) In coordinate geometry, suppose that A and B are two points on a given straight line, and let M be the point where the line through A parallel to the x-axis meets the line through B parallel to the y-axis. Then the **gradient** of the straight line is equal to MB/AM. (Notice that here MB is the *measure* of $\overrightarrow{MB}$ where the line through M and B has positive direction upwards. In other words, MB equals the length $|MB|$ if B is above M, and equals $-|MB|$ if B is below M. Similarly, $AM = |AM|$ if M is to the right of A, and $AM = -|AM|$ if M is to the left of A. Two cases are illustrated in the figure.)

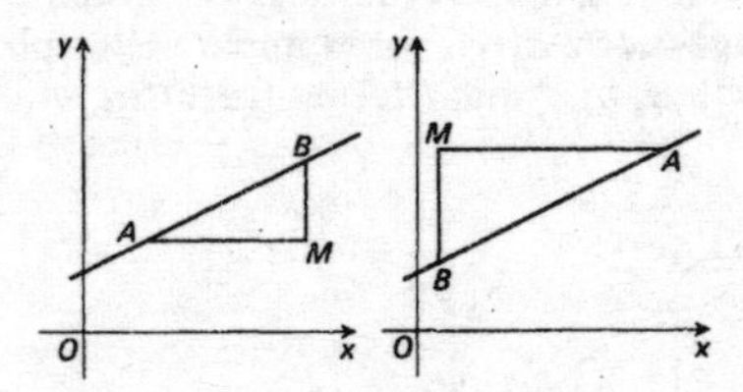

The gradient of the line through A and B may be denoted by m_{AB}, and, if A and B have coordinates (x_1, y_1) and (x_2, y_2), with $x_1 \neq x_2$, then

$$m_{AB} = \frac{y_2 - y_1}{x_2 - x_1}.$$

Though defined in terms of two points A and B on the line, the gradient of the line is independent of the choice of A and B. The line in the figure has gradient $\frac{1}{2}$.

Alternatively, the gradient may be defined as equal to $\tan\theta$, where either direction of the line makes an angle θ with the positive x-axis. (The different possible values for θ give the same value for tan θ.) If the line through A and B is vertical, that is, parallel to the y-axis, it is customary to say that the gradient is infinite. The following properties hold:

(i) Points A, B and C are collinear if and only if $m_{AB} = m_{AC}$. (This includes the case when m_{AB} and m_{AC} are both infinite.)

(ii) The lines with gradients m_1 and m_2 are parallel (to each other) if and only if $m_1 = m_2$. (This includes m_1 and m_2 both infinite.)

(iii) The lines with gradients m_1 and m_2 are perpendicular (to each other) if and only if $m_1 m_2 = -1$. (This must be reckoned to include the cases when $m_1 = 0$ and m_2 is infinite and vice versa.)

Graeco-Latin square The notion of *Latin square* can be extended to involve two sets of symbols. Suppose that one set of symbols consists of Roman letters and the other of Greek letters. A **Graeco-Latin square** is a square array in which each position contains one Roman letter and one Greek letter, such that the Roman letters form a Latin square, the Greek letters form a Latin square, and each Roman letter occurs with each Greek letter exactly once. An example of a 3×3 Graeco-Latin square is the following:

$$\begin{matrix} A\alpha & B\beta & C\gamma \\ B\gamma & C\alpha & A\beta \\ C\beta & A\gamma & B\alpha \end{matrix}$$

Such squares are used in the design of experiments.

gram In *SI units*, it is the *kilogram* that is the base unit for measuring mass. A **gram** is one-thousandth of a kilogram.

graph A number of **vertices** (or **points** or **nodes**), some of which are joined by **edges**. The edge joining the vertex u and the vertex v may be denoted by (u, v) or (v, u). The **vertex-set**, that is, the set of vertices, of a graph G may be denoted by $V(G)$ and the **edge-set** by $E(G)$. For example, the graph shown here on the left has $V(G) = \{u, v, w, x\}$ and $E(G) = \{(u, v), (u, w), (v, w), (w, x)\}$.

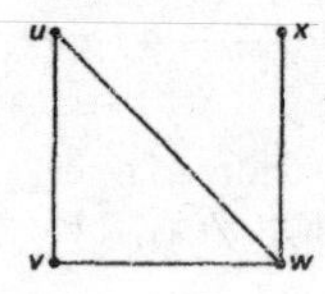

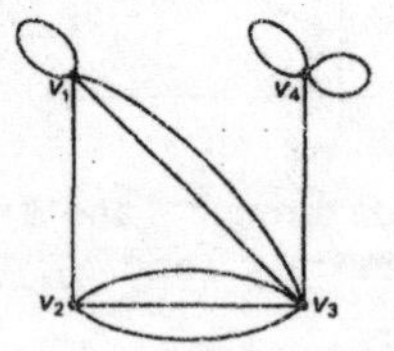

In general, a graph may have more than one edge joining a pair of vertices; when this occurs, these edges are called **multiple edges**. Also, a graph may have loops—a **loop** is an edge that joins a vertex to itself. In the other graph shown, there are 2 edges joining v_1 and v_3 and 3 edges joining v_2 and v_3; the graph also has three loops.

Normally, $V(G)$ and $E(G)$ are finite, but if this is not so, the result may also be called a graph, though some prefer to call this an **infinite graph**.

graph (of a function or mapping) For a *real function f*, the **graph** of f is the set of all pairs (x, y) in $\mathbf{R} \times \mathbf{R}$ such that $y = f(x)$ and x is in the domain of the function. For many real functions of interest, this gives a set of points that form a curve of some sort, possibly in a number of parts, that can be drawn in the plane. Such a curve defined by $y = f(x)$ is also called the graph of f. See also *mapping*.

graph (of a relation) Let R be a *binary relation* on a set S, so that, when a is related to b, this is written $a\,R\,b$. The **graph** of R is the corresponding subset of the *Cartesian product* $S \times S$, namely, the set of all pairs (a, b) such that $a\,R\,b$.

gravitational constant The constant of proportionality, denoted by G, that occurs in the *inverse square law of gravitation*. Its value is dependent on the decision to arrange for the *gravitational mass* and the *inertial mass* of a particle to have the same value. The dimensions of G are $L^3M^{-1}T^{-2}$, and its value is $6.672 \times 10^{-11}\ \mathrm{N\,m^2\,kg^{-2}}$.

gravitational force The force of attraction that exists between any two bodies, and described by the *inverse square law of gravitation*. See also *gravity*.

gravitational mass The parameter associated with a body that arises in the *inverse square law of gravitation*.

gravitational potential energy The *potential energy* associated with the gravitational force. When $\mathbf{F} = -\dfrac{GMm}{r^3}\mathbf{r}$, as in the *inverse square law of gravitation*, it can be shown that the gravitational potential energy $E_p = -GMm/r + \text{constant}$. When $\mathbf{F} = -mg\mathbf{k}$, as may be assumed near the Earth's surface, $E_p = mgz + \text{constant}$.

gravity Near the Earth's surface, a body experiences the gravitational force between the body and the Earth, which may be taken to be constant. The resulting acceleration due to gravity is $-g\mathbf{k}$, where $\mathbf{k}$ is a unit vector directed vertically upwards from the Earth's surface, assumed to be a horizontal plane. The constant g, which is the magnitude of the acceleration due to gravity, is equal to GM/R^2, where G is the *gravitational constant*, M is the mass of the Earth and R is the radius of the Earth. Near the Earth's surface, the value of g may be taken as $9.81\ \mathrm{m\,s^{-2}}$, though it varies between $9.78\ \mathrm{m\,s^{-2}}$ at the equator and $9.83\ \mathrm{m\,s^{-2}}$ at one of the poles.

great circle A circle on the surface of a sphere with its centre at the centre of the sphere (in contrast to a *small circle*). The shortest distance between two points on a sphere is along an arc of a great circle through the two points. This great circle is unique unless the two points are *antipodal*.

greatest common divisor For two non-zero integers a and b, any integer that is a divisor of both is a **common divisor**. Of all the common divisors, the greatest is the **greatest common divisor** (or **g.c.d.**), denoted by (a, b). The g.c.d. of a and b has the property of being divisible by every other common divisor of a and b. It is an important theorem that there are integers s and t such that the g.c.d. can be expressed as $sa + tb$. If the prime decompositions of a and b are known, the g.c.d. is easily found: for example, if $a = 168 = 2^3 \times 3 \times 7$ and $b = 180 = 2^2 \times 3^2 \times 5$, then the g.c.d. is $2^2 \times 3 = 12$. Otherwise, the g.c.d. can be found by the *Euclidean Algorithm*, which can be used also to find s and t to express the g.c.d. as $sa + tb$. Similarly, any finite set of non-zero integers $a_1, a_2, \ldots, a_n$ has a g.c.d., denoted by $(a_1, a_2, \ldots, a_n)$, and there are integers $s_1, s_2, \ldots, s_n$ such that this can be expressed as $s_1a_1 + s_2a_2 + \cdots + s_na_n$.

greatest lower bound = *infimum*.

greatest value Let f be a *real function* and D a subset of its domain. If there is a point c in D such that $f(c) \geq f(x)$ for all x in D, then $f(c)$ is the **greatest value** of f in D. There may be no such point: consider, for example, either the function f defined by $f(x) = 1/x$ or the function f defined by $f(x) = x$, with the open interval (0, 1) as D; or the function f defined by $f(x) = x - [x]$, with the closed interval [0, 1] as D. If the greatest value does exist, it may be attained at more than one point of D.

That a *continuous function* on a closed interval has a greatest value is ensured by the non-elementary theorem that such a function 'attains its bounds'. An important theorem states that a function, continuous on $[a, b]$ and *differentiable* in (a, b), attains its greatest value either at a *local maximum* (which is a *stationary point*) or at an end-point of the interval.

Green, George (1793–1841) British mathematician who developed the mathematical theory of electricity and magnetism. In his essay of 1828, following Poisson, he used the notion of potential and proved the result now known as Green's Theorem, which has wide applications in the subject. He had worked as a baker, and was self-taught in mathematics; he published other notable mathematical papers before beginning to study for a degree at Cambridge at the age of 40.

Gregory, James (1638–1675) Scottish mathematician who studied in Italy before returning to hold chairs at St Andrews and Edinburgh. He obtained infinite series for certain trigonometric functions such as $\tan^{-1}x$, and was one of the first to appreciate the difference between convergent and divergent series. A predecessor of Newton, he probably understood, in essence, the *Fundamental Theorem of Calculus* and knew of *Taylor series* forty years before Taylor's publication. He died at the age of 36.

Gregory–Newton forward difference formula Let $x_0, x_1, x_2, \ldots, x_n$ be equally spaced values, so that $x_i = x_0 + ih$, for $i = 1, 2, \ldots, n$. Suppose that the values $f_0, f_1, f_2, \ldots, f_n$ are known, where $f_i = f(x_i)$, for some function f.

The **Gregory–Newton forward difference formula** is a formula involving *finite differences* that gives an approximation for $f(x)$, where $x = x_0 + \theta h$, and $0 < \theta < 1$. It states that

$$f(x) \approx f_0 + \theta\Delta f_0 + \frac{\theta(\theta-1)}{2!}\Delta^2 f_0 + \frac{\theta(\theta-1)(\theta-2)}{3!}\Delta^3 f_0 + \cdots,$$

the series being terminated at some stage. The approximation $f(x) \approx f_0 + \theta\Delta f_0$ gives the result of *linear interpolation*. Terminating the series after one more term provides an example of quadratic interpolation.

Grelling's paradox Certain adjectives describe themselves and others do not. For example, 'short' describes itself and so does 'polysyllabic'. But 'long' does not describe itself, 'monosyllabic' does not, and nor does 'green'. The word 'heterological' means 'not describing itself'. The German mathematician K. Grelling pointed out what is known as **Grelling's paradox** that results from considering whether 'heterological' describes itself or not. The paradox has some similarities with *Russell's paradox*.

group An operation on a set is worth considering only if it has properties likely to lead to interesting and useful results. Certain basic properties recur in different parts of mathematics and, if these are recognized, use can be made of the similarities that exist in the different situations. One such set of basic properties is specified in the definition of a group. The following, then, are all examples of groups: the set of real numbers with addition, the set of non-zero real numbers with multiplication, the set of 2×2 real matrices with matrix addition, the set of vectors in 3-dimensional space with vector addition, the set of all bijective mappings from a set S onto itself with composition of mappings, the four numbers $1, i, -1, -i$ with multiplication. The definition is as follows: a **group** is a set G closed under an operation $\circ$ such that

1. for all a, b and c in G, $a \circ (b \circ c) = (a \circ b) \circ c$,
2. there is an identity element e in G such that $a \circ e = e \circ a = a$ for all a in G,
3. for each a in G, there is an inverse element a' in G such that $a \circ a' = a' \circ a = e$.

The group may be denoted by $\langle G, \circ\rangle$, or $(G, \circ)$, when it is necessary to specify the operation, but it may be called simply the group G when the intended operation is clear.

grouped data A set of data is said to be **grouped** when certain groups or categories are defined and the observations in each group are counted to give the frequencies. For numerical data, groups are often defined by means of *class intervals*.

Hadamard, Jacques (1865–1963) French mathematician, known for his proof in 1896 of the *Prime Number Theorem*. He also worked, amongst other things, on the *calculus of variations* and the beginnings of functional analysis.

half-life See *exponential decay*.

half-line The part of a straight line extending from a point indefinitely in one direction.

half-plane In coordinate geometry, if a line l has equation $ax + by + c = 0$, the set of points (x, y) such that $ax + by + c > 0$ forms the **open half-plane** on one side of l and the set of points (x, y) such that $ax + by + c < 0$ forms the open half-plane on the other side of l. When the line $ax + by + c = 0$ has been drawn, there is a useful method, if $c \neq 0$, of determining which half-plane is which: find out which of the two inequalities is satisfied by the origin. Thus the half-plane containing the origin is the one given by $ax + by + c > 0$ if $c > 0$, and is the one given by $ax + by + c < 0$ if $c < 0$.

A **closed half-plane** is a set of points (x, y) such that $ax + by + c \geq 0$ or such that $ax + by + c \leq 0$. The use of open and closed half-planes is the basis of elementary *linear programming*.

half-space Loosely, one of the two regions into which a plane divides 3-dimensional space. More precisely, if a plane has equation $ax + by + cz + d = 0$, the set of points (x, y, z) such that $ax + by + cz + d > 0$ and the set of points (x, y, z) such that $ax + by + cz + d < 0$ are **open** half-spaces lying on opposite sides of the plane. The set of points (x, y, z) such that $ax + by + cz + d \geq 0$, and the set of points (x, y, z) such that $ax + by + cz + d \leq 0$, are **closed** half-spaces.

half-turn symmetry See *symmetrical about a point*.

Hamilton, William Rowan (1805–1865) Ireland's greatest mathematician, whose main achievement was in the subject of geometrical optics, for which he laid a theoretical foundation that came close to anticipating quantum theory. His work is also of great significance for general mechanics. He is perhaps best known in pure mathematics for his algebraic theory of complex numbers, the invention of *quaternions* and the exploitation of non-commutative algebra. A child prodigy, he could, it is claimed, speak 13 languages at the age of 13. He became Professor of Astronomy at Dublin and Royal Astronomer of Ireland at the age of 22.

Hamiltonian graph In graph theory, one area of study has been concerned with the possibility of travelling around a *graph*, going along edges in such a way as to visit every vertex exactly once. For this purpose, the following definitions are made. A **Hamiltonian cycle** is a *cycle* that contains every vertex, and a graph is called **Hamiltonian** if it has a Hamiltonian cycle. The term arises from Hamilton's interest in the existence of such cycles in the graph of the dodecahedron—the graph with vertices and edges corresponding to the vertices and edges of a dodecahedron.

handshaking lemma The simple result that, in any *graph*, the sum of the degrees of all the vertices is even. (The name arises from its application to the total number of hands shaken when some members of a group of people shake hands.) It follows from the simple observation that the sum of the degrees of all the vertices of a graph is equal to twice the number of edges. A result which follows from it is that, in any graph, the number of vertices of odd degree is even.

Hardy, Godfrey Harold (1877–1947) British mathematician, a leading figure in mathematics in Cambridge. Often in collaboration with J. E. Littlewood, he published many papers on prime numbers, other areas of number theory and mathematical analysis.

harmonic mean See *mean*.

harmonic sequence A sequence $a_1, a_2, a_3, \ldots$ such that $1/a_1, 1/a_2, 1/a_3, \ldots$ is an *arithmetic sequence*. The most commonly occurring harmonic sequence is the sequence $1, \frac{1}{2}, \frac{1}{3}, \frac{1}{4}, \ldots$.

harmonic series A series $a_1 + a_2 + a_3 + \cdots$ in which $a_1, a_2, a_3, \ldots$ is a *harmonic sequence*. Often the term refers to the particular series $1 + \frac{1}{2} + \frac{1}{3} + \frac{1}{4} + \cdots$, where the n-th term a_n equals $1/n$. For this series, $a_n \to 0$. However, the series does not have a sum to infinity since, if s_n is the sum of the first n terms, then $s_n \to \infty$. For large values of n, $s_n \approx \ln n + \gamma$, where γ is *Euler's constant*.

h.c.f. = *highest common factor*.

hecto- Prefix used with *SI units* to denote multiplication by 10^2.

height (of a triangle) See *base* (of a triangle).

helix A curve on the surface of a (right-circular) cylinder that cuts the generators of the cylinder at a constant angle. Thus it is 'like a spiral staircase'.

hemisphere One half of a sphere cut off by a plane through the centre of the sphere.

hendecagon An eleven-sided polygon.

heptagon A seven-sided polygon.

Hermite, Charles (1822–1901) French mathematician who worked in algebra and analysis. In 1873, he proved that e is *transcendental*. Also notable is his proof that the general quintic equation can be solved using elliptic functions.

Hero (or **Heron**) **of Alexandria** (1st century AD) Greek scientist whose interests included optics, mechanical inventions and practical mathematics. A long-lost book, the *Metrica*, rediscovered in 1896, contains examples of mensuration, showing, for example, how to work out the areas of regular polygons and the volumes of different polyhedra. It also includes the earliest known proof of *Hero's formula.*

Hero's formula See *triangle.*

hertz The SI unit of *frequency*, abbreviated to 'Hz'. Normally used to measure the frequency of cycles, or oscillations, one hertz is equal to one cycle per second.

hexadecimal representation The representation of a number to *base* 16. In this system, 16 digits are required and it is normal to take 0, 1, 2, 3, 4, 5, 6, 7, 8, 9, A, B, C, D, E and F, where A to F represent the numbers that, in decimal notation, are denoted by 10, 11, 12, 13, 14 and 15. Then the hexadecimal representation of the decimal number 712, for example, is found by writing

$$712 = 2 \times 16^2 + 12 \times 16 + 8 = (2C8)_{16}.$$

It is particularly simple to change the representation of a number to base 2 (binary) to its representation to base 16 (hexadecimal) and vice versa: each block of 4 digits in base 2 (form blocks of 4, starting from the right-hand end) can be made to correspond to its hexadecimal equivalent. Thus

$$(101101001001101)_2 = (101\ 1010\ 0100\ 1101)_2 = (5A4D)_{16}.$$

Real numbers, not just integers, can also be written in hexadecimal notation, by using hexadecimal digits after a 'decimal' point, just as familiar *decimal* representations of real numbers are obtained to base 10. Hexadecimal notation is important in computing. It translates easily into binary notation, but is more concise and easier to read.

hexagon A six-sided polygon.

higher derivative If the function f is *differentiable* on an interval, its *derived function* f' is defined. If f' is also differentiable, then the derived function of this, denoted by f'', is the **second derived function** of f; its value at x, denoted by $f''(x)$, or d^2f/dx^2, is the **second derivative** of f at x. (The term 'second derivative' may be used loosely also for the second derived function f''.)

Similarly, if f'' is differentiable, then $f'''(x)$ or d^3f/dx^3, the third derivative of f at x, can be formed, and so on. The ***n*-th derivative** of f at x is denoted by $f^{(n)}(x)$ or d^nf/dx^n. The n-th derivatives, for $n \geq 2$, are called the **higher derivatives** of f. When $y = f(x)$, the higher derivatives may be denoted by $d^2y/dx^2, \ldots, d^ny/dx^n$ or $y'', y''', \ldots, y^{(n)}$. If, with a different notation, x is a function of t and the derivative dx/dt is denoted by $\dot{x}$, the second derivative d^2x/dt^2 is denoted by $\ddot{x}$.

higher-order partial derivative Given a function f of n variables $x_1, x_2, \ldots, x_n$, the *partial derivative* $\partial f/\partial x_i$, where $1 \leq i \leq n$, may also be reckoned to be a function of $x_1, x_2, \ldots, x_n$. So the partial derivatives of $\partial f/\partial x_i$ can be considered. Thus

$$\frac{\partial}{\partial x_i}\left(\frac{\partial f}{\partial x_i}\right) \quad \text{and} \quad \frac{\partial}{\partial x_j}\left(\frac{\partial f}{\partial x_i}\right) \quad (\text{for } j \neq i)$$

can be formed, and these are denoted, respectively, by

$$\frac{\partial^2 f}{\partial x_i^2} \quad \text{and} \quad \frac{\partial^2 f}{\partial x_j \partial x_i}.$$

These are the **second-order partial derivatives**. When $j \neq i$,

$$\frac{\partial^2 f}{\partial x_i \partial x_j} \quad \text{and} \quad \frac{\partial^2 f}{\partial x_j \partial x_i}$$

are different by definition, but the two are equal for most 'straightforward' functions f that are likely to be met. (It is not possible to describe here just what conditions are needed for equality.) Similarly, third-order partial derivatives such as

$$\frac{\partial^3 f}{\partial x_1^3}, \quad \frac{\partial^3 f}{\partial x_1 \partial x_2^2}, \quad \frac{\partial^3 f}{\partial x_1 \partial x_2 \partial x_3}, \quad \frac{\partial^3 f}{\partial x_2 \partial x_3 \partial x_1},$$

can be defined, as can fourth-order partial derivatives, and so on. Then the n-th-order partial derivatives, where $n \geq 2$, are called the **higher-order partial derivatives.**

When f is a function of two variables x and y, and the partial derivatives are denoted by f_x and f_y, then $f_{xx}, f_{xy}, f_{yx}, f_{yy}$ are used to denote

$$\frac{\partial^2 f}{\partial x^2}, \quad \frac{\partial^2 f}{\partial y \partial x}, \quad \frac{\partial^2 f}{\partial x \partial y}, \quad \frac{\partial^2 f}{\partial y^2}$$

respectively, noting particularly that f_{xy} means $(f_x)_y$ and f_{yx} means $(f_y)_x$. This notation can be extended to third-order (and higher) partial derivatives and to functions of more variables. With the value of f at (x, y) denoted by $f(x, y)$ and the partial derivatives denoted by f_1 and f_2, the second-order partial derivatives can be denoted by f_{11}, f_{12}, f_{21} and f_{22}, and this notation can also be extended to third-order (and higher) partial derivatives and to functions of more variables.

highest common factor = *greatest common divisor.*

Hilbert, David (1862–1943) German mathematician who was one of the founding fathers of twentieth-century pure mathematics, and in many ways the originator of the formalist school of mathematics which has been so dominant in the pure mathematics of this century. Born at Königsberg (Kaliningrad), he became professor at Göttingen in 1895, where he remained for the rest of his life. One of his fundamental contributions to formalism was his *Grundlagen der Geometrie* (Foundations of Geometry), published in 1899, which served to put geometry on a proper axiomatic basis, unlike the rather more intuitive 'axiomatization' of Euclid. He also made a major contribution to mathematical analysis. At the International Congress of Mathematics in 1900, he opened the new century by posing his famous list of 23 problems—problems that have kept mathematicians busy ever since and have generated a significant amount of the important work of this century. Hilbert is, for these reasons, often thought of as a thorough-going pure mathematician, but he was also the chairman of the famous atomic physics seminar at Göttingen that had a great influence on the development of quantum theory.

Hilbert's tenth problem The problem posed by Hilbert of finding an algorithm to determine whether or not a given Diophantine equation has solutions. It was proved by Y. Matijasevich in 1970 that no such algorithm exists.

histogram A diagram representing the *frequency distribution* of data grouped by means of class intervals. It consists of a sequence of rectangles, each of which has as its base one of the class intervals and is of a height taken so that the area is proportional to the frequency. If the class intervals are of equal lengths, then the heights of the rectangles are proportional to the frequencies. Some authors use the term 'histogram' when the data are discrete to describe a kind of *bar chart* in which the rectangles are shown touching.

The figure shows a histogram of a sample of 500 observations.

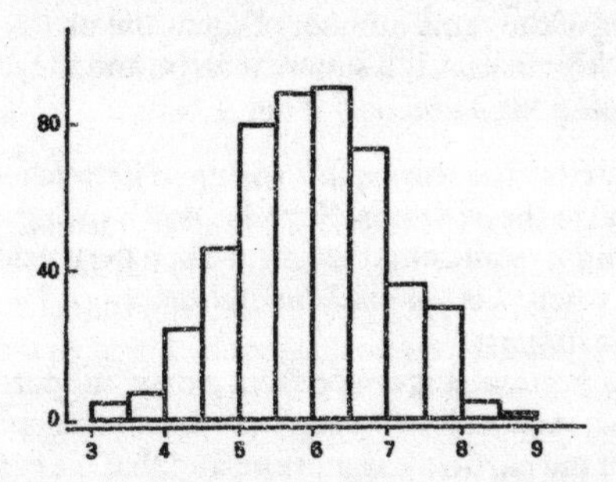

homogeneous first-order differential equation A first-order differential equation $dy/dx = f(x,y)$ in which the function f, of two variables, has the property that $f(kx, ky) = f(x,y)$ for all k. Examples of such functions are

$$\frac{x^2+3y^2}{2x^2-5xy}, \qquad 1+e^{x/y}, \qquad \frac{x}{\sqrt{x^2+y^2}}.$$

Any such function f can be written as a function of one variable v, where $v = y/x$. The method of solving homogeneous first-order differential equations is therefore to let $y = vx$ so that $dy/dx = x\,dv/dx + v$. The differential equation for v as a function of x that is obtained is always *separable.*

homogeneous linear differential equation See *linear differential equation with constant coefficients.*

homogeneous set of linear equations A set of m linear equations in n unknowns $x_1, x_2, \ldots, x_n$ that has the form

$$\begin{aligned} a_{11}x_1 + a_{12}x_2 + \cdots + a_{1n}x_n &= 0, \\ a_{21}x_1 + a_{22}x_2 + \cdots + a_{2n}x_n &= 0, \\ &\vdots \\ a_{m1}x_1 + a_{m2}x_2 + \cdots + a_{mn}x_n &= 0. \end{aligned}$$

Here, unlike the non-homogeneous case, the numbers on the right-hand sides of the equations are all zero. In matrix notation, this set of equations can be written $\mathbf{Ax} = \mathbf{0}$, where the unknowns form a column matrix $\mathbf{x}$. Thus $\mathbf{A}$ is the $m \times n$ matrix $[a_{ij}]$, and

$$\mathbf{x} = \begin{bmatrix} x_1 \\ \vdots \\ x_n \end{bmatrix}.$$

If $\mathbf{x}$ is a solution of a homogeneous set of linear equations, then so is any scalar multiple $k\mathbf{x}$ of it. There is always the trivial solution $\mathbf{x} = \mathbf{0}$. What is generally of concern is whether it has other solutions besides this one. For a homogeneous set consisting of the same number of equations as unknowns, the matrix of coefficients $\mathbf{A}$ is a square matrix, and the set of equations has non-trivial solutions if and only if $\det \mathbf{A} = 0$.

Hooke's law The law that says that the *tension* in a spring or a stretched elastic string is proportional to the *extension.* Suppose that a spring or elastic string has natural length l and actual length x. Then the tension T is given by $T = (\lambda/l)(x - l)$, where λ is the *modulus of elasticity,* or $T = k(x - l)$, where k is the *stiffness.*

Consider the motion of a particle suspended from a fixed support by a spring, the motion being in a vertical line through the equilibrium position of the particle. Suppose that the particle has mass m, and that the spring has stiffness k and natural length l. Let x be the length of the spring at time t. By using Hooke's law, the equation of motion $m\ddot{x} = mg - k(x - l)$ is obtained. The equilibrium position is given by $0 = mg - k(x - l)$; that is, $x = l + mg/k$. Letting $x = l + mg/k + X$ gives $\ddot{X} + \omega^2 X = 0$, where $\omega^2 = k/m$. Thus the particle performs *simple harmonic motion.*

l'Hôpital, Guillaume François Antoine, Marquis de (1661–1704) French mathematician who in 1696 produced the first textbook on differential calculus. This, and a subsequent book on analytical geometry, were standard texts for much of the eighteenth century. The first contains *l'Hôpital's rule*, known to be due to Jean Bernouilli, who is thought to have agreed to keep the Marquis l'Hôpital informed of his discoveries in return for financial support.

l'Hôpital's rule A rule for evaluating *indeterminate forms*. One form of the rule is the following:

THEOREM: Suppose that $f(x) \to 0$ and $g(x) \to 0$ as $x \to a$. Then

$$\lim_{x \to a} \frac{f(x)}{g(x)} = \lim_{x \to a} \frac{f'(x)}{g'(x)},$$

if the limit on the right-hand side exists.

For example,

$$\lim_{x \to 0} \frac{\sqrt{1+x}-1}{x} = \lim_{x \to 0} \frac{\frac{1}{2}(1+x)^{-1/2}}{1} = \frac{1}{2}.$$

The result also holds if $f(x) \to \infty$ and $g(x) \to \infty$ as $x \to a$. Moreover, the results apply if '$x \to a$' is replaced by '$x \to +\infty$' or '$x \to -\infty$'.

horsepower A unit of *power*, abbreviated to 'hp', once commonly used in Britain. 1 hp = 745.70 watts.

Huygens, Christiaan (1629–1695) Dutch mathematician, astronomer and physicist. In mathematics, he is remembered for his work on pendulum clocks and his contributions in the field of dynamics. These concerned, for example, the period of oscillation of a simple pendulum and matters such as the centrifugal force in uniform circular motion.

Hypatia (370–415) Greek philosopher who was the head of the neoplatonist school in Alexandria. It is known that she was consulted on scientific matters, and that she wrote commentaries on works of Diophantus and Apollonius. Her brutal murder by a fanatical mob is often taken to mark the beginning of Alexandria's decline as the outstanding centre of learning.

hyperbola A *conic* with eccentricity greater than 1. Thus it is the locus of all points P such that the distance from P to a fixed point F_1 (the **focus**) is equal to e (> 1) times the distance from P to a fixed line l_1 (the **directrix**). It turns out that there is another point F_2 and another line l_2 such that the same set of points would be obtained with these as focus and directrix. The hyperbola is also the conic section that results when a plane cuts a (double) cone in such a way that a section in two separate parts is obtained (see *conic*).

The line through F_1 and F_2 is the **transverse axis**, and the points V_1 and V_2 where it cuts the hyperbola are the **vertices**. The length $|V_1 V_2|$ is the **length of the transverse axis** and is usually taken to be $2a$. The midpoint of $V_1 V_2$ is the **centre** of the hyperbola. The line through the centre perpendicular to the transverse axis is the **conjugate axis**. It is usual to introduce b (> 0) defined by $b^2 = a^2(e^2 - 1)$, so that $e^2 = 1 + b^2/a^2$. It may be convenient to consider the points $(0, -b)$ and $(0, b)$ on the conjugate

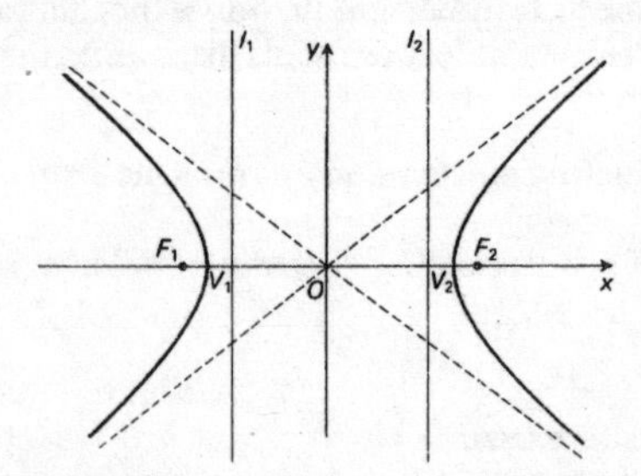

axis, despite the fact that the hyperbola does not cut the conjugate axis at all. The two separate parts of the hyperbola are the two *branches*.

If a coordinate system is taken with origin at the centre of the hyperbola, and x-axis along the transverse axis, the foci have coordinates $(ae, 0)$ and $(-ae, 0)$, the directrices have equations $x = a/e$ and $x = -a/e$, and the hyperbola has equation

$$\frac{x^2}{a^2} - \frac{y^2}{b^2} = 1.$$

Unlike the comparable equation for an ellipse, it is not necessarily the case here that $a > b$. When investigating the properties of a hyperbola, it is normal to choose this convenient coordinate system. It may be useful to take $x = a\sec\theta, y = b\tan\theta$ $(0 \leq \theta < 2\pi, \theta \neq \pi/2, 3\pi/2)$ as *parametric equations*. As an alternative, the parametric equations $x = a\cosh t$, $y = b\sinh t$ $(t \in \mathbf{R})$ (see *hyperbolic function*) may also be used, but give only one branch of the hyperbola.

A hyperbola with its centre at the origin and its transverse axis, of length $2a$, along the y-axis instead has equation $y^2/a^2 - x^2/b^2 = 1$ and has foci at $(0, ae)$ and $(0, -ae)$.

The hyperbola

$$\frac{x^2}{a^2} - \frac{y^2}{b^2} = 1$$

has two *asymptotes*, $y = (b/a)x$ and $y = (-b/a)x$. The shape of the hyperbola is determined by the eccentricity or, what is equivalent, by the ratio of b to a. The particular value $e = \sqrt{2}$ is important for this gives $b = a$. Then the asymptotes are perpendicular and the curve, which is of special interest, is a *rectangular hyperbola*.

hyperbolic cylinder A *cylinder* in which the fixed curve is a *hyperbola* and the fixed line to which the generators are parallel is perpendicular to the plane of the hyperbola. It is a *quadric*, and in a suitable coordinate system has equation

$$\frac{x^2}{a^2} - \frac{y^2}{b^2} = 1.$$

hyperbolic function Any of the functions cosh, sinh, tanh, sech, cosech and coth, defined as follows:

$$\cosh x = \tfrac{1}{2}(e^x + e^{-x}), \qquad \sinh x = \tfrac{1}{2}(e^x - e^{-x}),$$

$$\tanh x = \frac{\sinh x}{\cosh x}, \qquad \coth x = \frac{\cosh x}{\sinh x} \quad (x \neq 0),$$

$$\operatorname{sech} x = \frac{1}{\cosh x}, \qquad \operatorname{cosech} x = \frac{1}{\sinh x} \quad (x \neq 0).$$

The functions derive their name from the possibility of using $x = a\cosh t$, $y = b\sinh t$ ($t \in \mathbf{R}$) as *parametric equations* for (one branch of) a *hyperbola*. (The pronunciation of these functions causes difficulty. For instance, tanh may be pronounced as 'tansh' or 'than' (with the 'th' as in 'thing'); and sinh may be pronounced as 'shine' or 'sinch'. Some prefer to say 'hyperbolic tan' and 'hyperbolic sine'.) Many of the formulae satisfied by the hyperbolic functions are similar to corresponding formulae for the trigonometric functions, but some changes of sign must be noted. For example:

$$\cosh^2 x = 1 + \sinh^2 x,$$

$$\operatorname{sech}^2 x = 1 - \tanh^2 x,$$

$$\sinh(x + y) = \sinh x \cosh y + \cosh x \sinh y,$$

$$\cosh(x + y) = \cosh x \cosh y + \sinh x \sinh y,$$

$$\sinh 2x = 2\sinh x \cosh x,$$

$$\cosh 2x = \cosh^2 x + \sinh^2 x.$$

Since $\cosh(-x) = \cosh x$ and $\sinh(-x) = -\sinh x$, cosh is an *even function* and sinh is an *odd function*. The graphs of $\cosh x$ and $\sinh x$ are shown below. It is instructive to sketch both of them, together with the graphs of e^x and e^{-x}, on the same diagram. The graphs of the other hyperbolic functions follow on the next page.

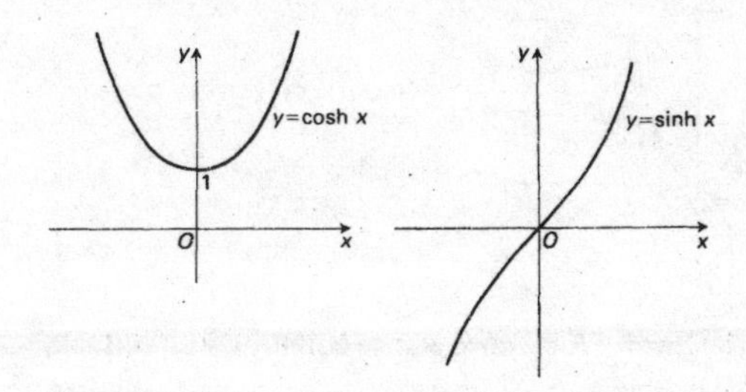

The following derivatives are easily established:

$$\frac{d}{dx}(\cosh x) = \sinh x, \quad \frac{d}{dx}(\sinh x) = \cosh x, \quad \frac{d}{dx}(\tanh x) = \operatorname{sech}^2 x.$$

See also *inverse hyperbolic function*.

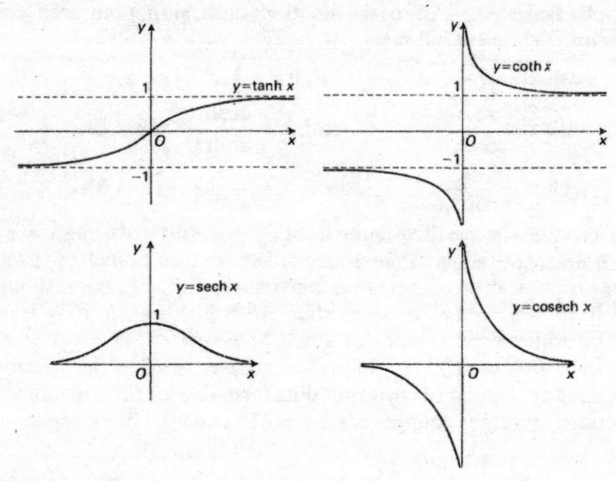

hyperbolic geometry See *non-Euclidean geometry.*

hyperbolic paraboloid A *quadric* whose equation in a suitable coordinate system is

$$\frac{x^2}{a^2} - \frac{y^2}{b^2} = \frac{2z}{c}.$$

The *yz*-plane and the *zx*-plane are planes of symmetry. Sections by planes parallel to the *xy*-plane are hyperbolas, the section by the *xy*-plane itself being a pair of straight lines. Sections by planes parallel to the other axial planes are parabolas. Planes through the *z*-axis cut the paraboloid in parabolas with vertex at the origin. The origin is a *saddle-point.*

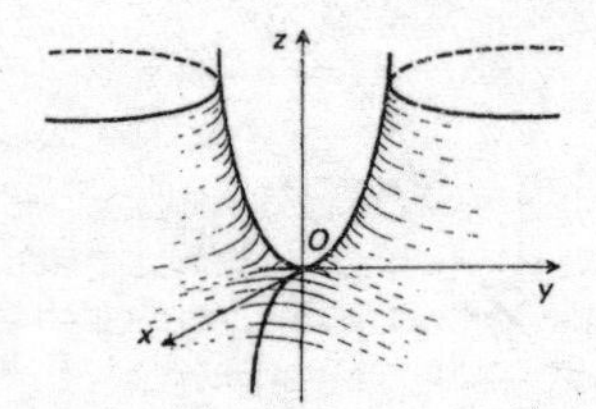

hyperboloid of one sheet A *quadric* whose equation in a suitable coordinate system is

$$\frac{x^2}{a^2} + \frac{y^2}{b^2} - \frac{z^2}{c^2} = 1.$$

The axial planes are planes of symmetry. The section by a plane $z = k$, parallel to the *xy*-plane, is an ellipse (a circle if $a = b$). The hyperboloid is

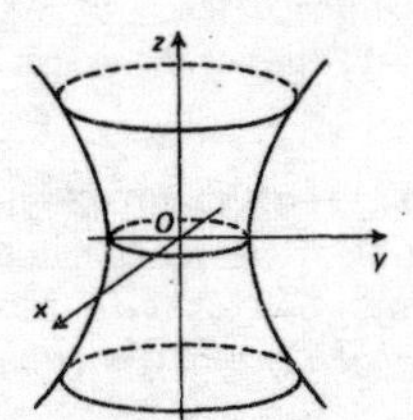

all in one piece or **sheet**. Sections by planes parallel to the other two axial planes are hyperbolas.

hyperboloid of two sheets A *quadric* whose equation in a suitable coordinate system is

$$\frac{x^2}{a^2}+\frac{y^2}{b^2}-\frac{z^2}{c^2}=-1.$$

The axial planes are planes of symmetry. The section by a plane $z=k$, parallel to the xy-plane, is, when non-empty, an ellipse (a circle if $a=b$). When k lies between $-c$ and c, the plane $z=k$ has no points of intersection with the hyperboloid, and the hyperboloid is thus in two pieces or **sheets**. Sections by planes parallel to the other two axial planes are hyperbolas. Planes through the z-axis cut the hyperboloid in hyperbolas with vertices at $(0,0,c)$ and $(0,0,-c)$.

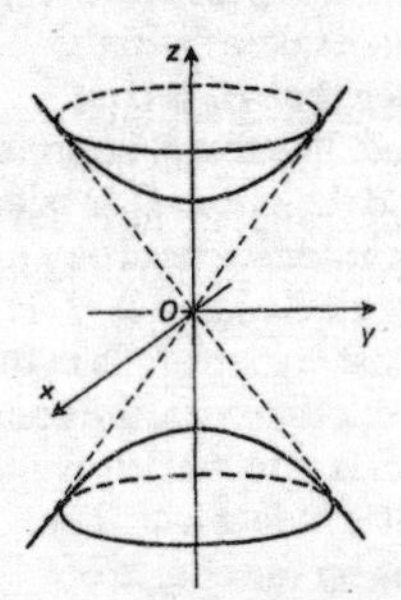

hypercube The generalization in n dimensions of a square in 2 dimensions and a cube in 3 dimensions. A geometrical description is difficult because of the problem of visualizing more than 3 dimensions. But the following approach describes the hypercube most frequently encountered, which has edges of unit length.

In the plane, the points with Cartesian coordinates (0, 0), (1, 0), (0, 1) and (1, 1) are the vertices of a square. In 3-dimensional space, the eight points with Cartesian coordinates (0, 0, 0), (1, 0, 0), (0, 1, 0), (0, 0, 1), (0, 1, 1), (1, 0, 1), (1, 1, 0) and (1, 1, 1) are the vertices of a cube. So, in *n-dimensional space*, the 2^n points with coordinates $(x_1, x_2, \ldots, x_n)$, where each $x_i = 0$ or 1, are the vertices of a hypercube. Two vertices are joined by an edge of the hypercube if they differ in exactly one of their coordinates. The vertices and edges of a hypercube in n dimensions form the vertices and edges of the graph known as the *n-cube*.

hypergeometric distribution Suppose that, in a set of N items, M items have a certain property and the remainder do not. A sample of n items is taken from the set. The discrete probability *distribution* for the number r of items in the sample having the property has *probability mass function* given by $\Pr(X = r) = \dfrac{{}^{M}C_r\,{}^{N-M}C_{n-r}}{{}^{N}C_n}$, where r runs from 0 up to the smaller of M and n. This is called a **hypergeometric distribution**. The mean is $\dfrac{nM}{N}$, and the variance is $\dfrac{nM(N-M)(N-n)}{N^2(N-1)}$.

hyperplane See *n-dimensional space*.

hypocycloid The curve traced out by a point on the circumference of a circle rolling round the inside of a fixed circle. A special case is the *astroid*.

hypotenuse The side of a right-angled triangle opposite the right angle.

hypothesis testing In statistics, a **hypothesis** is an assertion about a population. A **null hypothesis**, usually denoted by H_0, is a particular assertion that is to be accepted or rejected. The **alternative hypothesis** H_1 specifies some alternative to H_0. To decide whether H_0 is to be accepted or rejected, a **significance test** tests whether a sample taken from the population could have occurred by chance, given that H_0 is true.

From the sample, the *test statistic* is calculated. The test partitions the set of possible values into the **acceptance region** and the **critical** (or **rejection**) **region**. These depend upon the choice of the **significance level** α, which is the probability that the test statistic lies in the critical region if H_0 is true. Often α is chosen to be 5%, for example. If the test statistic falls in the critical region, the null hypothesis H_0 is rejected; otherwise, the conclusion is that there is no evidence for rejecting H_0 and it is said that H_0 is accepted.

A **Type I error** occurs if H_0 is rejected when it is in fact true. The probability of a Type I error is α, so this depends on the choice of significance level. A **Type II error** occurs if H_0 is accepted when it is in fact false. If the probability of a Type II error is β, the *power* of the test is defined as $1-\beta$. This depends upon the choice of alternative hypothesis. The null hypothesis H_0 normally specifies that a certain parameter has a certain value. If the alternative hypothesis H_1 says that the parameter is not equal to this value, then the test is said to be **two-tailed** (or **two-sided**). If H_1 says that the parameter is greater than this value (or less than this value), then the test is **one-tailed** (or **one-sided**).

i See *complex number.*

icosahedron (plural: icosahedra) A *polyhedron* with 20 faces, often assumed to be regular. The regular icosahedron is one of the *Platonic solids,* and its faces are equilateral triangles. It has 12 vertices and 30 edges.

icosidodecahedron One of the *Archimedean solids,* with 12 pentagonal faces and 20 triangular faces. It can be formed by cutting off the corners of a *dodecahedron* to obtain a polyhedron whose vertices lie at the midpoints of the edges of the original dodecahedron. It can also be formed by cutting off the corners of an *icosahedron* to obtain a polyhedron whose vertices lie at the midpoints of the edges of the original icosahedron.

identically distributed A number of random variables are **identically distributed** if they have the same *distribution.*

identity An equation which states that two expressions are equal for all values of any variables that occur, such as $x^2 - y^2 = (x + y)(x - y)$ and $x(x - 1)(x - 2) = x^3 - 3x^2 + 2x$. Sometimes the symbol $\equiv$ is used instead of $=$ to indicate that a statement is an identity.

identity element See *neutral element.*

identity mapping The **identity mapping** on a set S is the *mapping* $i_S\colon S \to S$ defined by $i_S(s) = s$ for all s in S. Identity mappings have the property that if $f\colon S \to T$ is a mapping, then $f \circ i_S = f$ and $i_T \circ f = f$.

identity matrix The $n \times n$ **identity matrix I**, or $\mathbf{I}_n$, is the $n \times n$ matrix

$$\begin{bmatrix} 1 & 0 & 0 & \dots & 0 \\ 0 & 1 & 0 & \dots & 0 \\ 0 & 0 & 1 & \dots & 0 \\ \vdots & \vdots & \vdots & \ddots & \vdots \\ 0 & 0 & 0 & \dots & 1 \end{bmatrix},$$

with each diagonal entry equal to 1 and all other entries equal to 0. It is the identity element for the set of $n \times n$ matrices with multiplication.

if and only if See *condition, necessary and sufficient.*

i.i.d. = *independent* and *identically distributed.*

image See *function* and *mapping.*

imaginary axis The *y*-axis in the *complex plane.* Its points represent *pure imaginary* numbers.

imaginary part A *complex number* z can be written $x + yi$, where x and y are real, and then y is the **imaginary part**. It is denoted by Im z or $\Im z$.

implication If p and q are statements, the statement 'p implies q' or 'if p then q' is an **implication** and is denoted by $p \Rightarrow q$. It is reckoned to be false only in the case when p is true and q is false. So its *truth table* is as follows:

p	q	$p \Rightarrow q$
T	T	T
T	F	F
F	T	T
F	F	T

The statement that 'p implies q and q implies p' may be denoted by $p \Leftrightarrow q$, to be read as 'p if and only if q'.

implicit differentiation When x and y satisfy a single equation, it may be possible to regard this as defining y as a function of x with a suitable domain, even though there is no explicit formula for y. In such a case, it may be possible to obtain the *derivative* of y by a method, called **implicit differentiation**, that consists of differentiating the equation as it stands and making use of the *chain rule*. For example, if $xy^2 + x^2y^3 - 1 = 0$, then

$$x2y\frac{dy}{dx} + y^2 + x^2 3y^2\frac{dy}{dx} + 2xy^3 = 0$$

and so, if $2xy + 3x^2y^2 \neq 0$,

$$\frac{dy}{dx} = -\frac{y^2 + 2xy^3}{2xy + 3x^2y^2}.$$

improper fraction See *fraction*.

improper integrals There are two kinds of **improper integral**. The first kind is one in which the interval of integration is infinite as, for example, in

$$\int_a^\infty f(x)\,dx.$$

It is said that this integral exists, and that its value is l, if the value of the integral from a to X tends to a limit l as $X \to \infty$. For example,

$$\int_1^X \frac{1}{x^2}\,dx = 1 - \frac{1}{X}$$

and, as $X \to \infty$, the right-hand side tends to 1. So

$$\int_1^\infty \frac{1}{x^2}\,dx = 1.$$

A similar definition can be given for the improper integral from $-\infty$ to a. If both the integrals

$$\int_{-\infty}^{a} f(x)\,dx \quad \text{and} \quad \int_{a}^{\infty} f(x)\,dx$$

exist and have values l_1 and l_2, then it is said that the integral from $-\infty$ to ∞ exists and that its value is $l_1 + l_2$.

The second kind of improper integral is one in which the function becomes infinite at some point. Suppose first that the function becomes infinite at one of the limits of integration as, for example, in

$$\int_0^1 \frac{1}{\sqrt{x}}\,dx.$$

The function $1/\sqrt{x}$ is not bounded on the closed interval $[0, 1]$, so, in the normal way, this integral is not defined. However, the function is bounded on the interval $[\delta, 1]$, where $0 < \delta < 1$, and

$$\int_\delta^1 \frac{1}{\sqrt{x}}\,dx = 2 - 2\sqrt{\delta}.$$

As $\delta \to 0$, the right-hand side tends to the limit 2. So the integral above, from 0 to 1, is taken, by definition, to be equal to 2; in the same way, any such integral can be given a value equal to the appropriate limit, if it exists. A similar definition can be made for an integral in which the function becomes infinite at the upper limit.

Finally, an integral in which the function becomes infinite at a point between the limits is dealt with as follows. It is written as the sum of two integrals, where the function becomes infinite at the upper limit of the first and at the lower limit of the second. If both these integrals exist, the original integral is said to exist and, in this way, its value can be obtained.

impulse The **impulse J** associated with a force **F** acting during the time interval from $t = t_1$ to $t = t_2$ is the definite integral of the force, with respect to time, over the interval. Thus **J** is the vector quantity given by

$$\mathbf{J} = \int_{t_1}^{t_2} \mathbf{F}\,dt.$$

Suppose that **F** is the total force acting on a particle of mass m. From Newton's second law of motion in the form $\mathbf{F} = (d/dt)(m\mathbf{v})$, it follows that the impulse **J** is equal to the change in the *linear momentum* of the particle.

If the force **F** is constant, then $\mathbf{J} = \mathbf{F}(t_2 - t_1)$. This can be stated as 'impulse = force × time'.

The concept of impulse is relevant in problems involving *collisions*.

Impulse has the dimensions MLT^{-1}, and the SI unit of measurement is the newton second, abbreviated to 'N s', or the kilogram metre per second, abbreviated to 'kg m s^{-1}'.

incentre The centre of the *incircle* of a triangle. It is the point at which the three internal bisectors of the angles of the triangle are concurrent.

incircle The circle that lies inside a triangle and touches the three sides. Its centre is the *incentre* of the triangle.

inclined plane A plane that is not horizontal. Its **angle of inclination** is the angle that a line of greatest slope makes with the horizontal.

include See *subset*.

inclusion–exclusion principle The number of elements in the union of 3 sets is given by

$$n(A \cup B \cup C) = n(A) + n(B) + n(C) - n(A \cap B) - n(A \cap C) - n(B \cap C) + n(A \cap B \cap C),$$

where $n(X)$ is the *cardinality* of the set X. This may be written

$$n(A \cup B \cup C) = \alpha_1 - \alpha_2 + \alpha_3,$$

where $\alpha_1 = n(A) + n(B) + n(C)$, $\alpha_2 = n(A \cap B) + n(A \cap C) + n(B \cap C)$ and $\alpha_3 = n(A \cap B \cap C)$. Now suppose that there are n sets instead of 3. Then the following, known as the **inclusion–exclusion principle**, holds:

$$n(A_1 \cup A_2 \cup \cdots \cup A_n) = \alpha_1 - \alpha_2 + \alpha_3 - \cdots + (-1)^{n-1}\alpha_n,$$

where α_i is the sum of the cardinalities of the intersections of the sets taken i at a time.

The principle is commonly used, when $A_1, A_2, \ldots, A_n$ are subsets of a universal set E, to find the number of elements not in any of the subsets. This number is equal to $n(E) - n(A_1 \cup A_2 \cup \cdots \cup A_n)$.

inconsistent A set of equations is **inconsistent** if its solution set is empty.

increasing function A *real function* f is **increasing** in or on an interval I if $f(x_1) \leq f(x_2)$ whenever x_1 and x_2 are in I with $x_1 < x_2$. Also, f is **strictly increasing** if $f(x_1) < f(x_2)$ whenever $x_1 < x_2$.

increasing sequence A sequence $a_1, a_2, a_3, \ldots$ is **increasing** if $a_i \leq a_{i+1}$ for all i, and **strictly increasing** if $a_i < a_{i+1}$ for all i.

increment The amount by which a certain quantity or variable is increased. An increment can normally be positive or negative, or zero, and is often required to be in some sense 'small'. If the variable is x, the increment may be denoted by Δx or δx.

indefinite integral See *integral*.

independent events Two events A and B are **independent** if the occurrence of either does not affect the probability of the other occurring. Otherwise, A and B are **dependent**. For independent events, the probability that they both occur is given by the product law $\Pr(A \cap B) = \Pr(A)\Pr(B)$.

For example, when an unbiased coin is tossed twice, the probability of obtaining 'heads' on the first toss is $\frac{1}{2}$, and the probability of obtaining 'heads' on the second toss is $\frac{1}{2}$. These two events are independent. So the probability of obtaining 'heads' on both tosses is equal to $\frac{1}{2} \times \frac{1}{2} = \frac{1}{4}$.

independent random variables Two random variables X and Y are **independent** if the value of either does not affect the value of the other. If two discrete random variables X and Y are independent, then $\Pr(X = x, Y = y) = \Pr(X = x)\Pr(Y = y)$. If two continuous random variables X and Y, with *joint probability density function* $f(x, y)$, are independent, then $f(x, y) = f_1(x)f_2(y)$, where f_1 and f_2 are the *marginal probability density functions.*

indeterminate form Suppose that $f(x) \to 0$ and $g(x) \to 0$ as $x \to a$. Then the limit of the quotient $f(x)/g(x)$ as $x \to a$ is said to give an **indeterminate form**, sometimes denoted by 0/0. It may be that the limit of $f(x)/g(x)$ can nevertheless be found by some method such as *l'Hôpital's rule.*

Similarly, if $f(x) \to \infty$ and $g(x) \to \infty$ as $x \to a$, then the limit of $f(x)/g(x)$ gives an indeterminate form, denoted by ∞/∞. Also, if $f(x) \to 0$ and $g(x) \to \infty$ as $x \to a$, then the limit of the product $f(x)g(x)$ gives an indeterminate form $0 \times \infty$.

index (plural: indices) Suppose that a is a real number. When the product $a \times a \times a \times a \times a$ is written as a^5, the number 5 is called the **index**. When the index is a positive integer p, then a^p means $a \times a \times \cdots \times a$, where there are p occurrences of a. It can then be shown that

(i) $a^p \times a^q = a^{p+q}$,
(ii) $a^p/a^q = a^{p-q}\ (a \neq 0)$,
(iii) $(a^p)^q = a^{pq}$,
(iv) $(ab)^p = a^p b^p$,

where, in (ii), for the moment, it is required that $p > q$. The meaning of a^p can however be extended so that p is not restricted to being a positive integer. This is achieved by giving a meaning for a^0, for a^{-p} and for $a^{m/n}$, where m is an integer and n is a positive integer. To ensure that (i) to (iv) hold when p and q are any rational numbers, it is necessary to take the following as definitions:

(v) $a^0 = 1$,
(vi) $a^{-p} = 1/a^p\ (a \neq 0)$,
(vii) $a^{m/n} = \sqrt[n]{a^m}$ (m an integer, n a positive integer).

Together, (i) to (vii) form the basic rules for indices.

The same notation is used in other contexts; for example, to define z^p, where z is a complex number, to define $\mathbf{A}^p$, where $\mathbf{A}$ is a square matrix, or to define g^p, where g is an element of a multiplicative group. In such cases, some of the above rules may hold and others may not.

index (in statistics) A figure used to show the variation in some quantity over a period of time, usually standardized relative to some base value. The index is given as a percentage with the base value equal to 100%. For example, the retail price index is used to measure changes in the cost of household items. Indices are often calculated by using a *weighted mean* of a number of constituent parts.

index set A set S may consist of elements, each of which corresponds to an element of a set I. Then I is the **index set**, and S is said to be **indexed** by I. For example, the set S may consist of elements a_i, where $i \in I$. This is written $S = \{a_i \mid i \in I\}$ or $\{a_i,\ i \in I\}$.

indirect proof A theorem of the form $p \Rightarrow q$ can be proved by establishing instead its *contrapositive*, by supposing $\neg q$ and showing that $\neg p$ follows. Such a method is called an **indirect proof**. Another example of an indirect proof is the method of *proof by contradiction*.

induction See *mathematical induction*.

inequality The following symbols have the meanings shown:

$\neq$ is not equal to,
$<$ is less than,
$\leq$ is less than or equal to,
$>$ is greater than,
$\geq$ is greater than or equal to.

An **inequality** is a statement of one of the forms: $a \neq b, a < b, a \leq b, a > b$ or $a \geq b$, where a and b are suitable quantities or expressions.

Given an inequality involving one real variable x, the problem may be to find the **solutions**, that is, the values of x that satisfy the inequality, in an explicit form. Often, the set of solutions, known as the **solution set**, may be given as an interval or as a union of intervals. For example, the solution set of the inequality

$$x - 2 > \frac{6}{x + 3}$$

is the set $(-4, -3) \cup (3, \infty)$.

Some authors use 'inequality' only for a statement, using one of the symbols above, that holds for all values of the variables that occur; for example, $x^2 + 2 > 2x$ is an inequality in this sense. Such authors use 'inequation' for a statement which holds only for some values of the variables involved. For them, an inequation has a solution set but an inequality does not; an inequality is comparable to an *identity*.

inertial frame of reference See *frame of reference*.

inertial mass The **inertial mass** of a particle is the constant of proportionality that occurs in the statement of Newton's second law of motion, linking the acceleration of the particle with the total force acting on the particle. Thus $m\mathbf{a} = \mathbf{F}$, where m is the inertial mass of the particle, $\mathbf{a}$ is the acceleration and $\mathbf{F}$ is the total force acting on the particle. See also *mass*.

inertia matrix See *angular momentum.*

inextensible string A string that has constant length, no matter how great the tension in the string.

inference The process of coming to some conclusion about a *population* based on a sample; or the conclusion about a population reached by such a process. The subject of statistical inference is concerned with the methods that can be used and the theory behind them.

infimum (plural: infima) See *bound.*

infinite product From an infinite sequence $a_1, a_2, a_3, \ldots$, an **infinite product** $a_1a_2a_3\ldots$ can be formed and denoted by

$$\prod_{r=1}^{\infty} a_r.$$

Let P_n be the n-th *partial product*, so that

$$P_n = \prod_{r=1}^{n} a_r.$$

If P_n tends to a limit P as $n \to \infty$, then P is the **value** of the infinite product. For example,

$$\prod_{r=2}^{\infty}\left(1 - \frac{1}{r^2}\right)$$

has the value $\frac{1}{2}$, since it can be shown that $P_n = (n+1)/2n$ and $P_n \to \frac{1}{2}$.

infinite sequence See *sequence.*

infinite series See *series.*

inflection = *inflexion.*

inflexion See *point of inflexion.*

initial conditions In applied mathematics, when differential equations with respect to time are to be solved, the known information about the system at a specified time, usually taken to be $t = 0$, may be called the **initial conditions**.

initial value The value of any physical quantity at a specified time reckoned as the starting-point, usually taken to be $t = 0$.

injection = *one-to-one mapping.*

injective mapping = *one-to-one mapping.*

inscribed circle (of a triangle) = *incircle.*

integer One of the 'whole' numbers: $\ldots, -3, -2, -1, 0, 1, 2, 3, \ldots$. The set of all integers is often denoted by **Z**. With the normal addition and multiplication, $\mathbb{Z}$ forms an *integral domain*. Kronecker said, 'God made the integers; everything else is the work of man.'

integer part For any real number x, there is a unique integer n such that $n \leq x < n + 1$. This integer n is the **integer part** of x, and is denoted by $[x]$. For example, $[\frac{9}{4}] = 2$ and $[\pi] = 3$, but notice that $[-\frac{9}{4}] = -3$. In a computer language, the function **INT(X)** may convert the real number **X** into an integer by truncating. If so, **INT(9/4) = 2** and **INT(PI) = 3**, but **INT(−9/4) = −2**. So **INT(X)** agrees with $[x]$ for $x \geq 0$ but not for $x < 0$.

The graph $y = [x]$ is shown below.

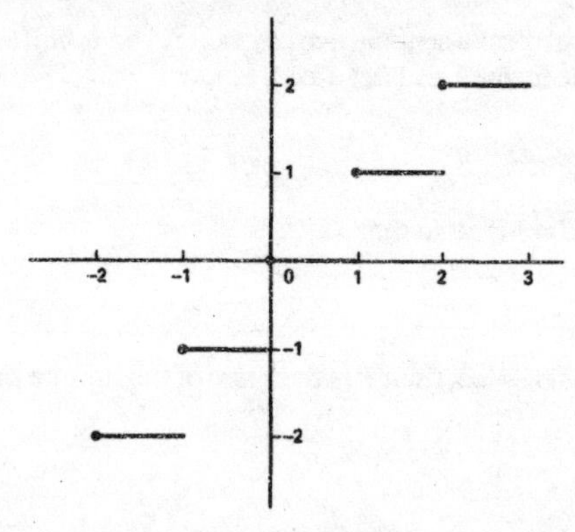

integral Let f be a function defined on the closed interval $[a, b]$. Take points $x_0, x_1, x_2, \ldots, x_n$ such that $a = x_0 < x_1 < x_2 < \cdots < x_{n-1} < x_n = b$, and in each subinterval $[x_i, x_{i+1}]$ take a point c_i. Form the sum

$$\sum_{i=0}^{n-1} f(c_i)(x_{i+1} - x_i);$$

that is, $f(c_0)(x_1 - x_0) + f(c_1)(x_2 - x_1) + \cdots + f(c_{n-1})(x_n - x_{n-1})$. Such a sum is called a **Riemann sum** for f over $[a, b]$. Geometrically, it gives the sum of the areas of n rectangles, and is an approximation to the area under the curve $y = f(x)$ between $x = a$ and $x = b$.

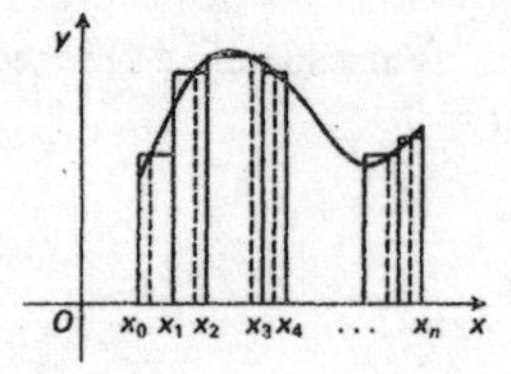

The **(Riemann) integral** of f over $[a, b]$ is defined to be the limit I (in a sense that needs more clarification than can be given here) of such a

Riemann sum as n, the number of points, increases and the size of the subintervals gets smaller. The value of I is denoted by

$$\int_a^b f(x)\,dx, \quad \text{or} \quad \int_a^b f(t)\,dt,$$

where it is immaterial what letter, such as x or t, is used in the integral. The intention is that the value of the integral is equal to what is intuitively understood to be the area under the curve $y = f(x)$. Such a limit does not always exist, but it can be proved that it does if, for example, f is a *continuous function* on $[a, b]$.

If f is continuous on $[a, b]$ and F is defined by

$$F(x) = \int_a^x f(t)\,dt,$$

then $F'(x) = f(x)$ for all x in $[a, b]$, so that F is an *antiderivative* of f. Moreover, if an antiderivative ϕ of f is known, the integral

$$\int_a^b f(t)\,dt$$

can be easily evaluated: the *Fundamental Theorem of Calculus* gives its value as $\phi(b) - \phi(a)$. Of the two integrals

$$\int_a^b f(x)\,dx \quad \text{and} \quad \int f(x)\,dx,$$

the first, with limits, is called a **definite integral** and the second, which denotes an antiderivative of f, is an **indefinite integral**.

integral calculus The subject that arose from the problem of trying to find the area of a region with a curved boundary. In general, this is calculated by a limiting process that yields gradually better approximations to the value. The *integral* is defined by a limiting process based on an intuitive idea of the area under a curve; the fundamental discovery was the link that exists between this and the *differential calculus*.

integral domain A commutative *ring* R with identity, with the additional property that

9. For all a and b in R, $ab = 0$ only if $a = 0$ or $b = 0$.

(The axiom numbering follows on from that used for ring.) Thus an integral domain is a commutative ring with identity with no *divisors of zero*. The natural example is the set $\mathbf{Z}$ of all integers with the usual addition and multiplication. Any *field* is an integral domain. Further examples of integral domains (these are not fields) are: the set $\mathbf{Z}[\sqrt{2}]$ of all real numbers of the form $a + b\sqrt{2}$, where a and b are integers, and the set $\mathbf{R}[x]$ of all polynomials in an indeterminate x, with real coefficients, each with the normal addition and multiplication.

integral part = *integer part*.

integrand The expression $f(x)$ in either of the integrals

$$\int_a^b f(x)\,dx, \qquad \int f(x)\,dx.$$

integrating factor See *linear first-order differential equation.*

integration The process of finding an *antiderivative* of a given function f. 'Integrate f' means 'find an antiderivative of f'. Such an antiderivative may be called an indefinite integral of f and be denoted by

$$\int f(x)\,dx.$$

The term 'integration' is also used for any method of evaluating a definite integral. The definite integral

$$\int_a^b f(x)\,dx$$

can be evaluated if an antiderivative ϕ of f can be found, because then its value is $\phi(b) - \phi(a)$. (This is provided that a and b both belong to an interval in which f is continuous.) However, for many functions f, there is no antiderivative expressible in terms of elementary functions, and other methods for evaluating the definite integral have to be sought, one such being so-called *numerical integration.*

What ways are there, then, of finding an antiderivative? If the given function can be recognized as the derivative of a familiar function, an antiderivative is immediately known. Some standard integrals are also given in the Table of Integrals (Appendix 3); more extensive tables of integrals are available. Certain **techniques of integration** may also be tried, among which are the following:

CHANGE OF VARIABLE: If it is possible to find a suitable function g such that the *integrand* can be written as $f(g(x))g'(x)$, it may be possible to find an indefinite integral using the **change of variable** $u = g(x)$; this is because

$$\int f(g(x))g'(x)\,dx = \int f(u)\,du,$$

a rule derived from the *chain rule* for differentiation. For example, in the integral

$$\int 2x(x^2+1)^8\,dx,$$

let $u = g(x) = x^2 + 1$. Then $g'(x) = 2x$ (this can be written '$du = 2x\,dx$'), and, using the rule above with $f(u) = u^8$, the integral equals

$$\int (x^2+1)^8 2x\,dx = \int u^8\,du = \tfrac{1}{9}u^9 = \tfrac{1}{9}(x^2+1)^9.$$

SUBSTITUTION: The rule above, derived from the chain rule for differentiation, can be written as

$$\int f(x)\,dx = \int f(g(u))g'(u)\,du.$$

It is used in this form to make the **substitution** $x = g(u)$. For example, in the integral

$$\int \frac{1}{(1+x^2)^{3/2}}\,dx,$$

let $x = g(u) = \tan u$. Then $g'(u) = \sec^2 u$ (this can be written '$dx = \sec^2 u\,du$'), and the integral becomes

$$\int \frac{\sec^2 u}{(1+\tan^2 u)^{3/2}}\,du = \int \frac{1}{\sec u}\,du = \int \cos u\,du = \sin u = \frac{x}{\sqrt{1+x^2}}.$$

INTEGRATION BY PARTS: The rule for **integration by parts**,

$$\int f(x)g'(x)\,dx = f(x)g(x) - \int g(x)f'(x)\,dx,$$

is derived from the rule for differentiating a product $f(x)g(x)$, and is useful when the integral on the right-hand side is easier to find than the integral on the left. For example, in the integral

$$\int x\cos x\,dx,$$

let $f(x) = x$ and $g'(x) = \cos x$. Then $g(x)$ can be taken as $\sin x$ and $f'(x) = 1$, so the method gives

$$\int x\cos x\,dx = x\sin x - \int \sin x\,1\,dx = x\sin x + \cos x.$$

See also *reduction formula* and *partial fractions*.

integration by parts See *integration.*

intercept See *straight line* (in the plane).

interior See *Jordan Curve Theorem.*

interior angle (of a polygon) The angle between two adjacent sides of a polygon that lies inside the polygon.

interior angle (with respect to a transversal of a pair of lines) See *transversal.*

interior point When the *real line* is being considered, the real number x is an **interior point** of the set S of real numbers if there is an open interval $(x-\delta, x+\delta)$, where $\delta > 0$, included in S.

Intermediate Value Theorem The following theorem stating an important property of *continuous functions*:

THEOREM: If the real function f is continuous on the closed interval $[a, b]$ and η is a real number between $f(a)$ and $f(b)$, then, for some c in (a, b), $f(c) = \eta$.

The theorem is useful for locating roots of equations. For example, suppose that $f(x) = x - \cos x$. Then f is continuous on [0, 1], and $f(0) < 0$ and $f(1) > 0$, so it follows from the Intermediate Value Theorem that the equation $f(x) = 0$ has a root in the interval (0, 1).

internal bisector Name sometimes given to the bisector of the interior angle of a triangle (or polygon).

internal force When a system of particles or a rigid body is being considered as a whole, an **internal force** is a force exerted by one particle of the system on another, or by one part of the rigid body on another. As far as the whole is concerned, the sum of the internal forces is zero. Compare *external force*.

interpolation Suppose that the values $f(x_0), f(x_1), \ldots, f(x_n)$ of a certain function f are known for the particular values $x_0, x_1, \ldots, x_n$. A method of finding an approximation for $f(x)$, for a given value of x somewhere between these particular values, is called **interpolation**. If $x_0 < x < x_1$, the method known as **linear interpolation** gives

$$f(x) \approx f(x_0) + \frac{x - x_0}{x_1 - x_0}(f(x_1) - f(x_0)).$$

This is obtained by supposing, as an approximation, that between x_0 and x_1 the graph of the function is a straight line joining the points $(x_0, f(x_0))$ and $(x_1, f(x_1))$. More complicated methods of interpolation use the values of the function at more than two values.

interquartile range A measure of *dispersion* equal to the difference between the first and third *quartiles* in a set of numerical data.

intersection The **intersection** of sets A and B (subsets of a *universal set*) is the set consisting of all objects that belong to A and belong to B, and it is denoted by $A \cap B$ (read as 'A **intersection** B'). Thus the term 'intersection' is used for both the resulting set and the operation, a *binary operation* on the set of all subsets of a universal set. The following properties hold:

(i) For all A, $A \cap A = A$ and $A \cap \emptyset = \emptyset$.
(ii) For all A and B, $A \cap B = B \cap A$; that is, $\cap$ is commutative.
(iii) For all A, B and C, $(A \cap B) \cap C = A \cap (B \cap C)$; that is, $\cap$ is associative.

In view of (iii), the intersection $A_1 \cap A_2 \cap \cdots \cap A_n$ of more than two sets can be written without parentheses, and it may also be denoted by

$$\bigcap_{i=1}^{n} A_i.$$

For the intersection of two events, see *event*.

interval A **finite interval** on the real line is a subset of **R** defined in terms of end-points a and b. Since each end-point may or may not belong to the subset, there are four types of finite interval:

(i) the closed interval $\{x \mid x \in \mathbf{R} \text{ and } a \leq x \leq b\}$, denoted by $[a, b]$,
(ii) the open interval $\{x \mid x \in \mathbf{R} \text{ and } a < x < b\}$, denoted by (a, b),
(iii) the interval $\{x \mid x \in \mathbf{R} \text{ and } a \leq x < b\}$, denoted by $[a, b)$,
(iv) the interval $\{x \mid x \in \mathbf{R} \text{ and } a < x \leq b\}$, denoted by $(a, b]$.

There are also four types of **infinite interval**:

(v) $\{x \mid x \in \mathbf{R} \text{ and } a \leq x\}$, denoted by $[a, \infty)$,
(vi) $\{x \mid x \in \mathbf{R} \text{ and } a < x\}$, denoted by (a, ∞),
(vii) $\{x \mid x \in \mathbf{R} \text{ and } x \leq a\}$, denoted by $(-\infty, a]$,
(viii) $\{x \mid x \in \mathbf{R} \text{ and } x < a\}$, denoted by (∞, a).

Here ∞ (read as 'infinity') and $-\infty$ (read as 'minus infinity') are not, of course, real numbers, but the use of these symbols provides a convenient notation.

If I is any of the intervals (i) to (iv), the **open interval determined by** I is (a, b); if I is (v) or (vi), it is (a, ∞) and, if I is (vii) or (viii), it is $(-\infty, a)$.

interval estimate See *estimate.*

invariant A property or quantity that is not changed by one or more specified operations or transformations. For example, a conic has equation

$$ax^2 + 2hxy + by^2 + 2gx + 2fy + c = 0$$

and, under a *rotation* of axes, the quantity $h^2 - ab$ is an invariant. The distance between two points and the property of perpendicularity between two lines are invariants under *translations* and rotations of the plane, but not under *dilatations.*

inverse element Suppose that, for the *binary operation* $\circ$ on the set S, there is a *neutral element* e. An element a' is an **inverse** (or **inverse element**) of the element a if $a \circ a' = a' \circ a = e$. If the operation is called multiplication, the neutral element is normally called the **identity element** and may be denoted by 1. Then the inverse a' may be called a **multiplicative inverse** of a and be denoted by a^{-1}, so that $aa^{-1} = a^{-1}a = 1$ (or e). If the operation is addition, the neutral element is denoted by 0, and the inverse a' may be called an **additive inverse** (or a **negative**) of a and be denoted by $-a$, so that $a + (-a) = (-a) + a = 0$. See also *group.*

inverse function For a *real function* f, its inverse function f^{-1} is to be a function such that if $y = f(x)$, then $x = f^{-1}(y)$. The conditions under which such a function exists need careful consideration. Suppose that f has domain S and range T, so that $f\colon S \to T$ is onto. The **inverse function** f^{-1}, with domain T and range S, can be defined (provided that f is a *one-to-one mapping*), as follows. For y in T, $f^{-1}(y)$ is the unique element x of S such that $f(x) = y$.

If the domain S is an interval I and f is *strictly increasing* on I or *strictly decreasing* on I, then f is certainly one-to-one. When f is *differentiable*, a sufficient condition can be given in terms of the sign of $f'(x)$. Thus, if f is continuous on an interval I and differentiable in the open interval (a, b)

determined by I (see *interval*), and $f'(x) > 0$ in (a, b) (or $f'(x) < 0$ in (a, b)), then, for f (with domain I) an inverse function exists.

When an inverse function is required for a given function f, it may be necessary to restrict the domain and obtain instead the inverse function of this *restriction* of f. For example, suppose that $f: \mathbf{R} \to \mathbf{R}$ is defined by $f(x) = x^2 - 4x + 5$. This function is not one-to-one. However, since $f'(x) = 2x - 4$, it can be seen that $f'(x) > 0$ for $x > 2$. Use f now to denote the function defined by $f(x) = x^2 - 4x + 5$ with domain $[2, \infty)$. The range is $[1, \infty)$ and the function f: $[2, \infty) \to [1, \infty)$ has an inverse function f^{-1}: $[1, \infty) \to [2, \infty)$. A formula for f^{-1} can be found by setting $y = x^2 - 4x + 5$ and, remembering that $x \in [2, \infty)$, obtaining $x = 2 + \sqrt{y - 1}$. So, with a change of notation, $f^{-1}(x) = 2 + \sqrt{x - 1}$ for $x \geq 1$.

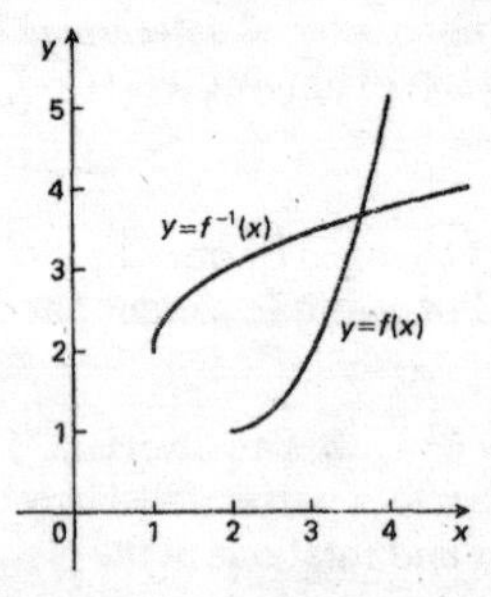

When the inverse function exists, the graphs $y = f(x)$ and $y = f^{-1}(x)$ are reflections of each other in the line $y = x$. The derivative of the inverse function can be found as follows. Suppose that f is differentiable and has inverse denoted now by g, so that if $y = f(x)$ then $x = g(y)$. Then, if $f'(x) \neq 0$, g is differentiable at y and

$$g'(y) = \frac{1}{f'(x)} = \frac{1}{f'(g(y))}.$$

When $f'(x)$ is denoted by dy/dx, then $g'(y)$ may be denoted by dx/dy and the preceding result says that, if $dy/dx \neq 0$, then

$$\frac{dx}{dy} = \frac{1}{dy/dx}.$$

This can be used safely only if it is known that the function in question has an inverse. For examples, see *inverse hyperbolic function* and *inverse trigonometric function*.

inverse hyperbolic function Each of the *hyperbolic functions* sinh and tanh is strictly increasing throughout the whole of its domain $\mathbf{R}$, so in each case an inverse function exists. In the case of cosh, the function has to be restricted to a suitable domain (see *inverse function*), taken to be $[0, \infty)$. The domain of the inverse function is, in each case, the range of the original function (after the restriction of the domain, in the case of cosh). The inverse functions obtained are: $\cosh^{-1}$: $[1,\infty) \to [0,\infty)$; $\sinh^{-1}$: $\mathbf{R} \to \mathbf{R}$; $\tanh^{-1}$: $(-1,1) \to \mathbf{R}$. These functions are given by the formulae:

$$\cosh^{-1} x = \ln\left(x + \sqrt{x^2 - 1}\right), \quad \text{for } x \geq 1,$$

$$\sinh^{-1} x = \ln\left(x + \sqrt{x^2 + 1}\right), \quad \text{for all } x,$$

$$\tanh^{-1} x = \ln\sqrt{\frac{1+x}{1-x}}, \quad \text{for } -1 < x < 1.$$

It is not so surprising that the inverse functions can be expressed in terms of the logarithmic function, since the original functions were defined in terms of the exponential function. The following derivatives can be obtained:

$$\frac{d}{dx}(\cosh^{-1} x) = \frac{1}{\sqrt{x^2 - 1}} \quad (x \neq 1),$$

$$\frac{d}{dx}(\sinh^{-1} x) = \frac{1}{\sqrt{x^2 + 1}}, \quad \frac{d}{dx}(\tanh^{-1} x) = \frac{1}{1 - x^2}.$$

inversely proportional See *proportion*.

inverse mapping Let $f: S \to T$ be a bijection, that is, a mapping that is both a *one-to-one mapping* and an *onto mapping*. Then a mapping, denoted by f^{-1}, from T to S may be defined as follows: for t in T, $f^{-1}(t)$ is the unique element s of S such that $f(s) = t$. The mapping f^{-1}: $T \to S$, which is also a bijection, is the **inverse mapping** of f. It has the property that $f \circ f^{-1} = i_T$ and $f^{-1} \circ f = i_S$, where i_S and i_T are the identity mappings on S and T, and $\circ$ denotes composition.

inverse matrix An **inverse** of a square matrix **A** is a matrix **X** such that **AX** = **I** and **XA** = **I**. (A matrix that is not square cannot have an inverse.) A square matrix **A** may or may not have an inverse, but if it has then that inverse is unique and **A** is said to be **invertible**. A matrix is invertible if and only if it is *non-singular*. Consequently, the term 'non-singular' is sometimes used for 'invertible'.

When det **A** $\neq 0$, the matrix $(1/\det \mathbf{A})$ adj **A** is the inverse of **A**, where adj **A** is the *adjoint* of **A**. For example, the 2×2 matrix **A** below is invertible if $ad - bc \neq 0$, and its inverse $\mathbf{A}^{-1}$ is as shown:

$$\mathbf{A} = \begin{bmatrix} a & b \\ c & d \end{bmatrix}, \quad \mathbf{A}^{-1} = \frac{1}{ad - bc}\begin{bmatrix} d & -b \\ -c & a \end{bmatrix}.$$

inverse of a complex number If z is a non-zero complex number and $z = x + yi$, the **(multiplicative) inverse** of z, denoted by z^{-1} or $1/z$, is

$$\frac{x}{x^2 + y^2} - \frac{y}{x^2 + y^2}\, i.$$

When z is written in *polar form*, so that $z = re^{i\theta} = r(\cos\theta + i\sin\theta)$, where $r \neq 0$, the inverse of z is $(1/r)e^{-i\theta} = (1/r)(\cos\theta - i\sin\theta)$. If z is represented

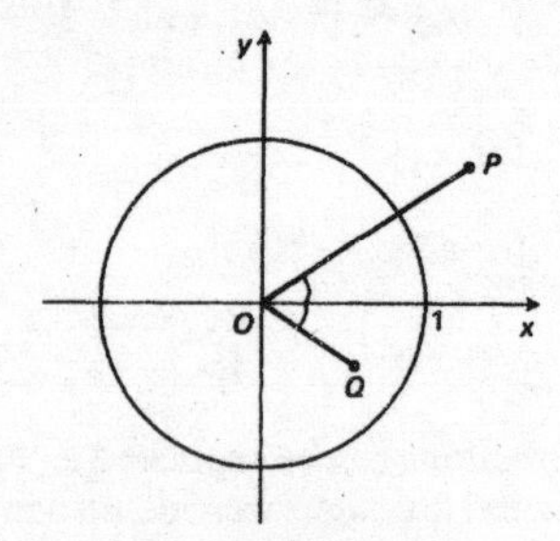

by P in the *complex plane*, then z^{-1} is represented by Q, where $\angle xOQ = -\angle xOP$ and $|OP| \,.\, |OQ| = 1$.

inverse square law of gravitation The following law that describes the force of attraction that exists between two particles, assumed to be isolated from all other bodies. Consider two particles P and S, with gravitational masses m and M, respectively. Let $\mathbf{r}$ be the position vector of P relative to S (that is, the vector represented by $\overrightarrow{SP}$), and let $r = |\mathbf{r}|$. Then the force $\mathbf{F}$ experienced by P is given by

$$\mathbf{F} = -\frac{GMm}{r^3}\mathbf{r},$$

where G is the *gravitational constant*. Since $\mathbf{r}/r$ is a unit vector, the magnitude of the gravitational force experienced by P is inversely proportional to the square of the distance between S and P. The force experienced by S is equal and opposite to that experienced by P.

inverse trigonometric function The inverse sine function $\sin^{-1}$ is, to put it briefly, the inverse function of sine, so that $y = \sin^{-1} x$ if $x = \sin y$. Thus $\sin^{-1} \frac{1}{2} = \pi/6$ because $\sin(\pi/6) = \frac{1}{2}$. However, $\sin(5\pi/6) = \frac{1}{2}$ also, so it might be thought that $\sin^{-1} \frac{1}{2} = \pi/6$ or $5\pi/6$. It is necessary to avoid such ambiguity so it is normally agreed that the value to be taken is the one lying in the interval $[-\pi/2, \pi/2]$. Similarly, $y = \tan^{-1} x$ if $x = \tan y$, and the value y is taken to lie in the interval $(-\pi/2, \pi/2)$. Also, $y = \cos^{-1} x$ if $x = \cos y$ and the value y is taken to lie in the interval $[0, \pi]$.

A more advanced approach provides more explanation. The inverse function of a trigonometric function exists only if the original function is restricted to a suitable domain. This can be an interval I in which the function is strictly increasing or strictly decreasing (see *inverse function*). So, to obtain an inverse function for $\sin x$, the function is restricted to a domain consisting of the interval $[-\pi/2, \pi/2]$; $\tan x$ is restricted to the interval $(-\pi/2, \pi/2)$; and $\cos x$ is restricted to $[0, \pi]$. The domain of the inverse function is, in each case, the range of the restricted function. Hence the following inverse functions are obtained: $\sin^{-1}$: $[-1, 1] \to [-\pi/2, \pi/2]$; $\tan^{-1}$: $\mathbf{R} \to (-\pi/2, \pi/2)$; $\cos^{-1}$: $[-1, 1] \to [0, \pi]$. The notation arcsin,

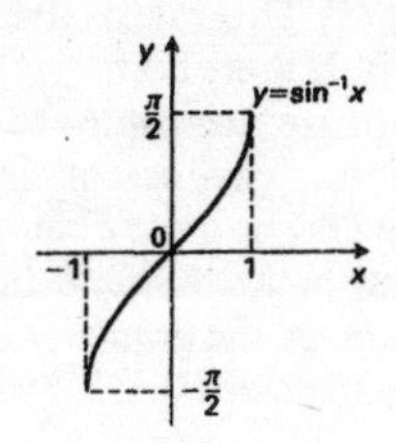

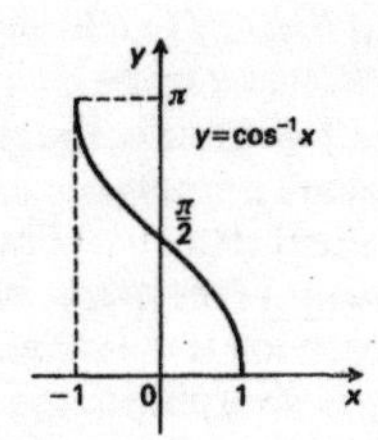

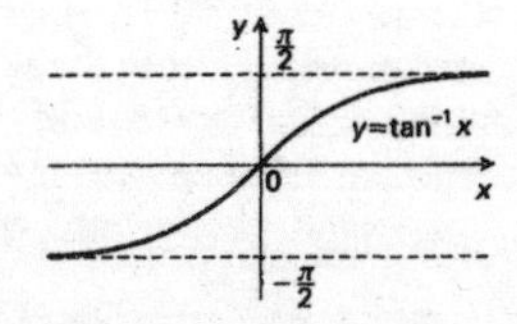

arctan and arccos, for $\sin^{-1}$, $\tan^{-1}$ and $\cos^{-1}$, is also used. The following derivatives can be obtained:

$$\frac{d}{dx}(\sin^{-1}x) = \frac{1}{\sqrt{1-x^2}} \quad (x \neq \pm 1),$$

$$\frac{d}{dx}(\cos^{-1}x) = -\frac{1}{\sqrt{1-x^2}} \quad (x \neq \pm 1), \quad \frac{d}{dx}(\tan^{-1}x) = \frac{1}{1+x^2}.$$

invertible matrix See *inverse matrix*.

irrational number A real number that is not *rational*. A famous proof, sometimes attributed to Pythagoras, shows that $\sqrt{2}$ is irrational; the method can also be used to show that numbers such as $\sqrt{3}$ and $\sqrt{7}$ are also irrational. It follows that numbers like $1 + \sqrt{2}$ and $1/(1 + \sqrt{2})$ are irrational. The proof that e is irrational is reasonably easy, and in 1761 Lambert showed that π is irrational.

isometry If P and Q are points in the plane, $|PQ|$ denotes the distance between P and Q. An **isometry** is a *transformation* of the plane that preserves the distance between points: it is a transformation with the property that, if P and Q are mapped to P' and Q', then $|P'Q'| = |PQ|$. Examples of isometries are *translations*, *rotations* and *reflections*. It can be shown that all the isometries of the plane can be obtained from translations, rotations and reflections, by composition. Two figures are *congruent* if there is an isometry that maps one onto the other.

isomorphic See *isomorphism*.

isomorphism (of groups) Let $\langle G, \circ \rangle$ and $\langle G', * \rangle$ be *groups*, so that $\circ$ is the operation on G and $*$ is the operation on G'. An **isomorphism** between $\langle G, \circ \rangle$ and $\langle G', * \rangle$ is a *one-to-one onto mapping* f from the set G to the set G'

such that, for all a and b in $G, f(a \circ b) = f(a) * f(b)$. This means that, if f maps a to a' and b to b', then f maps $a \circ b$ to $a' * b'$. If there is an isomorphism between two groups, the two groups are **isomorphic** to each other. Two groups that are isomorphic to one another have essentially the same structure; the actual elements of one group may be quite different objects from the elements of the other, but the way in which they behave with respect to the operation is the same. For example, the group $1, i, -1, -i$ with multiplication is isomorphic to the group of elements $0, 1, 2, 3$ with addition modulo 4.

isomorphism (of rings) Suppose that $\langle R, +, \times \rangle$ and $\langle R', \oplus, \otimes \rangle$ are *rings*. An **isomorphism** between them is a *one-to-one onto mapping* f from the set R to the set R' such that, for all a and b in $R, f(a + b) = f(a) \oplus f(b)$ and $f(a \times b) = f(a) \otimes f(b)$. If there is an isomorphism between two rings, the rings are **isomorphic** to each other and, as with isomorphic groups, the two have essentially the same structure.

isosceles trapezium A *trapezium* in which the two non-parallel sides have the same length.

isosceles triangle A triangle in which two sides have the same length. Sometimes the third side (which may be called the **base**) is required to be not of the same length. The two angles opposite the sides of equal length (which may be called the **base angles**) are equal.

iteration A method uses **iteration** if it yields successive approximations to a required value by repetition of a certain procedure. Examples are *fixed-point iteration* and *Newton's method* for finding a root of an equation $f(x) = 0$.

j In the notation for *complex numbers*, some authors, especially engineers, use *j* instead of *i*.

Jacobi, Carl Gustav Jacob (1804–1851) German mathematician responsible for notable developments in the theory of elliptic functions, a class of functions defined by, as it were, inverting certain integrals. Applying them to number theory, he was able to prove Fermat's conjecture that every integer is the sum of four perfect squares. He also published contributions on *determinants* and mechanics.

joint cumulative distribution function For two *random variables* X and Y, the **joint cumulative distribution function** is the function F defined by $F(x, y) = \Pr(X \leq x, Y \leq y)$. The term applies also to the generalization of this to more than two random variables. For two variables, it may be called the **bivariate** and, for more than two, the **multivariate cumulative distribution function**.

joint distribution The **joint distribution** of two or more *random variables* is concerned with the way in which the probability of their taking certain values, or values within certain intervals, varies. It may be given by the *joint cumulative distribution function*. More commonly, the joint distribution of some number of discrete random variables is given by their *joint probability mass function*, and that of some number of continuous random variables by their *joint probability density function*. For two variables it may be called the **bivariate** and, for more than two, the **multivariate distribution**.

joint probability density function For two continuous *random variables* X and Y, the **joint probability density function** is the function f such that

$$\Pr(a \leq X \leq b, c \leq Y \leq d) = \int_a^b \int_c^d f(x, y)\, dy\, dx.$$

The term applies also to the generalization of this to more than two random variables. For two variables, it may be called the **bivariate** and, for more than two, the **multivariate probability density function**.

joint probability mass function For two discrete *random variables* X and Y, the **joint probability mass function** is the function p such that $p(x_i, y_j) = \Pr(X = x_i, Y = y_j)$, for all i and j. The term applies also to the generalization of this to more than two random variables. For two variables, it may be called the **bivariate** and, for more than two, the **multivariate probability mass function**.

Jordan, (Marie-Ennemond-)Camille (1838–1922) French mathematician whose treatise on groups of permutations and the theory of equations drew attention to the work of Galois. In a later work on analysis, there appears his formulation of what is now called the *Jordan Curve Theorem*.

Jordan curve A simple closed curve. Thus a Jordan curve is a continuous plane curve that has no ends (or, in other words, that begins and ends at the same point) and does not intersect itself.

Jordan Curve Theorem An important theorem which says that a *Jordan curve* divides the plane into two regions, the **interior** and the **exterior** of the curve. The result is intuitively obvious but the proof is not elementary.

Josephus problem The story is told by the Jewish historian Josephus of a band of 41 rebels who, preferring suicide to capture, decided to stand in a circle and kill every third remaining person until no one was left. Josephus, one of the number, quickly calculated where he and his friend should stand so that they would be the last two remaining and so avoid being a part of the suicide pact. In its most general form, the **Josephus problem** is to find the position of the one who survives when there are n people in a circle and every m-th remaining person is eliminated.

joule The SI unit of *work*, abbreviated to 'J'. One joule is the work done when the point of application of a force of one *newton* moves a distance of one metre in the direction of the force.

Julia set The boundary of the set of points z_0 in the complex plane for which the application of the function $f(z) = z^2 + c$ repeatedly to the point z_0 produces a bounded sequence. The term may be used similarly for other functions as well. Julia sets are almost always *fractals*, and vary enormously in structure for different values of c.

Kelvin, Lord (1824–1907) The British mathematician, physicist and engineer, William Thomson, who, mostly in the earlier part of his career, made contributions to mathematics in the theories of electricity and magnetism and hydrodynamics.

Kepler, Johannes (1571–1630) German astronomer and mathematician, best known for his three laws of planetary motion, including the discovery that the planets travel in elliptical orbits round the Sun. He proposed the first two laws in 1609 after observations had convinced him that the orbit of Mars was an ellipse, and the third followed ten years later. He also developed methods used by Archimedes to determine the volumes of many solids of revolution by regarding them as composed of infinitely many infinitesimal parts.

Kepler's laws of planetary motion The following three laws about the motion of the planets round the Sun:

(i) Each planet travels in an elliptical *orbit* with the Sun at one focus.
(ii) The line joining a planet and the Sun sweeps out equal areas in equal times.
(iii) The squares of the times taken by the planets to complete an orbit are proportional to the cubes of the lengths of the major axes of their orbits.

These laws were formulated by Kepler, based on observations. Subsequently, Newton formulated his *inverse square law of gravitation* and his second law of motion, from which Kepler's laws can be deduced.

al-Khwārizmī, Muhammad ibn Mūsā (about AD 800) Mathematician and astronomer of Baghdad, the author of two important Arabic works on arithmetic and algebra. The word 'algorithm' is derived from the occurrence of his name in the title of the work that includes a description of the Hindu–Arabic number system. The word 'algebra' comes from the title *Al-jabr wa'l muqābalah* of his other work, which contains, among other things, methods of solving quadratic equations by a method akin to *completing the square*.

kilo- Prefix used with *SI units* to denote multiplication by 10^3.

kilogram In *SI units*, the base unit used for measuring *mass*, abbreviated to 'kg'. At present, it is defined as the mass of a certain platinum–iridium cylinder kept at Sèvres near Paris. One kilogram is approximately the mass of 1000 cm^3 of water.

kinematics The study of the motion of bodies without reference to the forces causing the motion. It is concerned with such matters as the positions, the velocities and the accelerations of the bodies.

kinetic energy The energy of a body attributable to its motion. The kinetic energy may be denoted by E_k or T. For a particle of mass m with speed v, $E_k = \frac{1}{2}mv^2$. For a rigid body rotating about a fixed axis with angular speed ω, and with moment of inertia I about the fixed axis, $E_k = \frac{1}{2}I\omega^2$.

Kinetic energy has the dimensions ML^2T^{-2}, and the SI unit of measurement is the *joule.*

kite A quadrilateral that has two adjacent sides of equal length and the other two sides of equal length. If the kite $ABCD$ has $AB = AD$ and $CB = CD$, the diagonals AC and BD are perpendicular, and AC bisects BD.

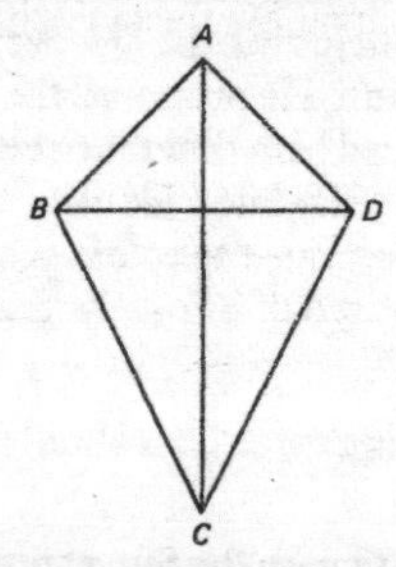

Klein, (Christian) Felix (1849–1925) German mathematician, who in his influential *Erlanger Programm* launched a unification of geometry, in which different geometries are classified by means of group theory. He is responsible for applying the terms 'elliptic' and 'hyperbolic' to describe *non-Euclidean geometries* and, in topology, he is remembered in the name of the *Klein bottle.*

Klein bottle The *Möbius band* has just one surface and has one edge. A sphere, on the other hand, has no edges, but it has two surfaces, the outside and the inside. An example with no edges and only one surface is the **Klein bottle.**

Consider, in the plane, the square of all points with Cartesian coordinates (x, y) such that $-1 \leq x \leq 1$ and $-1 \leq y \leq 1$. The operation of 'identifying' the point $(x, 1)$ with the point $(x, -1)$, for all x, can be thought of as forming a cylinder in 3 dimensions, with two surfaces, the outside and the inside, and two edges. The operation of 'identifying' the point $(1, y)$ with $(-1, -y)$ is like forming a cylinder but including a twist, so this can be thought of as resulting in a Möbius band. The Klein bottle is constructed by doing these two operations simultaneously. This is, of course, not practically possible with a square sheet of flexible material in 3 dimensions, but the result is nevertheless valid mathematically and it has no edges and just one surface.

Klein four-group There are essentially two *groups* with four elements. In other words, any group with four elements is *isomorphic* to one of these two. One is the *cyclic* group with four elements. The other is the **Klein**

four-group. It is *abelian*, and may be written as a multiplicative group with elements e, a, b and c, where e is the identity element, $a^2 = b^2 = c^2 = e$, $ab = c$, $bc = a$ and $ca = b$.

Koch curve Take an equilateral triangle as shown in the first diagram. Replace the middle third of each side by two sides of an equilateral triangle pointing outwards. This forms a six-pointed star, as shown in the second diagram. Repeat this construction to obtain the figure shown in the third diagram. The **Koch curve**, named after the Swedish mathematician Helge von Koch (1870–1924), is the curve obtained when this process is continued indefinitely. The interior of the curve has finite area, but the curve has infinite length.

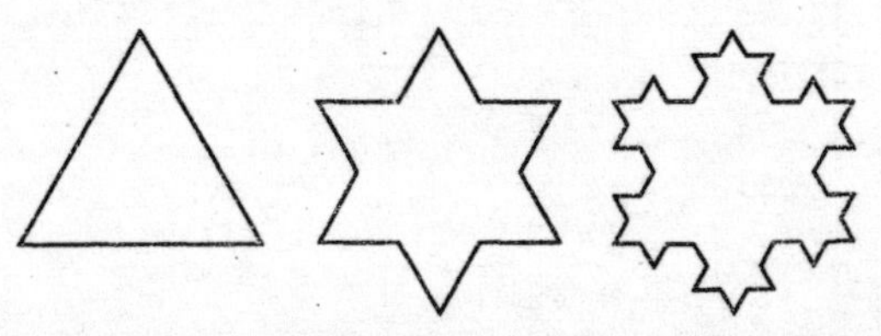

Kolmogorov, Andrei Nikolaevich (1903–1987) Russian mathematician who worked in a number of areas of mathematics, including probability. In this, his important contribution was to give the subject a rigorous foundation by using the language and notation of set theory.

Kolmogorov–Smirnov test A *non-parametric* test for testing the *null hypothesis* that a given sample has been selected from a population with a specified *cumulative distribution function* F. Let $x_1, x_2, \ldots, x_n$ be the sample values in ascending order, and let

$$z = \max_i \left| \frac{i}{n} - F(x_i) \right|.$$

If the null hypothesis is true, then z should be less than a value that can be obtained, for different significance levels, from tables.

Königsberg See *bridges of Königsberg*.

Kronecker, Leopold (1823–1891) German mathematician responsible for considerable contributions to number theory and other fields. But he is more often remembered as the first to cast doubts on non-constructive existence proofs, over which he disputed with Weierstrass and others. See *integer*.

Lagrange, Joseph-Louis (1736–1813) Arguably the greatest, alongside Euler, of eighteenth-century mathematicians. Although he was born in Turin and spent the early part of his life there, he eventually settled in Paris and is normally deemed to be French. Much of his important work was done in Berlin, where he was Euler's successor at the Academy. His work, in common with that of most important mathematicians of the time, covers the whole range of mathematics. He is known for results in number theory and algebra but is probably best remembered as a leading figure in the development of theoretical mechanics. His work *Mécanique analytique*, published in 1788, is a comprehensive account of the subject. In particular, he was mainly responsible for the methods of the *calculus of variations* and the consequent Lagrangian method in mechanics.

Lambert, Johann Heinrich (1728–1777) Swiss–German mathematician, scientist, philosopher and writer on many subjects, who in 1761 proved that π is an *irrational number*. He introduced the standard notation for *hyperbolic functions* and appreciated the difficulty of proving the *parallel postulate*.

lamina An object considered as having a 2-dimensional shape and density but having no thickness. It is used in a *mathematical model* to represent an object such as a thin plate or sheet in the real world.

Lami's Theorem The following theorem in mechanics, named after Bernard Lami, or Lamy (1640–1715):

THEOREM: Suppose that three forces act on a particle and that the resultant of the three forces is zero. Then the magnitude of each force is proportional to the sine of the angle between the other two forces.

In the given situation, the three forces are coplanar and the corresponding *polygon of forces* is a triangle. The lengths of the sides of this triangle are proportional to the magnitudes of the three forces and the result of the theorem is obtained by an application of the *sine rule*.

Laplace, Pierre-Simon, Marquis de (1749–1827) French mathematician, best known for his work on planetary motion, enshrined in his five-volume *Mécanique céleste*, and for his fundamental contributions to the theory of probability. It was Laplace who extended Newton's gravitational theory to the study of the whole solar system. He developed the strongly deterministic view that, once the starting conditions of a closed dynamical system such as the universe are known, its future development is then totally determined. Napoleon asked him where God fitted into all this. 'I have no need of that hypothesis,' replied Laplace. He briefly held the post of Minister of the Interior under Napoleon, but seems to have got on just as well with the restored monarchy.

Latin square A square array of symbols arranged in rows and columns in such a way that each symbol occurs exactly once in each row and once in each column. Two examples are given below. Latin squares are used in the design of experiments involving, for example, trials of different kinds of seed or fertilizer.

```
A B C D     A B C D
B A D C     D C B A
C D A B     C D A B
D C B A     B A D C
```

latitude Suppose that the *meridian* through a point P on the Earth's surface meets the equator at P'. (The meridian is taken here as a semicircle.) Let O be the centre of the Earth. The **latitude** of P is then the angle $P'OP$, measured in degrees north or south. The latitude of any point may be up to 90° north or up to 90° south. Latitude and *longitude* uniquely fix the position of a point on the Earth's surface.

latus rectum The chord through the focus of a *parabola* and perpendicular to the axis. For the parabola $y^2 = 4ax$, the latus rectum has length $4a$.

l.c.m. = *least common multiple.*

leading diagonal = *main diagonal.*

least common multiple For two non-zero integers a and b, an integer that is a multiple of both is a **common multiple.** Of all the positive common multiples, the least is the **least common multiple (l.c.m.)**, denoted by $[a, b]$. The l.c.m. of a and b has the property of dividing any other common multiple of a and b. If the prime decompositions of a and b are known, the l.c.m. is easily obtained: for example, if $a = 168 = 2^3 \times 3 \times 7$ and $b = 180 = 2^2 \times 3^2 \times 5$, then the l.c.m. is $2^3 \times 3^2 \times 5 \times 7 = 2520$. For positive integers a and b, the l.c.m. is equal to $ab/(a, b)$, where (a, b) is the *greatest common divisor* (g.c.d.).

Similarly, any finite set of non-zero integers, $a_1, a_2, \ldots, a_n$ has an l.c.m. denoted by $[a_1, a_2, \ldots, a_n]$. However, when $n > 2$ it is not true, in general, that $[a_1, a_2 \ldots, a_n]$ is equal to $a_1a_2 \ldots a_n/(a_1, a_2 \ldots, a_n)$, where $(a_1, a_2 \ldots, a_n)$ is the g.c.d.

least squares The method of **least squares** is used to estimate *parameters* in statistical models such as those that occur in *regression.* Estimates for the parameters are obtained by minimizing the sum of the squares of the differences between the observed values and the predicted values under the model. For example, suppose that, in a case of linear regression, where $E(Y) = \alpha + \beta X$, there are n paired observations $(x_1, y_1), (x_2, y_2), \ldots, (x_n, y_n)$. Then the method of least squares gives a and b as *estimators* for α and β, where a and b are chosen so as to minimize $\sum(y_i - a - bx_i)^2$.

least upper bound = *supremum.*

least value Let f be a *real function* and D a subset of its domain. If there is a point c in D such that $f(c) \leq f(x)$ for all x in D, then $f(c)$ is the **least value** of f in D. There may be no such point: consider, for example, either the function f defined by $f(x) = -1/x$, or the function f defined by $f(x) = x$, with the open interval $(0, 1)$ as D; or the function f defined by $f(x) = [x] - x$, with the closed interval $[0, 1]$ as D. If the least value does exist, it may be attained at more than one point of D.

That a *continuous function* on a closed interval has a least value is ensured by the non-elementary theorem that such a function 'attains its bounds'. An important theorem states that a function, continuous on $[a, b]$ and *differentiable* in (a, b), attains its least value either at a *local minimum* (which is a *stationary point*) or at an end-point of the interval.

Lebesgue, Henri-Léon (1875–1941) French mathematician who revolutionized the theory of integration. Building on the work of earlier French mathematicians on the notion of measure, he developed the Lebesgue integral, one of the outstanding concepts in modern analysis. It generalizes the *Riemann integral*, which can deal only with continuous functions and other functions not too unlike them. His two major texts on the subject were published in the first few years of this century.

left and **right derivative** For the *real function* f, if

$$\lim_{h \to 0-} \frac{f(a+h) - f(a)}{h}$$

exists, this limit is the **left derivative** of f at a. Similarly, if

$$\lim_{h \to 0+} \frac{f(a+h) - f(a)}{h}$$

exists, this limit is the **right derivative** of f at a. (See *limit from the left and right*.) The *derivative* $f'(a)$ exists if and only if the left derivative and the right derivative of f at a exist and are equal. An example where the left and right derivatives both exist but are not equal is provided by the function f, where $f(x) = |x|$ for all x. At 0, the left derivative equals -1 and the right derivative equals $+1$.

left-handed system See *right-handed system*.

Legendre, Adrien-Marie (1752–1833) French mathematician who, with Lagrange and Laplace, formed a trio associated with the period of the French Revolution. He was well known in the nineteenth century for his highly successful textbook on the geometry of Euclid. But his real work was concerned with calculus. He was responsible for the classification of elliptic integrals into their standard forms. The so-called Legendre polynomials, solutions of a certain differential equation, are among the most important of the special functions. In an entirely different area, he along with Euler conjectured and partially proved an important result in number theory known as the law of quadratic reciprocity.

Leibniz, Gottfried Wilhelm (1646–1716) German mathematician, philosopher, scientist and writer on a wide range of subjects, who was, with Newton, the founder of the calculus. Newton's discovery of differential calculus was perhaps ten years earlier than Leibniz's, but Leibniz was the first to publish his account, written independently of Newton, in 1684. Soon after, he published an exposition of integral calculus that included the *Fundamental Theorem of Calculus*. He also wrote on other branches of mathematics, making significant contributions to the development of symbolic logic, a lead which was not followed up until the end of the nineteenth century.

Leibniz's Theorem The following result for the n-th derivative of a product:

THEOREM: If $h(x) = f(x)g(x)$ for all x, the n-th derivative of h is given by

$$h^{(n)}(x) = \sum_{r=0}^{n} \binom{n}{r} f^{(r)}(x) g^{(n-r)}(x),$$

where the coefficients $\binom{n}{r}$ are *binomial coefficients*.

For example, to find $h^{(8)}(x)$, when $h(x) = x^2 \sin x$, let $f(x) = x^2$ and $g(x) = \sin x$. Then $f'(x) = 2x$ and $f''(x) = 2$; and $g^{(8)}(x) = \sin x, g^{(7)}(x) = -\cos x$ and $g^{(6)}(x) = -\sin x$. So

$$\begin{aligned} h^{(8)}(x) &= x^2 \sin x + \binom{8}{1} 2x(-\cos x) + \binom{8}{2} 2(-\sin x) \\ &= x^2 \sin x - 16x \cos x - 56 \sin x. \end{aligned}$$

lemma A mathematical theorem proved primarily for use in the proof of a subsequent theorem.

length (of a binary code) See *binary code*.

length (of a binary word) See *binary word*.

length (of a line segment) The **length** $|AB|$ of the *line segment AB*, and the length $|\overrightarrow{AB}|$ of the *directed line-segment*, is equal to the distance between A and B. It is equal to zero when A and B coincide, and is otherwise always positive.

length (of a sequence) See *sequence*.

length (of a series) See *series*.

length (of a vector) See *vector*.

length of the major and **minor axis** See *ellipse*.

Leonardo of Pisa See *Fibonacci*.

lever A rigid bar or rod free to rotate about a point or axis, called the **fulcrum**, normally used to transmit a force at one point to a force at

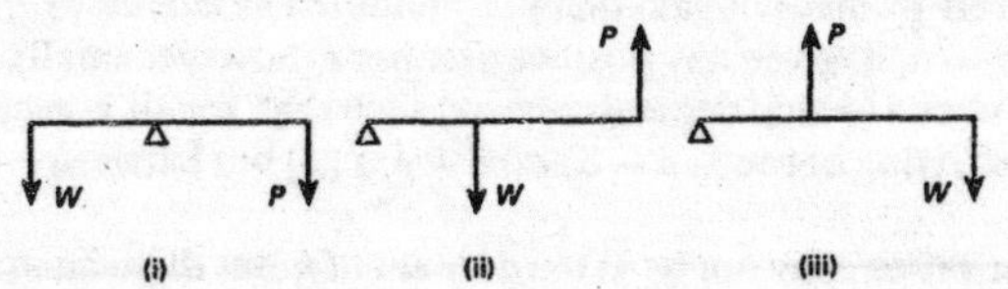

another. In diagrams the fulcrum is usually denoted by a small triangle. In the three examples shown in the figure above, the lever is being used as a *machine* in which the effort is the force **P** and the load is **W**. In each case, the *mechanical advantage* can be found by using the *principle of moments*. Everyday examples corresponding to the three kinds of lever shown are (i) an oar as used in rowing, (ii) a stationary wheelbarrow and (iii) the jib of a certain kind of crane.

l'Hôpital, l'Hôpital's rule See under H.

liar paradox The paradox that arises from considering the statement 'This statement is false.'

Lie, (Marius) Sophus (1842–1899) Norwegian mathematician responsible for advances in differential equations and differential geometry, and primarily remembered for a three-volume treatise on transformation groups. Like Klein, he brought group theory to bear on geometry.

lift See *aerodynamic drag*.

light In mechanics, an object such as a string or a rod is said to be **light** if its weight may be considered negligible compared with that of other objects involved.

likelihood function For a sample from a population with an unknown *parameter*, the **likelihood function** is the probability that the sample could have occurred at random; it is a function of the unknown parameter. The **method of maximum likelihood** selects the value of the parameter that maximizes the likelihood function. For algebraic convenience, it is common to work with the logarithm of the likelihood function. The concept applies equally well when there are two or more unknown parameters.

limit (of $f(x)$) The **limit**, if it exists, of $f(x)$ as x tends to a is a number l with the property that, as x gets closer and closer to a, $f(x)$ gets closer and closer to l. This is written

$$\lim_{x \to a} f(x) = l.$$

Such an understanding may be adequate at an elementary level. It is important to realize that this limit may not equal $f(a)$; indeed, $f(a)$ may not necessarily be defined.

At a more advanced level, the definition just given needs to be made more precise: a formal statement says that $f(x)$ can be made as close to l as we please by restricting x to a sufficiently small neighbourhood of a (but excluding a itself). Thus, it is said that $f(x)$ tends to l as x tends to a, written $f(x) \to l$ as $x \to a$, if, given any positive number ϵ (however small), there is a positive number δ (which depends upon ϵ) such that, for all x, except possibly a itself, lying between $a - \delta$ and $a + \delta$, $f(x)$ lies between $l - \epsilon$ and $l + \epsilon$.

Notice that a itself may not be in the domain of f; but there must be a neighbourhood of a, the whole of which (apart possibly from a itself) is in the domain of f. For example, let f be the function defined by

$$f(x) = \frac{\sin x}{x} \quad (x \neq 0).$$

Then 0 is not in the domain of f, but it can be shown that

$$\lim_{x \to 0} \frac{\sin x}{x} = 1.$$

In the above, l is a real number, but it is also possible that $f(x) \to \pm\infty$, where appropriate formal definitions can be given: it is said that $f(x) \to +\infty$ as $x \to a$ if, given any positive number K (however large), there is a positive number δ (which depends upon K) such that, for all x, except possibly a itself, lying between $a - \delta$ and $a + \delta$, $f(x)$ is greater than K. It is clear, for example, that $1/x^2 \to +\infty$ as $x \to 0$. There is a similar definition for $f(x) \to -\infty$ as $x \to a$.

Sometimes the behaviour of $f(x)$ as x gets indefinitely large is of interest and importance: it is said that $f(x) \to l$ as $x \to \infty$ if, given any positive number ϵ (however small), there is a number X (which depends upon ϵ) such that, for all $x > X$, $f(x)$ lies between $l - \epsilon$ and $l + \epsilon$. This means that $y = l$ is a horizontal *asymptote* for the graph $y = f(x)$. Similar definitions can be given for $f(x) \to l$ as $x \to -\infty$, and for $f(x) \to \pm\infty$ as $x \to \infty$ and $f(x) \to \pm\infty$ as $x \to -\infty$.

limit (of a sequence) The **limit**, if it exists, of an infinite sequence $a_1, a_2, a_3, \ldots$ is a number l with the property that a_n gets closer and closer to l as n gets indefinitely large. Such an understanding may be adequate at an elementary level.

At a more advanced level, the statement just given needs to be made more precise: it is said that the sequence $a_1, a_2, a_3, \ldots$ has the limit l if, given any positive number ϵ (however small), there is a number N (which depends upon ϵ) such that, for all $n > N$, a_n lies between $l - \epsilon$ and $l + \epsilon$. This can be written $a_n \to l$.

The sequence $0, \frac{1}{2}, \frac{3}{4}, \frac{7}{8}, \frac{15}{16}, \ldots$, for example, has the limit 1. To take another example, the sequence $-1, \frac{1}{2}, -\frac{1}{3}, \frac{1}{4}, -\frac{1}{5}, \ldots$ has the limit 0; since this is the sequence whose n-th term is $(-1)^n/n$, this fact can be stated as $(-1)^n/n \to 0$.

There are, of course, sequences that do not have a limit. These can be classified into different kinds.

(i) A formal definition says that $a_n \to \infty$ if, given any positive number K (however large), there is an integer N (which depends upon K) such that, for all $n > N$, a_n is greater than K. For example, $a_n \to \infty$ for the sequence 1, 4, 9, 16, ..., in which $a_n = n^2$.

(ii) There is a similar definition for $a_n \to -\infty$, and an example is the sequence −4, −5, −6, ..., in which $a_n = -n - 3$.

(iii) If a sequence does not have a limit but is *bounded*, such as the sequence $-\frac{1}{2}, \frac{2}{3}, -\frac{3}{4}, \frac{4}{5}, \ldots$, in which $a_n = (-1)^n n/(n+1)$, it may be said to **oscillate finitely**.

(iv) If a sequence is not bounded, but it is not the case that $a_n \to \infty$ or $a_n \to -\infty$ (the sequence 1, 2, 1, 4, 1, 8, 1, ... is an example), the sequence may be said to **oscillate infinitely**.

limit from the left and **right** The statement that $f(x)$ tends to l as x tends to a **from the left** can be written: $f(x) \to l$ as $x \to a-$. Another way of writing this is

$$\lim_{x \to a-} f(x) = l.$$

The formal definition says that this is so if, given $\epsilon > 0$, there is a number $\delta > 0$ such that, for all x strictly between $a - \delta$ and a, $f(x)$ lies between $l - \epsilon$ and $l + \epsilon$. In place of $x \to a-$, some authors use $x \nearrow a$. In the same way, the statement that $f(x)$ tends to l as x tends to a **from the right** can be written: $f(x) \to l$ as $x \to a+$. Another way of writing this is

$$\lim_{x \to a+} f(x) = l.$$

The formal definition says that this is so if, given $\epsilon > 0$, there is a number $\delta > 0$ such that, for all x strictly between a and $a + \delta$, $f(x)$ lies between $l - \epsilon$ and $l + \epsilon$. In place of $x \to a+$, some authors use $x \searrow a$. For example, if $f(x) = x - \lfloor x \rfloor$, then

$$\lim_{x \to 1-} f(x) = 1, \quad \lim_{x \to 1+} f(x) = 0.$$

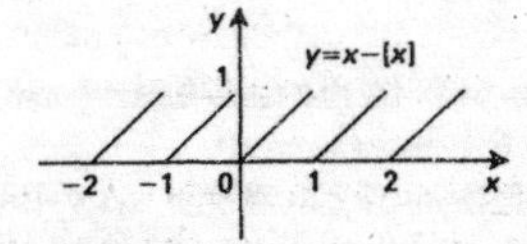

limit of integration In the definite integral

$$\int_a^b f(x)\,dx,$$

the **lower limit** (of integration) is a and the **upper limit** (of integration) is b.

Lindemann, (Carl Louis) Ferdinand von (1852–1939) German mathematician, professor at Königsberg and then at Munich, who proved in 1882 that π is a *transcendental number*.

line of action The **line of action** of a force is a straight line through the *point of application* of the force and parallel to the direction of the force.

line of symmetry See *symmetrical about a line.*

line segment If A and B are two points on a straight line, the part of the line between and including A and B is a **line segment**. This may be denoted by AB or BA. (The notation AB may also be used with a different meaning, as a real number, the *measure* of $\overrightarrow{AB}$ on a directed line.) Some authors use 'line segment' to mean a *directed line-segment.*

linear algebra The topics of linear equations, matrices, vectors, and of the algebraic structure known as a vector space, are intimately linked, and this area of mathematics is known as **linear algebra.**

linear congruence equation See *congruence equation.*

linear differential equation A *differential equation* of the form

$$a_n(x)y^{(n)}(x) + a_{n-1}(x)y^{(n-1)}(x) + \cdots + a_1(x)y'(x) + a_0(x)y(x) = f(x),$$

where $a_0, a_1, \ldots, a_n$ and f are given functions, and $y', \ldots, y^{(n)}$ are *derivatives* of y. See also *linear differential equation with constant coefficients* and *linear first-order differential equation.*

linear differential equation with constant coefficients For simplicity, consider such an equation of second-order,

$$a\frac{d^2y}{dx^2} + b\frac{dy}{dx} + cy = f(x), \qquad 1$$

where a, b and c are given constants and f is a given function. (Higher-order equations can be treated similarly.) Suppose that f is not the zero function. Then the equation

$$a\frac{d^2y}{dx^2} + b\frac{dy}{dx} + cy = 0 \qquad 2$$

is the **homogeneous** equation that corresponds to the **non-homogeneous** equation **1**. The two are connected by the following result:

THEOREM: If $y = G(x)$ is the general solution of **2** and $y = y_1(x)$ is a particular solution of **1**, then $y = G(x) + y_1(x)$ is the general solution of **1**.

Thus the problem of solving **1** is reduced to the problem of finding the **complementary function** (C.F.) $G(x)$, which is the general solution of **2**, and a particular solution $y_1(x)$ of **1**, usually known in this context as a **particular integral** (P.I.).

The complementary function is found by looking for solutions of **2** of the form $y = e^{mx}$ and obtaining the **auxiliary equation** $am^2 + bm + c = 0$. If this equation has distinct real roots m_1 and m_2, the C.F. is $y = Ae^{m_1x} + Be^{m_2x}$; if it has one (repeated) root m, the C.F. is $y = (A + Bx)e^{mx}$; if it has non-real roots $\alpha \pm \beta i$, the C.F. is $y = e^{\alpha x}(A\cos\beta x + B\sin\beta x)$.

The most elementary way of obtaining a particular integral is to try something similar in form to $f(x)$. Thus, if $f(x) = e^{kx}$, try as the P.I.

$y_1(x) = pe^{kx}$. If $f(x)$ is a polynomial in x, try a polynomial of the same degree. If $f(x) = \cos kx$ or $\sin kx$, try $y_1(x) = p\cos kx + q\sin kx$. In each case, the values of the unknown coefficients are found by substituting the possible P.I. into the equation 1. If $f(x)$ is the sum of two terms, a P.I. corresponding to each may be found and the two added together.

Using these methods, the general solution of $y'' - 3y' + 2y = 4x + e^{3x}$, for example, is found to be $y = Ae^x + Be^{2x} + 2x + 3 + \frac{1}{2}e^{3x}$.

linear equation Consider, in turn, **linear equations** in one, two and three variables. A linear equation in one variable x is an equation of the form $ax + b = 0$. If $a \neq 0$, this has the solution $x = -b/a$. A linear equation in two variables x and y is an equation of the form $ax + by + c = 0$. If a and b are not both zero, and x and y are taken as Cartesian coordinates in the plane, the equation is an equation of a *straight line.* A linear equation in three variables x, y and z is an equation of the form $ax + by + cz + d = 0$. If a, b and c are not all zero, and x, y and z are taken as Cartesian coordinates in 3-dimensional space, the equation is an equation of a *plane.* See also *simultaneous linear equations.*

linear first-order differential equation A *differential equation* of the form

$$\frac{dy}{dx} + P(x)y = Q(x),$$

where P and Q are given functions. One method of solution is to multiply both sides of the equation by an **integrating factor** $\mu(x)$, given by

$$\mu(x) = \exp\left(\int P(x)\,dx\right).$$

This choice is made because then the left-hand side of the equation becomes

$$\mu(x)\frac{dy}{dx} + \mu(x)P(x)y,$$

which is the exact derivative of $\mu(x)y$, and the solution can be found by integration.

linear function In real analysis, a **linear function** is a function f such that $f(x) = ax + b$ for all x in $\mathbf{R}$, where a and b are real numbers with, normally, $a \neq 0$.

linear interpolation See *interpolation.*

linearly dependent and **independent** A set of vectors $\mathbf{u}_1, \mathbf{u}_2, \ldots, \mathbf{u}_r$ is **linearly independent** if $x_1\mathbf{u}_1 + x_2\mathbf{u}_2 + \cdots + x_r\mathbf{u}_r = \mathbf{0}$ holds only if $x_1 = 0$, $x_2 = 0, \ldots, x_r = 0$. Otherwise, the set is **linearly dependent.** In 3-dimensional space, any set of four or more vectors must be linearly dependent. A set of three vectors is linearly independent if and only if the three are not coplanar. A set of two vectors is linearly independent if and only if the two are not parallel or, in other words, if and only if neither is a scalar multiple of the other.

linear momentum The **linear momentum** of a particle is the product of its mass and its velocity. It is a vector quantity, usually denoted by **p**, and $\mathbf{p} = m\mathbf{v}$. See also *conservation of linear momentum*.

Linear momentum has the dimensions MLT^{-1}, and the SI unit of measurement is the kilogram metre per second, abbreviated to 'kg m s^{-1}'.

linear programming The branch of mathematics concerned with maximizing or minimizing a linear function subject to a number of linear constraints. It has applications in economics, town planning, management in industry, and commerce, for example. In its simplest form, with two variables, the constraints determine a **feasible region**, which is the interior of a polygon in the plane. The **objective function** to be maximized or minimized attains its maximum or minimum value at a vertex of the feasible region. For example, consider the problem of maximizing $4x_1 - 3x_2$ subject to $x_1 - 2x_2 \geq -4, 2x_1 + 3x_2 \leq 13, x_1 - x_2 \leq 4$, and $x_1 \geq 0, x_2 \geq 0$. The feasible region is the interior of the polygon *OABCD* shown in the figure, and the objective function $4x_1 - 3x_2$ attains its maximum value of 17 at the point *B* with coordinates (5, 1).

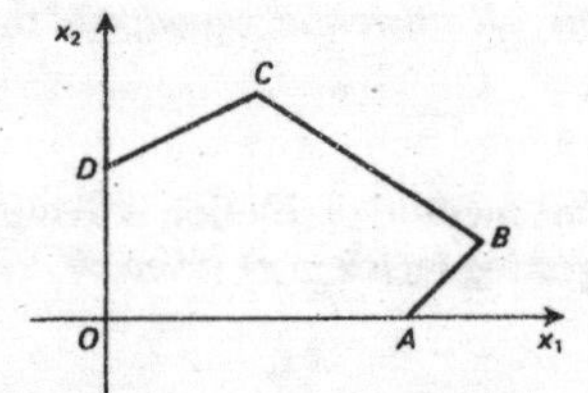

linear regression See *regression*.

linear transformation See *transformation* (of the plane).

Liouville, Joseph (1809–1882) Prolific and influential French mathematician, who was founder and editor of a notable French mathematical journal and proved important results in the fields of number theory, differential equations, differential geometry and complex analysis. In 1844, he proved the existence of *transcendental numbers* and constructed an infinite class of such numbers.

litre A unit, not an SI unit, used in certain contexts for measuring volume, abbreviated to 'l'. A litre is equal to one cubic decimetre, which equals 1000 cm^3. One millilitre, abbreviated to 1 ml, is equal to 1 cm^3. At one time, a litre was defined as the volume of one kilogram of water at 4 °C at standard atmospheric pressure.

load See *machine*.

Lobachevsky, Nikolai Ivanovich (1792–1856) Russian mathematician who in 1829 published his discovery, independently of Bolyai, of *hyperbolic* *geometry*. He continued to publicize his ideas, but his work received wide recognition only after his death.

local maximum (plural: maxima) For a function *f*, a **local maximum** is a point *c* that has a neighbourhood at every point of which, except for *c*, $f(x) < f(c)$. If *f* is *differentiable* at a local maximum *c*, then $f'(c) = 0$; that is, the local maximum is a *stationary point*.

local minimum (plural: minima) For a function *f*, a **local minimum** is a point *c* that has a neighbourhood at every point of which, except for *c*, $f(x) > f(c)$. If *f* is *differentiable* at a local minimum *c*, then $f'(c) = 0$; that is, the local minimum is a *stationary point*.

location In statistics, a measure of **location** is a single figure which gives a typical or, in some sense, central value for a distribution or sample. The most common measures of location are the *mean*, the *median* and, less usefully, the *mode*. For several reasons, the mean is usually the preferred measure of location, but when a distribution is *skew* the median may be more appropriate.

locus (plural: loci) A **locus** (in the plane) is the set of all points in the plane that satisfy some given condition or property. For example, the locus of all points that are a given distance from a fixed point is a circle; the locus of all points equidistant from two given points, *A* and *B*, is the perpendicular bisector of *AB*. If, in a given coordinate system, a locus can be expressed in the form $\{(x, y) \mid f(x, y) = 0\}$, the equation $f(x, y) = 0$ is called an **equation** of the locus (in the given coordinate system). The equation may be said to 'represent' the locus.

logarithm Let *a* be any positive number, not equal to 1. Then, for any real number *x*, the meaning of a^x can be defined (see approach 1 to the *exponential function to base a*). The **logarithmic function to base** *a*, denoted by $\log_a$, is defined as the inverse of this function. So $y = \log_a x$ if and only if $x = a^y$. Thus $\log_a x$ is the index to which *a* must be raised in order to get *x*. Since any power a^y of *a* is positive, *x* must be positive for $\log_a x$ to be defined, and so the domain of the function $\log_a$ is the set of positive real numbers. (See *inverse function* for a more detailed explanation about the domain of an inverse function.) The value $\log_a x$ is called simply the **logarithm** of *x* to base *a*. The notation $\log x$ may be used when the base intended is understood. The following properties hold, where *x*, *y* and *r* are real, with $x > 0$ and $y > 0$:

(i) $\log_a(xy) = \log_a x + \log_a y$.
(ii) $\log_a(1/x) = -\log_a x$.
(iii) $\log_a(x/y) = \log_a x - \log_a y$.
(iv) $\log_a(x^r) = r \log_a x$.
(v) Logarithms to different bases are related by the formula

$$\log_b x = \frac{\log_a x}{\log_a b}.$$

Logarithms to base 10 are called **common logarithms**. Tables of common logarithms were in the past used for arithmetical calculations. Logarithms to base e are called **natural** or **Napierian logarithms**, and the notation ln can be used instead of $\log_e$. However, this presupposes that the value of e has been defined independently. It is preferable to define the *logarithmic function* ln in quite a different way. From this, e can be defined and then the equivalence of ln and $\log_e$ proved.

logarithmic function The **logarithmic function** ln must be distinguished from the logarithmic function to base a (see *logarithm*). Here are two approaches:

1. Suppose that the value of the number e has already been obtained independently. Then the *logarithm* of x to base e can be defined and denoted by $\log_e x$, and the logarithmic function ln can be taken to be just this function $\log_e$. The problem with this approach is its reliance on a prior definition of e and the difficulty of subsequently proving some of the important properties of ln.

2. The following is more satisfactory. Let f be the function defined, for $t > 0$, by $f(t) = 1/t$. Then the logarithmic function ln is defined, for $x > 0$, by

$$\ln x = \int_1^x \frac{1}{t}\, dt.$$

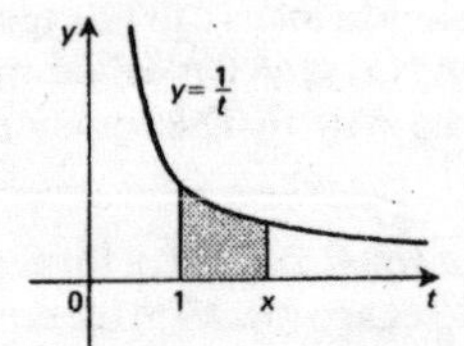

The intention is best appreciated in the case when $x > 1$, for then $\ln x$ gives the area under the graph of f in the interval $[1, x]$. This function ln is *continuous* and increasing; it is *differentiable* and, from the fundamental relationship between differentiation and integration, its *derived function* is the function f. Thus it has been established that

$$\frac{d}{dx}(\ln x) = \frac{1}{x} \quad (x > 0).$$

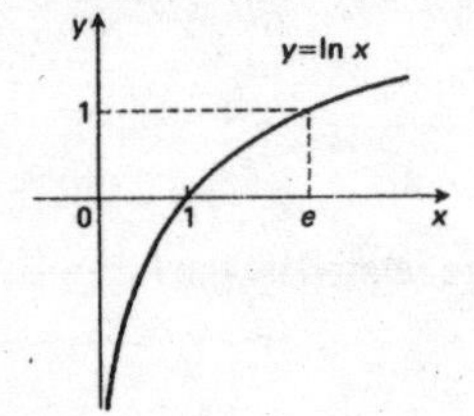

From the definition, the following properties can be obtained, where x, y and r are real, with $x > 0$ and $y > 0$:

(i) $\ln(xy) = \ln x + \ln y$.
(ii) $\ln(1/x) = -\ln x$.
(iii) $\ln(x/y) = \ln x - \ln y$.
(iv) $\ln(x^r) = r \ln x$.

With this approach, exp can be defined as the inverse function of ln, and the number e defined as exp 1. Finally, it is shown that $\ln x$ and $\log_e x$ are identical.

logarithmic function to base a See *logarithm*.

logarithmic plotting Two varieties of logarithmic graph paper are commonly available. One, known as **semi-log** or **single log** paper, has a standard scale on the x-axis and a *logarithmic scale* on the y-axis. If given data satisfy an equation $y = ba^x$ (where a and b are constants), exhibiting *exponential growth*, then on this special graph paper the corresponding points lie on a straight line. The other variety is known as **log–log** or **double log** paper and has a logarithmic scale on both axes. In this case, data satisfying $y = bx^m$ (where b and m are constants) produce points that lie on a straight line. Given some experimental data, such **logarithmic plotting** on one variety or the other of logarithmic graph paper can be used to determine what relationship might have given rise to such data.

logarithmic scale A method of representing numbers (certainly positive and usually greater than or equal to 1) by points on a line as follows. Suppose that one direction along the line is taken as positive, and that the point O is taken as origin. The number x is represented by the point P in such a way that OP is proportional to $\log x$, where logarithms are to base 10. Thus the number 1 is represented by O; and, if the point A represents 10, the point B that represents 100 is such that $OB = 2OA$.

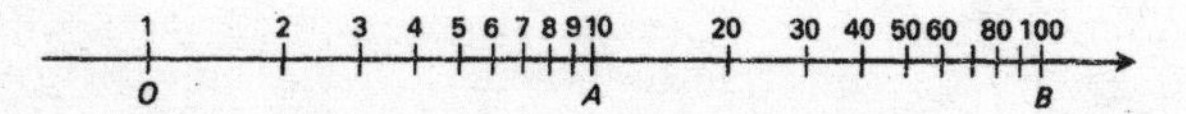

logarithmic spiral = *equiangular spiral*.

logically equivalent Two *compound statements* involving the same components are **logically equivalent** if they have the same *truth tables*. This means that, for all possible truth values of the components, the resulting truth values of the two statements are the same. For example, the truth table for the statement $(\neg p) \vee q$ can be completed as follows:

p	q	$\neg p$	$(\neg p) \vee q$
T	T	F	T
T	F	F	F
F	T	T	T
F	F	T	T

By comparing the last column here with the truth table for $p \Rightarrow q$ (see *implication*), it can be seen that $(\neg p) \vee q$ and $p \Rightarrow q$ are logically equivalent.

longitude The **longitude** of a point P on the Earth's surface is the angle, measured in degrees east or west, between the *meridian* through P and the *prime meridian* through Greenwich (the meridians being taken here as semicircles). If the meridian through P and the meridian through Greenwich meet the equator at P' and G' respectively, the longitude is the angle $P'OG'$, where O is the centre of the Earth. The longitude of any point may be up to 180° east or up to 180° west. Longitude and *latitude* uniquely fix the position of a point on the Earth's surface.

loop See *graph.*

lower bound See *bound.*

lower limit See *limit of integration.*

lower triangular matrix See *triangular matrix.*

machine A device that enables energy from one source to be modified and transmitted as energy in a different form or for a different purpose. Simple examples, in mechanics, are *pulleys* and *levers*, which may be used as components in a more complex machine. The force applied to the machine is the **effort**, and the force exerted by the machine is equal and opposite to the **load**.

If the effort has magnitude P and the load has magnitude W, the **mechanical advantage** of the machine is the ratio W/P. The **velocity ratio** is the ratio of the distance travelled by the effort to the distance travelled by the load. It is a constant for a particular machine, calculable from first principles. For example, in the case of a pulley system, the velocity ratio equals the number of ropes supporting the load or, equivalently, the number of pulleys in the system.

The **efficiency** is the ratio of the work obtained from the machine to the work put into the machine, often stated as a percentage. In the real world, the efficiency of a machine is less than 1 because of factors such as friction. In an ideal machine for which the efficiency is 1, the mechanical advantage equals the velocity ratio.

Maclaurin, Colin (1698–1746) Scottish mathematician who was the outstanding British mathematician of the generation following Newton's. He developed and extended the subject of calculus. His textbook on the subject contains important original results, but the *Maclaurin series*, which appears in it, is just a special case of the *Taylor series* known considerably earlier. He also obtained notable results in geometry and wrote a popular textbook on algebra.

Maclaurin series (or **expansion**) Suppose that f is a *real function*, all of whose *derived functions* $f^{(r)}$ $(r = 1, 2, \ldots)$ exist in some interval containing 0. It is then possible to write down the power series

$$f(0) + \frac{f'(0)}{1!}x + \frac{f''(0)}{2!}x^2 + \cdots + \frac{f^{(n)}(0)}{n!}x^n + \cdots.$$

This is the **Maclaurin series** (or **expansion**) for f. For many important functions, it can be proved that the Maclaurin series is convergent, either for all x or for a certain range of values of x, and that for these values the sum of the series is $f(x)$. For these values it is said that the Maclaurin series is a 'valid' expansion of $f(x)$. The Maclaurin series for some common functions, with the values of x for which they are valid, are given in the Table of Series (Appendix 4). The function f, defined by $f(0) = 0$ and $f(x) = e^{-1/x^2}$ for all $x \neq 0$, is notorious in this context. It can be shown that all of its derived functions exist and that $f^{(r)}(0) = 0$ for all r. Consequently, its Maclaurin series is convergent and has sum 0, for all x. This shows, perhaps contrary to expectation, that, even when the Maclaurin series for a function f is convergent, its sum is not necessarily $f(x)$. See also *Taylor's Theorem*.

magic square A square array of numbers arranged in rows and columns in such a way that all the rows, all the columns and both diagonals add up to the same total. Often the numbers used are $1, 2, \ldots, n^2$, where n is the number of rows and columns. The example on the left below is said to be of ancient Chinese origin; that on the right is featured in an engraving by Albrecht Dürer.

$$\begin{matrix} 4 & 9 & 2 \\ 3 & 5 & 7 \\ 8 & 1 & 6 \end{matrix} \qquad \begin{matrix} 16 & 3 & 2 & 13 \\ 5 & 10 & 11 & 8 \\ 9 & 6 & 7 & 12 \\ 4 & 15 & 14 & 1 \end{matrix}$$

magnitude (of a vector) See *vector*. Also, if the vector $\mathbf{a}$ is given in terms of its components (with respect to the standard vectors $\mathbf{i}$, $\mathbf{j}$ and $\mathbf{k}$) in the form $\mathbf{a} = a_1\mathbf{i} + a_2\mathbf{j} + a_3\mathbf{k}$, the magnitude of $\mathbf{a}$ is given by the formula

$$|\mathbf{a}| = \sqrt{a_1^2 + a_2^2 + a_3^2}.$$

main diagonal In the $n \times n$ matrix $[a_{ij}]$, the entries $a_{11}, a_{22}, \ldots, a_{nn}$ form the **main diagonal**.

major arc See *arc*.

major axis See *ellipse*.

Mandelbrot set The set of points c in the complex plane for which the application of the function $f(z) = z^2 + c$ repeatedly to the point $z = 0$ produces a bounded sequence. The set is an extremely complicated object whose boundary is a *fractal*. It can also be defined as the set of points c for which the *Julia set* of the above function is a connected set. The Mandelbrot set is very frequently used as an illustration of a geometrical fractal.

mantissa See *floating-point notation*.

mapping A **mapping** (or **function**) f from S to T, where S and T are non-empty sets, is a rule which associates, with each element of S, a unique element of T. The set S is the **domain** and T is the **codomain** of f. The phrase 'f from S to T' is written '$f: S \to T$'. For s in S, the unique element of T that f associates with s is the **image of s under f** and is denoted by $f(s)$. If $f(s) = t$, it is said that f **maps** s to t, written $f: s \mapsto t$. The subset of T consisting of those elements that are images of elements of S under f, that is, the subset $\{ t \mid t = f(s), \text{ for some } s \text{ in } S \}$, is the **image** (or **range**) of f, denoted by $f(S)$. For the mapping $f: S \to T$, the subset $\{ (s, f(s)) \mid s \in S \}$ of the *Cartesian product* $S \times T$ is the *graph* of f. The graph of a mapping has the property that, for each s in S, there is a unique element (s, t) in the graph. Some authors define a mapping from S to T to be a subset of $S \times T$ with this property; then the image of s under this mapping is defined to be the unique element t such that (s, t) is in this subset.

marginal distribution Suppose that X and Y are discrete random variables with *joint probability mass function* p. Then the **marginal distribution** of X is the distribution with probability mass function p_1, where

$$p_1(x) = \sum_y p(x, y).$$

A similar formula gives the marginal distribution of Y.

Suppose that X and Y are continuous random variables with joint probability density function f. Then the marginal distribution of X is the distribution with probability density function f_1, where

$$f_1(x) = \int_{-\infty}^{\infty} f(x, t)\, dt.$$

marginal probability density function The *probability density function* of a *marginal distribution*.

Markov, Andrei Andreevich (1856–1922) Russian mathematician who, mostly after 1900, proved important results in probability, developing the notion of a *Markov chain* and initiating the study of *stochastic processes*.

Markov chain Consider a *stochastic process* $X_1, X_2, X_3, \ldots$ in which the state space is discrete. This is a **Markov chain** if the probability that X_{n+1} takes a particular value depends only on the value of X_n and not on the values of $X_1, X_2, \ldots, X_{n-1}$. (This definition can be adapted so as to apply to a stochastic process with a continuous state space, or to a more general stochastic process $\{X(t), t \in T\}$, to give what is called a **Markov process**.) In most Markov chains, the probability that $X_{n+1} = j$ given that $X_n = i$, denoted by p_{ij}, does not depend on n. In that case, if there are N states, these values p_{ij} are called the **transition probabilities** and form the **transition matrix** $[p_{ij}]$, an $N \times N$ *row stochastic matrix*.

mass With any body there are associated two parameters: the *gravitational mass*, which occurs in the *inverse square law of gravitation*, and the *inertial mass*, which occurs in Newton's second law of motion. By suitable scaling, the two values can be taken to be equal, and their common value is the **mass** of the particle.

The SI unit of mass is the *kilogram*.

mathematical induction The method of proof 'by **mathematical induction**' is based on the following principle:

PRINCIPLE OF MATHEMATICAL INDUCTION: Let there be associated, with each positive integer n, a proposition $P(n)$, which is either true or false. If

(i) $P(1)$ is true,
(ii) for all k, $P(k)$ implies $P(k + 1)$,

then $P(n)$ is true for all positive integers n.

This essentially describes a property of the positive integers; either it is accepted as a principle that does not require proof or it is proved as a theorem from some agreed set of more fundamental axioms. The following are typical of results that can be proved by induction:

1. For all positive integers n, $\sum_{r=1}^{n} r^2 = \frac{1}{6}n(n+1)(2n+1)$.
2. For all positive integers n, the n-th derivative of $\frac{1}{x}$ is $(-1)^n \frac{n!}{x^{n+1}}$.
3. For all positive integers n, $(\cos\theta + i\sin\theta)^n = \cos n\theta + i\sin n\theta$.

In each case, it is clear what the proposition $P(n)$ should be, and that (i) can be verified. The method by which the so-called induction step (ii) is proved depends upon the particular result to be established.

A modified form of the principle is this. Let there be associated, with each integer $n \geq n_0$, a proposition $P(n)$, which is either true or false. If (i) $P(n_0)$ is true, and (ii) for all $k \geq n_0$, $P(k)$ implies $P(k+1)$, then $P(n)$ is true for all integers $n \geq n_0$. This may be used to prove, for example, that $3^n > n^3$ for all integers $n \geq 4$.

mathematical model A problem of the **real world** is a problem met by a physicist, an economist, an engineer, or indeed anyone in their normal working conditions or everyday life. Mathematics may help to solve the problem. But to apply the mathematics it is often necessary to develop an abstract mathematical problem, called a **mathematical model** of the original, that approximately corresponds to the real world problem. Developing such a model may involve making assumptions and simplifications. The mathematical problem can then be investigated and perhaps solved. When interpreted in terms of the real world, this may provide an appropriate solution to the original problem.

matrix (plural: matrices) A rectangular array of **entries** displayed in rows and columns and enclosed in brackets. The entries are elements of some suitable set, either specified or understood. They are often numbers, perhaps integers, real numbers or complex numbers, but they may be, say, polynomials or other expressions. An $m \times n$ matrix has m rows and n columns and can be written

$$\begin{bmatrix} a_{11} & a_{12} & \cdots & a_{1n} \\ a_{21} & a_{22} & \cdots & a_{2n} \\ \vdots & \vdots & \ddots & \vdots \\ a_{m1} & a_{m2} & \cdots & a_{mn} \end{bmatrix}.$$

Round brackets may be used instead of square brackets. The subscripts are read as though separated by commas: for example, a_{23} is read as 'a, two, three'. The matrix above may be written in abbreviated form as $[a_{ij}]$, where the number of rows and columns is understood and a_{ij} denotes the entry in the i-th row and j-th column. See also *addition* (of matrices), *multiplication* (of matrices) and *inverse matrix*.

matrix game A *game* in which a *matrix* contains numerical data giving information about what happens according to the strategies chosen by two opponents or players. By convention, the matrix $[a_{ij}]$ gives the pay-off to one of the players, R: when R chooses the i-th row and the other player, C, chooses the j-th column, then C pays to R an amount of a_{ij} units. (If a_{ij} is negative, then in fact R pays C a certain amount.) This is an example of a **zero-sum game** because the total amount received by the two players is zero: R receives a_{ij} units and C receives $-a_{ij}$ units.

If the game is a *strictly determined game*, and R and C use *conservative strategies*, C will always pay R a certain amount, which is the value of the game. If the game is not strictly determined, then by the *Fundamental Theorem of Game Theory*, there is again a value for the game, being the expectation (loosely, the average pay-off each time when the game is played many times) when R and C use *mixed strategies* that are optimal.

matrix of coefficients For a set of m linear equations in n unknowns $x_1, x_2, \ldots, x_n$:

$$\begin{aligned} a_{11}x_1 + a_{12}x_2 + \cdots + a_{1n}x_n &= b_1, \\ a_{21}x_1 + a_{22}x_2 + \cdots + a_{2n}x_n &= b_2, \\ &\vdots \\ a_{m1}x_1 + a_{m2}x_2 + \cdots + a_{mn}x_n &= b_m, \end{aligned}$$

the **matrix of coefficients** is the $m \times n$ matrix $[a_{ij}]$.

matrix of cofactors For $\mathbf{A}$, a square matrix $[a_{ij}]$, the **matrix of cofactors** is the matrix, of the same order, that is obtained by replacing each entry a_{ij} by its *cofactor* A_{ij}. It is used to find the *adjoint* of $\mathbf{A}$, and hence the inverse $\mathbf{A}^{-1}$.

Maupertuis, Pierre-Louis Moreau de (1698–1759) French scientist and mathematician known as the first to formulate a proposition called the principle of least action. He was an active supporter in France of Newton's gravitational theory. He led an expedition to Lapland to measure the length of a degree along a meridian, which confirmed that the Earth is an *oblate spheroid*.

maximin strategy See *conservative strategy*.

maximum For **maximum value**, see *greatest value*. See also *local maximum*.

maximum likelihood estimator The *estimator* for an unknown parameter given by the method of maximum likelihood. See *likelihood*.

Maxwell, James Clerk (1831–1879) British mathematical physicist, who to physicists probably ranks second to Newton in stature. Of considerable importance in applied mathematics are the differential equations known as

Maxwell's equations, which are fundamental to the field theory of electromagnetism.

mean The **mean** of the numbers $a_1, a_2, \dots, a_n$ is equal to

$$\frac{a_1 + a_2 + \cdots + a_n}{n}.$$

This is the number most commonly used as the average. It may be called the **arithmetic mean** to distinguish it from other means such as those described below. When each number a_i is to have weight w_i, the **weighted mean** is equal to

$$\frac{w_1 a_1 + w_2 a_2 + \cdots + w_n a_n}{w_1 + w_2 + \cdots + w_n}.$$

The **geometric mean** of the positive numbers $a_1, a_2, \dots, a_n$ is $\sqrt[n]{a_1 a_2 \dots a_n}$. Given two positive numbers a and b, suppose that $a < b$. The arithmetic mean m is then equal to $\frac{1}{2}(a + b)$, and a, m, b is an arithmetic sequence. The geometric mean g is equal to $\sqrt{ab}$, and a, g, b is a geometric sequence. The arithmetic mean of 3 and 12 is $7\frac{1}{2}$, and the geometric mean is 6. It is a theorem of elementary algebra that, for any positive numbers $a_1, a_2, \dots, a_n$, the arithmetic mean is greater than or equal to the geometric mean; that is,

$$\frac{1}{n}(a_1 + a_2 + \cdots + a_n) \geq \sqrt[n]{a_1 a_2 \dots a_n}.$$

The **harmonic mean** of positive numbers $a_1, a_2, \dots, a_n$ is the number h such that $1/h$ is the arithmetic mean of $1/a_1, 1/a_2, \dots, 1/a_n$. Thus

$$h = \frac{n}{(1/a_1) + \cdots + (1/a_n)}.$$

In statistics, the mean of a set of observations $x_1, x_2, \dots, x_n$, called a **sample mean**, is denoted by $\bar{x}$, so that

$$\bar{x} = \frac{x_1 + x_2 + \cdots + x_n}{n}.$$

The mean of a random variable X is equal to the *expected value* $E(X)$. This may be called the **population mean** and denoted by μ. A sample mean $\bar{x}$ may be used to estimate μ.

mean absolute deviation For the sample $x_1, x_2, \dots, x_n$, with mean $\bar{x}$, the **mean absolute deviation** is equal to

$$\frac{\sum |x_i - \bar{x}|}{n}.$$

It is a measure of *dispersion*, but is not often used.

mean squared deviation For the sample $x_1, x_2, \dots, x_n$, with mean $\bar{x}$, the **mean squared deviation (about the mean)** is the second moment about the mean and is equal to

$$\frac{\sum(x_i - \bar{x})^2}{n}.$$

It is in fact the *variance*, in the form with n as the denominator.

mean squared error For an *estimator* X of a parameter θ, the **mean squared error** is equal to $E((X - \theta)^2)$. For an unbiased estimator this equals $\mathrm{Var}(X)$ (see *expected value* and *variance*).

mean value (of a function) Let f be a function that is continuous on the closed interval $[a, b]$. The **mean value** of f over the interval $[a, b]$ is equal to

$$\frac{1}{b-a}\int_a^b f(x)\,dx.$$

The mean value $\bar{y}$ has the property that the area under the curve $y = f(x)$ between $x = a$ and $x = b$ is equal to the area under the horizontal line $y = \bar{y}$ between $x = a$ and $x = b$, which is the area of a rectangle with width $b - a$ and height $\bar{y}$.

For example, the function may represent the temperature measured at a certain place over a period of 24 hours. The mean value gives what may be considered an 'average' value for the temperature over that period.

Mean Value Theorem The following theorem, which has very important consequences in differential calculus:

THEOREM: Let f be a function that is continuous on $[a, b]$ and differentiable in (a, b). Then there is a number c with $a < c < b$ such that

$$f'(c) = \frac{f(b) - f(a)}{b - a}.$$

The result stated in the theorem can be expressed as a statement about the graph of f: if A, with coordinates $(a, f(a))$, and B, with coordinates $(b, f(b))$, are the points on the graph corresponding to the end-points of the interval, there must be a point C on the graph between A and B at which the tangent is parallel to the chord AB.

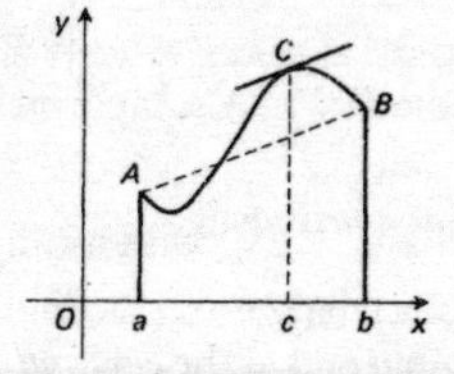

The theorem is normally deduced from *Rolle's Theorem*, which is in fact the special case of the Mean Value Theorem in which $f(a) = f(b)$. A rigorous proof of either theorem relies on the non-elementary result that a *continuous function* on a closed interval attains its bounds. The Mean Value Theorem has immediate corollaries, such as the following. With the appropriate conditions on f,

(i) if $f'(x) = 0$ for all x, then f is a *constant function*,
(ii) if $f'(x) > 0$ for all x, then f is *strictly increasing*.

The important *Taylor's Theorem* can also be seen as an extension of the Mean Value Theorem.

measure Let A and B be two points on a *directed line*, and let $\overrightarrow{AB}$ be the *directed line-segment* from A to B. The **measure** AB is defined by

$$AB = \begin{cases} |AB|, & \text{if } \overrightarrow{AB} \text{ is in the positive direction,} \\ -|AB|, & \text{if } \overrightarrow{AB} \text{ is in the negative direction.} \end{cases}$$

Suppose, for example, that a horizontal line has positive direction to the right (like the usual x-axis). Then the definition of measure gives $AB = 2$ if B is 2 units to the right of A, and $AB = -2$ if B is 2 units to the left of A.

The following properties are easily deduced from the definition:

(i) For points A and B on a directed line, $AB = -BA$.
(ii) For points A, B and C on a directed line, $AB + BC = AC$ (established by considering the different possible relative positions of A, B and C).

measure of dispersion See *dispersion*.

measure of location See *location*.

mechanical advantage See *machine*.

median (in statistics) Suppose that the observations in a set of numerical data are ranked in ascending order. Then the **(sample) median** is the middle observation if there are an odd number of observations, and is the average of the two middlemost observations if there are an even number. Thus, if there are n observations, the median is the $\frac{1}{2}(n+1)$-th in ascending order when n is odd, and is the average of the $\frac{1}{2}n$-th and the $(\frac{1}{2}n+1)$-th when n is even.

The **median** of a continuous *distribution* with *probability density function* f is the number m such that

$$\int_{-\infty}^{m} f(x)\,dx = 0.5.$$

It is the value that divides the distribution into two halves.

median (of a triangle) A line through a vertex of a triangle and the midpoint of the opposite side. The three medians are concurrent at the *centroid*.

mega- Prefix used with *SI units* to denote multiplication by 10^6.

member = *element*.

Menelaus of Alexandria (about AD 100) Greek mathematician whose only surviving work is an important treatise on spherical geometry with applications to astronomy. In this, he was the first to use *great circles* extensively and to consider the properties of *spherical triangles*. He extended to spherical triangles the theorem about plane triangles that is named after him.

Menelaus' Theorem The following theorem, due to Menelaus of Alexandria:

THEOREM: Let L, M and N be points on the sides BC, CA and AB of a triangle (possibly extended). Then L, M and N are collinear if and only if

$$\frac{BL}{LC}\cdot\frac{CM}{MA}\cdot\frac{AN}{NB}=-1.$$

(Note that BC, for example, is considered to be directed from B to C so that LC, for example, is positive if LC is in the same direction as BC and negative if it is in the opposite direction; see *measure*.) See *Ceva's Theorem*.

mensuration The measurement or calculation of lengths, angles, areas and volumes associated with geometrical figures.

meridian Strictly, a *great circle* on the Earth, assumed to be a sphere, through the north and south poles. However, the meridian through a point P often means not the circle but the semicircle through P with NS as diameter, where N is the north pole and S is the south pole.

Mersenne, Marin (1588–1648) French monk, philosopher and mathematician who provided a valuable channel of communication between such contemporaries as Descartes, Fermat, Galileo and Pascal: 'To inform Mersenne of a discovery is to publish it throughout the whole of Europe.' In an attempt to find a formula for prime numbers, he considered the numbers $2^p - 1$, where p is prime. Not all such numbers are prime: $2^{11} - 1$ is not. Those that are prime are called *Mersenne primes*.

Mersenne prime A *prime* of the form $2^p - 1$, where p is a prime. The number of known primes of this form is over 30, and keeps increasing as they are discovered by using computers. In 1995, the largest known was $2^{1257787} - 1$. Each Mersenne prime gives rise to an even *perfect number*.

method of least squares See *least squares*.

method of maximum likelihood See *likelihood*.

method of moments See *moment* (in statistics).

metre In *SI units*, the base unit used for measuring length, abbreviated to 'm'. It was once defined as one ten-millionth of the distance from one of the poles to the equator, and later as the length of a platinum–iridium bar kept under specified conditions in Paris. Now it is defined in terms of the wavelength of a certain light wave produced by the krypton-86 atom.

micro- Prefix used with *SI units* to denote multiplication by 10^{-6}.

midpoint Let A and B be points in the plane with Cartesian coordinates (x_1, y_1) and (x_2, y_2). Then, as a special case of the *section formulae*, the midpoint of AB has coordinates $(\frac{1}{2}(x_1 + x_2)\ \frac{1}{2}(y_1 + y_2))$. For points A and B in 3-dimensional space, with Cartesian coordinates (x_1, y_1, z_1) and (x_2, y_2, z_2), the midpoint of AB has coordinates

$$(\tfrac{1}{2}(x_1 + x_2), \tfrac{1}{2}(y_1 + y_2), \tfrac{1}{2}(z_1 + z_2)).$$

If points A and B have position vectors $\mathbf{a}$ and $\mathbf{b}$, the midpoint of AB has position vector $\frac{1}{2}(\mathbf{a} + \mathbf{b})$.

milli- Prefix used with *SI units* to denote multiplication by 10^{-3}.

minimax strategy See *conservative strategy*.

Minimax Theorem = *Fundamental Theorem of Game Theory*.

minimum For **minimum value** see *least value*. See also *local minimum*.

Minkowski, Hermann (1864–1909) Mathematician, born in Lithuania of German parents, whose concept of space and time as a four-dimensional entity was a significant contribution to the development of relativity. He also obtained important results in number theory and was a close friend of Hilbert.

minor arc See *arc*.

minor axis See *ellipse*.

minute (angular measure) See *degree* (angular measure).

mirror-image See *reflection* (of the plane).

mixed fraction See *fraction*.

mixed strategy In a *matrix game*, if a player chooses the different rows (or columns) with certain probabilities, the player is using a **mixed strategy**. If the game is given by an $m \times n$ matrix, a mixed strategy for one player, R, is given by an m-tuple $\mathbf{x}$, where $\mathbf{x} = (x_1, x_2, \ldots, x_m)$, in which $x_i \geq 0$, for $i = 1, 2, \ldots, m$, and $x_1 + x_2 + \cdots + x_m = 1$. Here x_i is the probability that R chooses the i-th row. For example, if $\mathbf{x} = (\frac{1}{3}, \frac{1}{3}, \frac{1}{3})$, it means that R chooses each row with equal probability or, loosely, that R chooses each row equally often. If $\mathbf{x} = (0, \frac{1}{4}, \frac{3}{4})$, then R never chooses the first row, and chooses the third row three times as often as the second row. A mixed strategy for the other player, C, is defined similarly.

Möbius, August Ferdinand (1790–1868) German astronomer and mathematician, whose work in mathematics was mainly in geometry and topology. One of his contributions was his conception of the *Möbius band*.

Möbius band (or strip) A continuous flat loop with one twist in it. Between any two points on it, a continuous line can be drawn on the surface without

crossing an edge. Thus the band has only one surface and likewise only one edge.

mode The most frequently occurring observation in a sample or, for grouped data, the group with the highest frequency. For a continuous random variable, a value at which the probability density function has a local maximum is called a mode.

model See *mathematical model*.

modulo *n*, addition and multiplication The word 'modulo' means 'to the modulus'. For any positive integer n, let S be the *complete set of residues* $\{0, 1, 2, \ldots, n-1\}$. Then **addition modulo** n on S is defined as follows. For a and b in S, take the usual sum of a and b as integers, and let r be the element of S to which the result is *congruent* (modulo n); the sum $a + b$ (mod n) is equal to r. Similarly, **multiplication modulo** n is defined by taking $ab \pmod{n}$ to be equal to s, where s is the element of S to which the usual product of a and b is congruent (modulo n). For example, addition and multiplication modulo 5 are given by the following tables:

+	0	1	2	3	4
0	0	1	2	3	4
1	1	2	3	4	0
2	2	3	4	0	1
3	3	4	0	1	2
4	4	0	1	2	3

×	0	1	2	3	4
0	0	0	0	0	0
1	0	1	2	3	4
2	0	2	4	1	3
3	0	3	1	4	2
4	0	4	3	2	1

modulus of a complex number (plural: moduli) If z is a *complex number* and $z = x + yi$, the **modulus** of z, denoted by $|z|$ (read as 'mod z'), is equal to $\sqrt{x^2 + y^2}$. (As always, the sign $\sqrt{\ }$ means the non-negative square root.) If z is represented by the point P in the *complex plane*, the modulus of z equals the distance $|OP|$. Thus $|z| = r$, where (r, θ) are the polar coordinates of P. If z is real, the modulus of z equals the *absolute value* of the real number, so the two uses of the same notation $|\ |$ are consistent, and the term 'modulus' may be used for 'absolute value'.

modulus of a congruence See *congruence*.

modulus of elasticity A parameter, usually denoted by λ, associated with the material of which an elastic string is made. It occurs in the constant of proportionality in *Hooke's law*.

moment (in mechanics) A means of measuring the turning effect of a force about a point. For a system of coplanar forces, the moment of one of the forces **F** about any point A in the plane can be defined as the product of the magnitude of **F** and the distance from A to the line of action of **F**, and is considered to be acting either clockwise or anticlockwise. For example, suppose that forces with magnitudes F_1 and F_2 act at B and C, as shown in the figure. The moment of the first force about A is $F_1 d_1$ clockwise, and the moment of the second force about A is $F_2 d_2$ anticlockwise. The *principle of moments* considers when a system of coplanar forces produces a state of equilibrium.

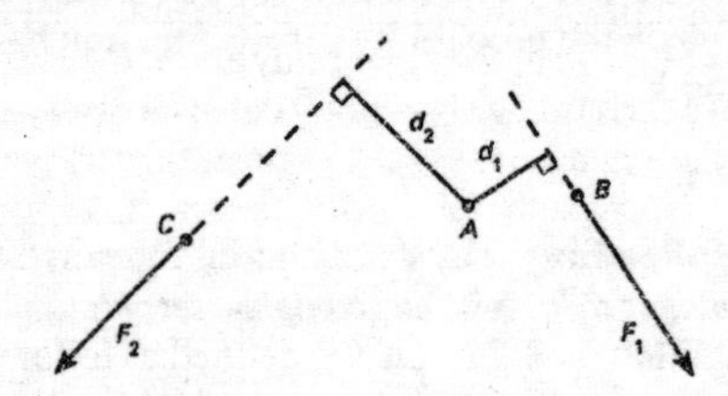

However, a better approach to measuring the turning effect is to define the moment of a force about a point as a vector, as follows. The moment of the force **F**, acting at a point B, about the point A is the vector $(\mathbf{r}_B - \mathbf{r}_A) \times \mathbf{F}$, where this involves a *vector product*. The use of vectors not only eliminates the need to distinguish between clockwise and anticlockwise directions, but facilitates the measuring of the turning effects of non-coplanar forces acting on a 3-dimensional body.

Similarly, for a particle P with position vector **r** and *linear momentum* **p**, the moment of the linear momentum of P about the point A is the vector $(\mathbf{r} - \mathbf{r}_A) \times \mathbf{p}$. This is the *angular momentum* of the particle P about the point A.

Suppose that a *couple* consists of a force **F** acting at B and a force $-\mathbf{F}$ acting at C. Let $\mathbf{r}_B$ and $\mathbf{r}_C$ be the position vectors of B and C, and let $\mathbf{r}_A$ be the position vector of the point A. The moment of the couple about A is equal to $(\mathbf{r}_B - \mathbf{r}_A) \times \mathbf{F} + (\mathbf{r}_C - \mathbf{r}_A) \times (-\mathbf{F}) = (\mathbf{r}_B - \mathbf{r}_C) \times \mathbf{F}$, which is independent of the position of A.

moment (in statistics) For a set of observations $x_1, x_2, \ldots, x_n$, the j-th **(sample) moment** about p is equal to

$$\frac{\sum (x_i - p)^j}{n}.$$

For a random variable X, the j-th **(population) moment** about p is equal to $\mathrm{E}((X - p)^j)$. (See *expected value*.) The first moment about 0 is the *mean*. For the second moment about the mean, see *variance*.

Suppose that a sample is taken from a population with k unknown parameters. In the **method of moments**, the first k population moments are equated to the first k sample moments to find k equations in k unknowns. These can be solved to find the **moment estimates** of the parameters.

moment estimate See *moment* (in statistics).

moment estimator The *estimator* found by the method of moments. See *moment* (in statistics).

moment of inertia A quantity relating to a rigid body and a given axis, derived from the way in which the mass of the rigid body is distributed relative to the axis. It arises in the calculation of the *kinetic energy* and the *angular momentum* of the rigid body. With Cartesian coordinates, let I_{xx}, I_{yy} and I_{zz} denote the moments of inertia of the rigid body about the x-axis, the y-axis and the z-axis. Suppose then that a rigid body consists of a system of n particles $P_1, P_2, \ldots, P_n$, where P_i has mass m_i and position vector $\mathbf{r}_i$, and $\mathbf{r}_i = x_i\mathbf{i} + y_i\mathbf{j} + z_i\mathbf{k}$. Then

$$I_{xx} = \sum_{i=1}^{n} m_i(y_i^2 + z_i^2), \qquad I_{yy} = \sum_{i=1}^{n} m_i(z_i^2 + x_i^2),$$

$$I_{zz} = \sum_{i=1}^{n} m_i(x_i^2 + y_i^2).$$

For solid bodies, a moment of inertia is given by an integral. For example,

$$I_{xx} = \int_V \rho(\mathbf{r})(y^2 + z^2)\,dV, \quad I_{yy} = \int_V \rho(\mathbf{r})(z^2 + x^2)\,dV,$$

$$I_{zz} = \int_V \rho(\mathbf{r})(x^2 + y^2)\,dV,$$

where $\rho(\mathbf{r})$ is the density at the point with position vector $\mathbf{r}$ and $\mathbf{r} = x\mathbf{i} + y\mathbf{j} + z\mathbf{k}$.

For a uniform rod of length l and mass m, the moment of inertia about an axis running along the length of the rod is zero. The moment of inertia about an axis perpendicular to the rod, through the centre of mass, equals $\frac{1}{12}ml^2$. The moment of inertia about an axis perpendicular to the rod, through one end of the rod, equals $\frac{1}{3}ml^2$.

For a uniform circular plate of radius a and mass m, the moment of inertia about an axis perpendicular to the plate, through the centre of the circle, equals $\frac{1}{2}ma^2$.

For a uniform rectangular plate of length a, width b and mass m, the moment of inertia about an axis along a side of length a equals $\frac{1}{3}mb^2$.

Moments of inertia about other axes may be calculated by using the *Parallel Axis Theorem* and the *Perpendicular Axis Theorem*.

moment of momentum = *angular momentum*.

momentum See *linear momentum* and *angular momentum*. Often 'momentum' is used to mean linear momentum.

Monge, Gaspard (1746–1818) French mathematician who was a high-ranking minister in Napoleon's government and influential in the founding of the École Polytechnique in Paris. The courses in geometry that he taught there were concerned to a great extent with the subjects of his research: he was the originator of what is called descriptive geometry and the founder of modern analytical solid geometry.

monic polynomial A *polynomial* in one variable in which the coefficient of the highest power is 1. The polynomial $a_nx^n + a_{n-1}x^{n-1} + \cdots + a_1x + a_0$ is monic if $a_n = 1$.

monotonic function A *real function* f is **monotonic** on or over an interval I if it is either increasing on I ($f(x_1) \leq f(x_2)$ whenever $x_1 < x_2$) or decreasing on I ($f(x_1) \geq f(x_2)$ whenever $x_1 < x_2$). Also, f is **strictly monotonic** if it is either strictly increasing or strictly decreasing.

monotonic sequence A sequence $a_1, a_2, a_3, \ldots$ is **monotonic** if it is either increasing ($a_i \leq a_{i+1}$ for all i) or decreasing ($a_i \geq a_{i+1}$ for all i), and **strictly monotonic** if it is either strictly increasing or strictly decreasing.

Monte Carlo method The use of sampling involving the generation of *random numbers*, normally by computer, to solve mathematical problems such as the approximation of certain integrals.

Morley's Theorem The following theorem due to Frank Morley (1860–1937):

THEOREM: In any triangle, the points of intersection of adjacent trisectors of the angles form the vertices of an equilateral triangle.

It is remarkable that this elegant fact of Euclidean geometry was not discovered until so long after Euclid's time.

moving average For a sequence of observations $x_1, x_2, x_3, \ldots$, the **moving average** of order n is the sequence of *arithmetic means*

$$\frac{x_1 + \cdots + x_n}{n}, \quad \frac{x_2 + \cdots + x_{n+1}}{n}, \quad \frac{x_3 + \cdots + x_{n+2}}{n}, \quad \ldots.$$

A weighted moving average such as

$$\frac{x_1 + 2x_2 + x_3}{4}, \quad \frac{x_2 + 2x_3 + x_4}{4}, \quad \frac{x_3 + 2x_4 + x_5}{4}, \quad \ldots$$

may also be used.

Müller, Johann See *Regiomontanus.*

multiple See *divides.*

multiple edges See *graph.*

multiple regression See *regression.*

multiple root See *root.*

multiplication (of complex numbers) For complex numbers z_1 and z_2, given by $z_1 = a + bi$ and $z_2 = c + di$, the **product** is defined by $z_1 z_2 = (ac - bd) + (ad + bc)i$. In some circumstances it may be more convenient to write z_1 and z_2 in *polar form*, thus:

$$z_1 = r_1 e^{i\theta_1} = r_1(\cos\theta_1 + i\sin\theta_1),$$
$$z_2 = r_2 e^{i\theta_2} = r_2(\cos\theta_2 + i\sin\theta_2).$$

Then $z_1 z_2 = r_1 r_2 e^{i(\theta_1+\theta_2)} = r_1 r_2(\cos(\theta_1 + \theta_2) + i\sin(\theta_1 + \theta_2))$. Thus 'you multiply the moduli and add the arguments'.

multiplication (of matrices) Suppose that matrices **A** and **B** are *conformable* for multiplication, so that **A** has order $m \times n$ and **B** has order $n \times p$. Let $\mathbf{A} = [a_{ij}]$ and $\mathbf{B} = [b_{ij}]$. Then **(matrix) multiplication** is defined by taking the **product AB** to be the $m \times p$ matrix **C**, where $\mathbf{C} = [c_{ij}]$ and

$$c_{ij} = a_{i1}b_{1j} + a_{i2}b_{2j} + \cdots + a_{in}b_{nj} = \sum_{r=1}^{n} a_{ir}b_{rj}.$$

The product **AB** is not defined if **A** and **B** are not conformable for multiplication. Matrix multiplication is not *commutative*; for example, if

$$\mathbf{A} = \begin{bmatrix} 0 & 1 \\ 0 & 0 \end{bmatrix} \quad \text{and} \quad \mathbf{B} = \begin{bmatrix} 1 & 0 \\ 0 & 0 \end{bmatrix},$$

then $\mathbf{AB} \neq \mathbf{BA}$. Moreover, it is not true that $\mathbf{AB} = \mathbf{O}$ implies that either $\mathbf{A} = \mathbf{O}$ or $\mathbf{B} = \mathbf{O}$, as the same example of **A** and **B** shows. However, matrix multiplication is associative: $\mathbf{A}(\mathbf{BC}) = (\mathbf{AB})\mathbf{C}$, and the distributive laws hold: $\mathbf{A}(\mathbf{B} + \mathbf{C}) = \mathbf{AB} + \mathbf{AC}$ and $(\mathbf{A} + \mathbf{B})\mathbf{C} = \mathbf{AC} + \mathbf{BC}$. Strictly, in each case, what should be said is that, if **A**, **B** and **C** are such that one side of the equation exists, then so does the other side and the two sides are equal.

multiplication modulo *n* See *modulo n, addition and multiplication.*

multiplicative group A *group* in which the operation is called multiplication, usually denoted by '.' but with the product $a\,.\,b$ normally written as ab.

multiplicative inverse See *inverse element.*

multiplicity See *root.*

multivariate Relating to more than two random variables. See *joint cumulative distribution function, joint distribution, joint probability density function* and *joint probability mass function.*

mutually disjoint = *pairwise disjoint.*

mutually exclusive events Events A and B are **mutually exclusive** if A and B cannot both occur; that is, if $A \cap B = \emptyset$. For mutually exclusive events, the probability that either one event or the other occurs is given by the addition law, $\Pr(A \cup B) = \Pr(A) + \Pr(B)$.

For example, when a die is thrown, the probability of obtaining a 'one' is $\frac{1}{6}$, and the probability of obtaining a 'two' is $\frac{1}{6}$. These events are mutually exclusive. So the probability of obtaining a 'one' or a 'two' is equal to $\frac{1}{6}+\frac{1}{6}=\frac{1}{3}$.

mutually prime = *relatively prime.*

N See *natural number.*

nano- Prefix used with *SI units* to denote multiplication by 10^{-9}.

Napier, John (1550–1617) Scottish mathematician responsible for the publication in 1614 of the first tables of logarithms. His mathematical interests, pursued in spare time from church and state affairs, were in spherical trigonometry and, notably, in computation. His logarithms were defined geometrically rather than in terms of a base. In effect, he took the logarithm of x to be y, with

$$x = 10^7 \times \left(1 - \frac{1}{10^7}\right)^y.$$

Napierian logarithm See *logarithm.*

Napier's bones An invention by Napier for carrying out multiplication. It consisted of a number of rods made of bone or ivory, which were marked with numbers from the multiplication tables. Multiplication using the rods involved some addition of digits and was very similar to the familiar method of long multiplication. Logarithms were not involved at all.

natural logarithm See *logarithm.*

natural number One of the numbers 1, 2, 3, ... Some authors also include 0. The set of natural numbers is often denoted by **N**.

***n*-cube** A *simple graph*, denoted by Q_n, whose vertices and edges correspond to the vertices and edges of an n-dimensional *hypercube.* Thus, there are 2^n vertices that can be labelled with the *binary words* of length n, and there is an edge between two vertices if they are labelled with

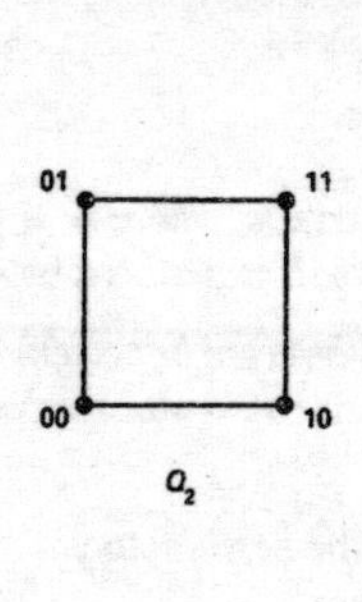

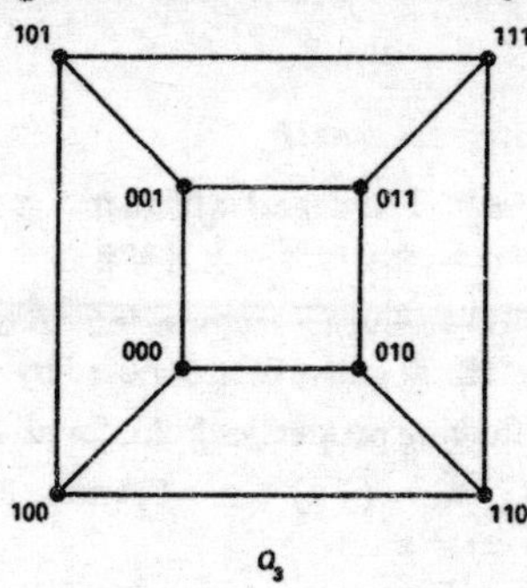

words that differ in exactly one digit. The graphs Q_2 and Q_3 are shown in the figure.

n-dimensional space Points in the plane can be specified by means of Cartesian coordinates (x, y), and points in 3-dimensional space may be assigned Cartesian coordinates (x, y, z). It can be useful, similarly, to consider space of n dimensions, for general values of n, by defining a point to be given by n coordinates. Many familiar ideas from geometry of 2 and 3 dimensions can be generalized to space of higher dimensions. For example, if the point P has coordinates $(x_1, x_2, \ldots, x_n)$, and the point Q has coordinates $(y_1, y_2, \ldots, y_n)$, then the distance PQ can be defined to be equal to

$$\sqrt{(x_1 - y_1)^2 + (x_2 - y_2)^2 + \cdots + (x_n - y_n)^2}.$$

Straight lines can be defined, and so can the notion of angle between them. The set of points whose coordinates satisfy a linear equation $a_1x_1 + a_2x_2 + \cdots + a_nx_n = b$ is called a **hyperplane**, which divides the space into two *half-spaces*, as a plane does in 3 dimensions. There are other so-called subspaces of different dimensions, a straight line being a subspace of dimension 1 and a hyperplane being a subspace of dimension $n - 1$. Other curves and surfaces can also be considered. In n dimensions, the generalization of a square in 2 dimensions and a cube in 3 dimensions is a *hypercube*.

negation If p is a statement, then the statement 'not p', denoted by $\neg p$, is the **negation** of p. It states, in some suitable wording, the opposite of p. For example, if p is 'It is raining', then $\neg p$ is 'It is not raining'; if p is '2 is not an integer', then $\neg p$ is 'It is not the case that 2 is not an integer' or, in other words, '2 is an integer'. If p is true then $\neg p$ is false, and vice versa. So the *truth table* of $\neg p$ is:

p	$\neg p$
T	F
F	T

negative See *inverse element*.

negative (of a matrix) The **negative** of an $m \times n$ matrix $\mathbf{A}$, where $\mathbf{A} = [a_{ij}]$, is the $m \times n$ matrix $\mathbf{C}$, where $\mathbf{C} = [c_{ij}]$ and $c_{ij} = -a_{ij}$. It is denoted by $-\mathbf{A}$.

negative (of a vector) Given a vector $\mathbf{a}$, let $\overrightarrow{AB}$ be a *directed line-segment* representing $\mathbf{a}$. The **negative** of $\mathbf{a}$, denoted by $-\mathbf{a}$, is the vector represented by $\overrightarrow{BA}$. The following properties hold, for all $\mathbf{a}$:

(i) $\mathbf{a} + (-\mathbf{a}) = (-\mathbf{a}) + \mathbf{a} = \mathbf{0}$, where $\mathbf{0}$ is the zero vector,
(ii) $-(-\mathbf{a}) = \mathbf{a}$.

negative direction See *directed line*.

neighbourhood On the *real line*, a **neighbourhood** of the real number a is an open interval $(a - \delta, a + \delta)$, where $\delta > 0$, with its centre at a.

nested multiplication A polynomial $f(x)$, such as $2x^3 - 7x^2 + 5x + 11$, can be evaluated for $x = h$ by calculating h^2 and h^3, multiplying by appropriate coefficients and summing the terms. But fewer operations are required if the polynomial is rewritten as $((2x - 7)x + 5)x + 11$ and then evaluated. This method, known as **nested multiplication**, is therefore more efficient and is recommended when evaluation is to be carried out by hand or by computer. A polynomial $a_5x^5 + a_4x^4 + a_3x^3 + a_2x^2 + a_1x + a_0$ of degree 5, for example, would be rewritten for this purpose as $((((a_5x + a_4)x + a_3)x + a_2)\,x + a_1)x + a_0$. The steps involved in this evaluation correspond exactly to the calculations that are made in the process of *synthetic division*, in which the remainder gives $f(h)$.

network A *digraph* in which every arc is assigned a **weight** (normally some non-negative number). In some applications, something may be thought of as flowing or being transported between the vertices of a network, with the weight of each arc giving its 'capacity'. In some other cases, the vertices of a network may represent steps in a process and the weight of the arc joining u and v may give the time that must elapse between step u and step v.

neutral element An element e is a **neutral element** for a *binary operation* $\circ$ on a set S if, for all a in S, $a \circ e = e \circ a = a$. If the operation is called multiplication, a neutral element is normally called an **identity element** and may be denoted by 1. If the operation is called addition, such an element is normally denoted by 0, and is often called a zero element. However, there is a case for preferring the term 'neutral element', as there is an alternative definition for the term 'zero element' (see *zero element*).

neutral equilibrium See *equilibrium*.

Newton, Isaac (1642–1727) English physicist and mathematician who dominated and revolutionized the mathematics and physics of the seventeenth century. Gauss and Einstein seem to have conceded to him the place of top mathematician of all time. He was responsible for the essentials of the calculus, the theory of mechanics, the law of gravity, the theory of planetary motion, the *binomial series*, *Newton's method* in numerical analysis, and many important results in the theory of equations. A morbid dislike of criticism held him back from publishing much of his work. For example, in 1684, Edmond Halley suggested to Newton that he investigate the law of attraction that would yield Kepler's laws of planetary motion. Newton replied immediately that it was the inverse square law—he had worked that out many years earlier. Rather startled by this, Halley set to work to persuade Newton to publish his results. This Newton eventually did in the form of his *Principia*, published in 1687, perhaps the single most powerful work in the long history of mathematics.

newton The SI unit of *force*, abbreviated to 'N'. One newton is the force required to give a mass of one kilogram an acceleration of one metre per second per second.

Newton quotient The quotient $(f(a+h)-f(a))/h$, which is used to determine whether f is *differentiable* at a, and if so to find the *derivative*.

Newton–Raphson method = *Newton's method*.

Newton's inverse square law of gravitation See *inverse square law of gravitation*.

Newton's law of restitution See *coefficient of restitution*.

Newton's laws of motion Three laws of motion, applicable to particles of constant mass, formulated by Newton in his *Principia*. They can be stated as follows:

(i) A particle continues to be at rest or moving with constant velocity unless the total force acting on the particle is non-zero.
(ii) The rate of change of linear momentum of a particle is proportional to the total force acting on the particle.
(iii) Whenever a particle exerts a force on a second particle, the second particle exerts an equal and opposite force on the first.

The first law can be interpreted as saying that a force is required to alter the velocity of a particle. The linear momentum $\mathbf{p}$ of a particle, mentioned in the second law, is given by $\mathbf{p}=m\mathbf{v}$, where m is the mass and $\mathbf{v}$ is the velocity. So, when m is constant, $d\mathbf{p}/dt=m(d\mathbf{v}/dt)=m\mathbf{a}$, where $\mathbf{a}$ is the acceleration. The second law is therefore often stated as $m\mathbf{a}=\mathbf{F}$, where $\mathbf{F}$ is the total force acting on the particle. The third law states that 'to every action there is an equal and opposite reaction'.

Newton's method The following method of finding successive approximations to a root of an equation $f(x)=0$. Suppose that x_0 is a first approximation, known to be quite close to a root. If the root is in fact x_0+h, where h is 'small', taking the first two terms of the Taylor series gives $f(x_0+h)\approx f(x_0)+hf'(x_0)$. Since $f(x_0+h)=0$, it follows that $h\approx -f(x_0)/f'(x_0)$. Thus x_1, given by

$$x_1=x_0-\frac{f(x_0)}{f'(x_0)},$$

is likely to be a better approximation to the root. To see the geometrical significance of the method, suppose that P_0 is the point $(x_0, f(x_0))$ on the curve $y=f(x)$, as shown in the left-hand figure. The value x_1 is given by the point at which the tangent to the curve at P_0 meets the x-axis. It may be possible to repeat the process to obtain successive approximations $x_0, x_1, x_2, \ldots$, where

$$x_{n+1}=x_n-\frac{f(x_n)}{f'(x_n)}.$$

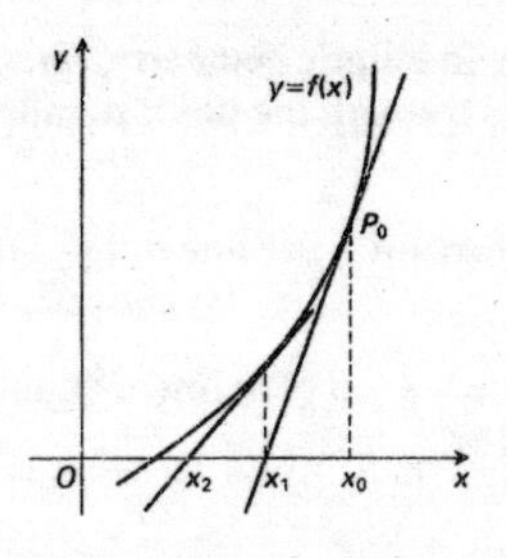

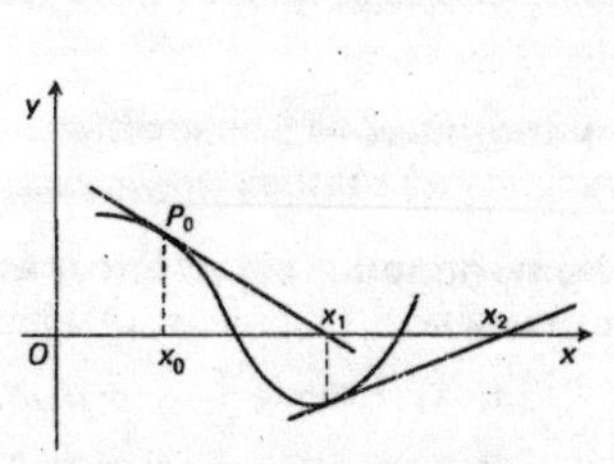

These may be successively better approximations to the root as required, but in the right-hand figure is shown the graph of a function, with a value x_0 close to a root, for which x_1 and x_2 do not give successively better approximations.

Neyman, Jerzy (1894–1981) Statistician known for his work in the theories of *hypothesis testing* and *estimation*. He introduced the notion of *confidence interval*. Born in Romania, he spent some years at University College, London before going to the University of California at Berkeley in 1938.

nine-point circle Consider a triangle with vertices A, B and C. Let A', B' and C' be the midpoints of the three sides. Let D, E and F be the feet of the perpendiculars from A, B and C, respectively, to the opposite sides. Then AD, BE and CF are concurrent at the *orthocentre* H. Let P, Q and R be the midpoints of AH, BH and CH. The points A', B', C', D, E, F, P, Q and R lie on a circle called the **nine-point circle**. The centre of this circle lies on the *Euler line* of the triangle at the midpoint of OH, where O is the *circumcentre*. **Feuerbach's Theorem** states that the nine-point circle touches the *incircle* and the three *excircles* of the triangle. This was proved in 1822 by Karl Wilhelm Feuerbach (1800–1834).

node See *graph* and *tree*.

Noether, Amalie ('Emmy') (1882–1935) German mathematician, known for her highly creative work in the theory of *rings*, non-commutative algebras and other areas of *abstract algebra*. She was responsible for the growth of an active group of algebraists at Göttingen in the period up to 1933.

nominal See *data*.

nonagon = *enneagon*.

non-Euclidean geometry After unsuccessful attempts had been made at proving that the *parallel postulate* could be deduced from the other postulates of Euclid's, the matter was settled by the discovery of **non-Euclidean geometries** by Gauss, Lobachevsky and Bolyai. In these, all Euclid's axioms hold except the parallel postulate. In the so-called **hyperbolic geometry**, given a point not on a given line, there are at least two lines through the point parallel to the line. In **elliptic geometry**, given a point not on a given line, there are no lines through the point parallel to the line.

non-homogeneous linear differential equation See *linear differential equation with constant coefficients.*

non-homogeneous set of linear equations A set of m linear equations in n unknowns $x_1, x_2, \ldots, x_n$ that has the form

$$\begin{aligned} a_{11}x_1 + a_{12}x_2 + \cdots + a_{1n}x_n &= b_1, \\ a_{21}x_1 + a_{22}x_2 + \cdots + a_{2n}x_n &= b_2, \\ &\vdots \\ a_{m1}x_1 + a_{m2}x_2 + \cdots + a_{mn}x_n &= b_m, \end{aligned}$$

where $b_1, b_2, \ldots, b_m$ are not all zero. (Compare with *homogeneous set of linear equations.*) In matrix notation, this set of equations may be written as $\mathbf{Ax} = \mathbf{b}$, where $\mathbf{A}$ is the $m \times n$ matrix $[a_{ij}]$, and $\mathbf{b}$ ($\neq \mathbf{0}$) and $\mathbf{x}$ are column matrices:

$$\mathbf{b} = \begin{bmatrix} b_1 \\ b_2 \\ \vdots \\ b_m \end{bmatrix}, \qquad \mathbf{x} = \begin{bmatrix} x_1 \\ x_2 \\ \vdots \\ x_m \end{bmatrix}.$$

Such a set of equations may be inconsistent, have a unique solution, or have infinitely many solutions (see *simultaneous linear equations*). For a set consisting of the same number of linear equations as unknowns, the matrix of coefficients $\mathbf{A}$ is a square matrix, and the set of equations has a unique solution, namely $\mathbf{x} = \mathbf{A}^{-1}\mathbf{b}$, if and only if $\mathbf{A}$ is invertible.

non-parametric methods Methods of *inference* that make no assumptions about the underlying population *distribution*. Non-parametric tests are often concerned with hypotheses about the median of a population and use the ranks of the observations. Examples of non-parametric tests are the *Wilcoxon rank-sum test*, the *Kolmogorov–Smirnov test* and the *sign test*.

non-singular A square matrix $\mathbf{A}$ is **non-singular** if it is not *singular*; that is, if $\det \mathbf{A} \neq 0$, where $\det \mathbf{A}$ is the determinant of $\mathbf{A}$. See also *inverse matrix*.

norm See *partition* (of an interval).

normal (to a curve) Suppose that P is a point on a curve in the plane. Then the **normal** at P is the line through P perpendicular to the tangent at P.

normal (to a plane) A line perpendicular to the plane. A normal is perpendicular to any line that lies in the plane.

normal (to a surface) See *tangent plane*.

normal distribution The continuous probability *distribution* with *probability density function* f given by

$$f(x) = \frac{1}{\sqrt{2\pi\sigma^2}} \exp\left(-\frac{(x-\mu)^2}{2\sigma^2}\right),$$

denoted by $N(\mu, \sigma^2)$. It has mean μ and variance σ^2. The distribution is widely used in statistics because many experiments produce data that are approximately normally distributed, the sum of random variables from non-normal distributions is approximately normally distributed (see *Central Limit Theorem*), and it is the limiting distribution of distributions such as the *binomial*, *Poisson* and *chi-squared distributions.* It is called the **standard normal distribution** when $\mu = 0$ and $\sigma^2 = 1$.

If X has the distribution $N(\mu, \sigma^2)$ and $Z = (X - \mu)/\sigma$, then Z has the distribution N(0, 1). The diagram shows the graph of the probability density function of N(0, 1).

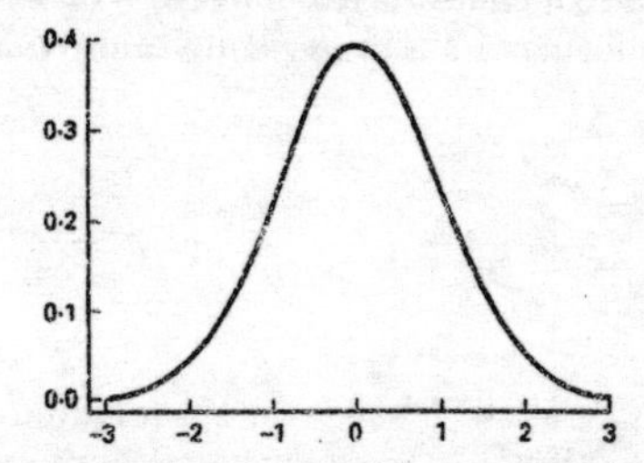

The following table gives, for each value z, the percentage of observations which exceed z, for the standard normal distribution N(0, 1). Thus the values are to be used for *one-tailed tests.* Interpolation may be used for values of z not included.

z	0.0	0.5	1.0	1.28	1.5	1.64	1.96	2.33	2.57	3.0	3.5
%	50	30.9	15.9	10.0	6.7	5.0	2.5	1.0	0.5	0.14	0.02

normal reaction See *contact force.*

normal vector (to a plane) A vector whose direction is perpendicular to the plane. A normal vector is perpendicular to any vector whose direction lies in the plane.

not See *negation.*

n-th derivative See *higher derivative.*

n-th-order partial derivative See *higher-order partial derivative.*

n-th root For any real number a, an ***n*-th root** of a is a number x such that $x^n = a$. (When $n = 2$, it is called a *square root*, and when $n = 3$ a **cube root**.)

First consider n even. If a is negative, there is no real number x such that $x^n = a$. If a is positive, there are two such numbers, one positive and one negative. For $a \geq 0$, the notation $\sqrt[n]{a}$ is used to denote quite specifically the non-negative n-th root of a. For example, $\sqrt[4]{16} = 2$, and 16 has two fourth roots, namely 2 and -2.

Next consider n odd. For all values of a, there is a unique number x such that $x^n = a$, and it is denoted by $\sqrt[n]{a}$. For example, $\sqrt[3]{-8} = -2$.

n-th root of unity A complex number z such that $z^n = 1$. The n distinct n-th roots of unity are $e^{i2k\pi/n}$ $(k = 0, 1, \ldots, n-1)$, or

$$\cos\frac{2k\pi}{n} + i\sin\frac{2k\pi}{n} \quad (k = 0, 1, \ldots, n-1).$$

They are represented in the *complex plane* by points that lie on the unit circle and are vertices of a regular n-sided polygon. The n-th roots of unity for $n = 5$ and $n = 6$ are shown in the figure. The n-th roots of unity always include the real number 1, and also include the real number -1 if n is even. The non-real n-th roots of unity form pairs of *conjugates*. See also *cube root of unity* and *fourth root of unity*.

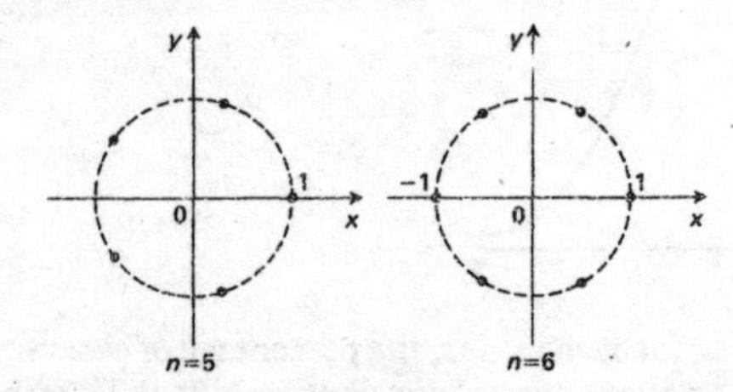

n-tuple An **n-tuple** consists of n objects normally taken in a particular order, denoted, for example, by $(x_1, x_2, \ldots, x_n)$. The term is a generalization of a *triple* when $n = 3$, and a *quadruple* when $n = 4$. When $n = 2$ it is an *ordered pair*.

null hypothesis See *hypothesis testing*.

nullity The **nullity** of a matrix $\mathbf{A}$ is a number relating to the set of solutions of $\mathbf{Ax} = \mathbf{0}$, the corresponding *homogeneous set of linear equations*: it can be defined as the number of *parameters* required when the solutions of the equations are expressed in terms of parameters. In other words, it is the number of unknowns that are free to take arbitrary values. A way of obtaining the solutions in terms of parameters is by transforming the matrix to one in *reduced echelon form*. There is an important connection between the nullity and the *rank* of a matrix: for an $m \times n$ matrix $\mathbf{A}$, the nullity of $\mathbf{A}$ equals n minus the rank of $\mathbf{A}$.

null matrix = *zero matrix*.

null sequence A sequence whose limit is 0.

null set = *empty set*.

number line = *real line*.

number systems The early Egyptian number system used different symbols for 1, 10, 100, and so on, with each symbol repeated the required number of times. Later, the Babylonians had symbols for 1 and 10 repeated similarly, but for larger numbers they used a positional notation with base 60, so that groups of symbols were positioned to indicate the number of different powers of 60.

The Greek number system used letters to stand for numbers. For example, α, β, γ and δ represented 1, 2, 3 and 4, and ι, κ, λ and μ represented 10, 20, 30 and 40. The Roman number system is still known today and used for some special purposes. Roughly speaking, each Roman numeral is repeated as often as necessary to give the required total, with the larger numerals appearing before the smaller, except that if a smaller precedes a larger its value is subtracted. For example, IX, XXVI and CXLIV represent 9, 26 and 144.

The Hindu–Arabic number system, in which numbers are generally written today, uses the Arabic numerals and a positional notation with base 10. It originated in India, where records of its use go back to the sixth century. It was introduced to Europe in the twelfth century, promoted by Fibonacci and others.

numeral A symbol used to denote a number. The **Roman numerals** I, V, X, L, C, D and M represent 1, 5, 10, 50, 100, 500 and 1000 in the Roman number system. The **Arabic numerals** 0, 1, 2, 3, 4, 5, 6, 7, 8 and 9 are used in the Hindu–Arabic number system to give numbers in the form generally familiar today. See *number system*.

numerator See *fraction*.

numerical analysis Many mathematical problems to which answers are required in practical situations cannot be solved in all generality. The only way may be to use a numerical method; that is, to consider the problem in such a way that, from given data, a solution—or an approximation to one—is obtained essentially by numerical calculations. The subject dealing with the derivation and analysis of such methods is called **numerical analysis**.

numerical integration The methods of **numerical integration** are used to find approximate values for *definite integrals* and are useful for the following reasons. It may be that there is no analytical method of finding an *antiderivative* of the *integrand*. Possibly, such a method may exist but be very complicated. In another case, the integrand may not be known explicitly; it might be that only its values at certain points within the interval of integration are known. Among the elementary methods of numerical integration are the *trapezium rule* and *Simpson's rule*.

numerical value = *absolute value*.

objective function See *linear programming*.

oblique A pair of intersecting lines, to be taken, for example, as coordinate axes, are **oblique** if they are not (or at least not necessarily) perpendicular. Three concurrent lines, similarly, are oblique if they are not mutually perpendicular.

observation A particular value taken by a *random variable*. Normally, n observations of the random variable X are denoted by $x_1, x_2, \ldots, x_n$.

observer With a *frame of reference* it is convenient to associate an **observer** who is thought of as viewing the motion with respect to this frame of reference and having the capability to measure positions in space and the duration of time intervals.

obtuse angle An angle that is greater than a *right angle* and less than two right angles. An **obtuse-angled triangle** is one in which one angle is obtuse.

octagon An eight-sided polygon.

octahedron (plural: octahedra) A *polyhedron* with 8 faces, often assumed to be regular. The regular octahedron is one of the *Platonic solids* and its faces are equilateral triangles. It has 6 vertices and 12 edges.

octant In a Cartesian coordinate system in 3-dimensional space, the axial planes divide the rest of the space into eight regions called **octants**. The set of points $\{(x, y, z) \mid x > 0, y > 0, z > 0\}$ may be called the **positive** (or possibly the **first**) **octant**.

odd function The *real function* f is an **odd function** if $f(-x) = -f(x)$ for all x (in the domain of f). Thus the graph $y = f(x)$ of an odd function is symmetrical about the origin; that is, it has a half-turn symmetry about the origin, because whenever (x, y) lies on the graph then so does $(-x, -y)$. For example, f is an odd function when $f(x)$ is defined as any of the following: $2x$, x^3, $x^7 - 8x^3 + 5x$, $1/(x^3 - x)$, $\sin x$, $\tan x$.

ogive A graph of the *cumulative frequency distribution* of a sample or of the *cumulative distribution function* of a random variable.

one-sided test See *hypothesis testing*.

one-tailed test See *hypothesis testing*.

one-to-one correspondence A *one-to-one mapping* between two sets that is also an *onto mapping*. Thus each element of the first set is made to correspond with exactly one element of the second set, and vice versa. Such a correspondence is also called a *bijection*.

one-to-one mapping A *mapping* $f: S \to T$ is **one-to-one** if, whenever s_1 and s_2 are distinct elements of S, their images $f(s_1)$ and $f(s_2)$ are distinct elements of T. So f is one-to-one if $f(s_1) = f(s_2)$ implies that $s_1 = s_2$.

only if See *condition, necessary and sufficient.*

onto mapping A *mapping* $f: S \to T$ is **onto** if every element of the codomain T is the image under f of at least one element of the domain S. So f is onto if the image (or range) $f(S)$ is the whole of T.

open disc See *disc.*

open half-plane See *half-plane.*

open half-space See *half-space.*

open interval The **open interval** (a, b) is the set

$$\{ x \mid x \in \mathbf{R} \text{ and } a < x < b \}$$

For **open interval determined by** I, see *interval.*

operation An **operation** on a set S is a rule that associates with some number of elements of S a resulting element. If this resulting element is always also in S, then S is said to be **closed under** the operation. An operation that associates with one element of S a resulting element is called a *unary operation*; one that associates with two elements of S a resulting element is a *binary operation.*

optimal A value or solution is **optimal** if it is, in a specified sense, the best possible. For example, in a *linear programming* problem the maximum or minimum value of the objective function, whichever is required, is the optimal value and a point at which it is attained is an optimal solution.

optimal strategy See *Fundamental Theorem of Game Theory.*

optimization The process of finding the best possible solution to a problem. In mathematics, this often consists of maximizing or minimizing the value of a certain function, perhaps subject to given constraints.

or See *disjunction.*

orbit The path traced out by a body experiencing a *central force*, such as the gravitational force associated with a second body. When the force is given by the *inverse square law of gravitation*, the orbit is either elliptic, parabolic or hyperbolic, with the second body at a focus of the conic.

order (of a differential equation) See *differential equation.*

order (of a group) The **order** of a *group* G is the number of elements in G.

order (of a matrix) An $m \times n$ matrix is said to have **order** $m \times n$ (read as 'm by n'). An $n \times n$ matrix may be called a square matrix of order n.

order (of a partial derivative) See *higher-order partial derivative*.

order (of a root) See *root*.

ordered pair An **ordered pair** consists of two objects considered in a particular order. Thus, if $a \neq b$, the ordered pair (a, b) is not the same as the ordered pair (b, a). See *Cartesian product*.

ordinary differential equation See *differential equation*.

ordinate The y-coordinate in a Cartesian coordinate system in the plane.

origin See *coordinates* (on a line), *coordinates* (in the plane) and *coordinates* (in 3-dimensional space).

orthocentre The point at which the *altitudes* of a triangle are concurrent. The orthocentre lies on the *Euler line* of the triangle.

orthogonal curves Two curves, or straight lines, are **orthogonal** if they intersect at right angles.

orthogonal matrix A square matrix $\mathbf{A}$ is **orthogonal** if $\mathbf{A}^T\mathbf{A} = \mathbf{I}$, where $\mathbf{A}^T$ is the *transpose* of $\mathbf{A}$. The following properties hold:

- (i) If $\mathbf{A}$ is orthogonal, $\mathbf{A}^{-1} = \mathbf{A}^T$ and so $\mathbf{A}\mathbf{A}^T = \mathbf{I}$.
- (ii) If $\mathbf{A}$ is orthogonal, $\det \mathbf{A} = \pm 1$.
- (iii) If $\mathbf{A}$ and $\mathbf{B}$ are orthogonal matrices of the same order, then $\mathbf{AB}$ is orthogonal.

orthogonal vectors Two non-zero vectors are **orthogonal** if their directions are perpendicular. Thus non-zero vectors $\mathbf{a}$ and $\mathbf{b}$ are orthogonal if and only if $\mathbf{a} \cdot \mathbf{b} = 0$.

orthonormal An **orthonormal** set of vectors is a set of mutually orthogonal unit vectors. In 3-dimensional space, three mutually orthogonal unit vectors form an **orthonormal basis**. The standard orthonormal basis in a Cartesian coordinate system consists of the unit vectors $\mathbf{i}$, $\mathbf{j}$ and $\mathbf{k}$ along the three coordinate axes.

oscillate finitely and **infinitely** See *limit* (of a sequence).

oscillations A particle or rigid body performs **oscillations** if it travels to and fro in some way about a central position, usually a point of stable *equilibrium*. Examples include the swings of a simple or compound pendulum, the bobbing up and down of a particle suspended by a spring, and the vibrations of a violin string. The oscillations are *damped* when there is a resistive force and *forced* when there is an applied force.

outlier An observation that is deemed to be unusual and possibly erroneous because it does not follow the general pattern of the data in the sample.

pairwise disjoint Sets $A_1, A_2, \ldots, A_n$ are said to be **pairwise disjoint** if $A_i \cap A_j = \emptyset$ for all i and j, with $i \neq j$. The term can also be applied to an infinite collection of sets.

Pappus of Alexandria (about AD 320) Considered to be the last great Greek geometer. He wrote commentaries on Euclid and Ptolemy, but most valuable is his *Synagoge* ('Collection'), which contains detailed accounts of much Greek mathematics, some of which would otherwise be unknown.

Pappus' Theorems The following two theorems about surfaces and solids of revolutions:

THEOREM: Suppose that an arc of a plane curve is rotated through one revolution about a line in the plane that does not cut the arc. Then the area of the surface of revolution obtained is equal to the length of the arc times the distance travelled by the centroid of the curve.

THEOREM: Suppose that a plane region is rotated through one revolution about a line in the plane that does not cut the region. Then the volume of the solid of revolution obtained is equal to the area of the region times the distance travelled by the centroid of the region.

The theorems can be used to find surface areas and volumes, such as those of a *torus*. They can also be used to find the positions of centroids. For example, using the second theorem and the known volume of a sphere, the position of the centroid of the region bounded by a semicircle and a diameter can be found.

parabola A *conic* with eccentricity equal to 1. Thus a parabola is the locus of all points P such that the distance from P to a fixed point F (the **focus**) is equal to the distance from P to a fixed line l (the **directrix**). It is obtained as a plane section of a cone in the case when the plane is parallel to a generator of the cone (see *conic*). A line through the focus perpendicular to the directrix is the **axis** of the parabola, and the point where the axis cuts the parabola is the **vertex**. It is possible to take the vertex as origin, the axis of the parabola as the x-axis, the tangent at the vertex as the y-axis and the focus as the point $(a, 0)$. In this coordinate system, the directrix has equation $x = -a$ and the parabola has equation $y^2 = 4ax$. When investigating the properties of a parabola, it is normal to choose this convenient coordinate system.

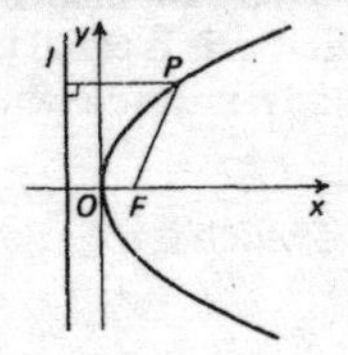

Different values of a give parabolas of different sizes, but all parabolas are the same shape. For any value of t, the point $(at^2, 2at)$ satisfies $y^2 = 4ax$, and conversely any point of the parabola has coordinates $(at^2, 2at)$ for some value of t. Thus $x = at^2, y = 2at$ may be taken as *parametric equations* for the parabola $y^2 = 4ax$.

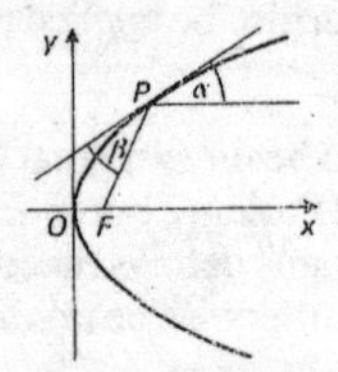

One important property of a parabola is this. For a point P on the parabola, let α be the angle between the tangent at P and a line through P parallel to the axis of the parabola, and let β be the angle between the tangent at P and a line through P and the focus, as shown in the figure; then $\alpha = \beta$. This is the basis of the parabolic reflector: if a source of light is placed at the focus of a parabolic reflector, each ray of light is reflected parallel to the axis, so producing a parallel beam of light.

parabolic cylinder A *cylinder* in which the fixed curve is a *parabola* and the fixed line to which the generators are parallel is perpendicular to the plane of the parabola. It is a *quadric*, and in a suitable coordinate system has equation

$$\frac{x^2}{a^2} = \frac{2y}{b}.$$

paraboloid See *elliptic paraboloid* and *hyperbolic paraboloid*.

paradox A situation in which an entirely reasonable assumption appears to lead to an unreasonable conclusion. It may seem that, if a certain statement is true, then it follows that it is false; and also that, if it is false, then it is true. Examples of this are *Grelling's paradox*, the *liar paradox* and *Russell's paradox*. See also *Zeno of Elea*.

Parallel Axis Theorem The following theorem about *moments of inertia* of a rigid body:

THEOREM: Let I_C be the moment of inertia of a rigid body about an axis through C, the centre of mass of the rigid body. Then the moment of inertia of the body about some other axis parallel to the first axis is equal to $I_C + md^2$, where m is the mass of the rigid body and d is the distance between the two parallel axes.

parallelepiped A *polyhedron* with six faces, each of which is a parallelogram. (The word is commonly misspelt.)

parallelogram A quadrilateral in which (i) both pairs of opposite sides are parallel, and (ii) the lengths of opposite sides are equal. Either property, (i) or (ii), in fact implies the other. The area of a parallelogram equals 'base times height'. That is to say, if one pair of parallel sides, of length b, are a distance h apart, the area equals bh. Alternatively, if the other pair of sides have length a and θ is the angle between adjacent sides, the area equals $ab \sin \theta$.

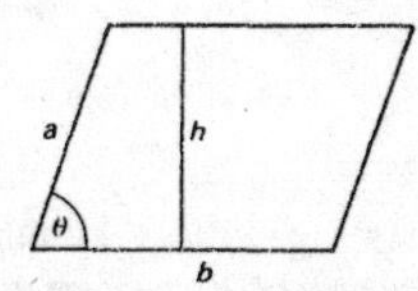

parallelogram law See *addition* (of vectors).

parallel postulate The axiom of Euclidean geometry which says that, if two straight lines are cut by a *transversal* and the interior angles on one side add up to less than two right angles, then the two lines meet on that side. It is equivalent to Playfair's axiom, which says that, given a point not on a given line, there is precisely one line through the point parallel to the line. The parallel postulate was shown to be independent of the other axioms of Euclidean geometry in the nineteenth century, when *non-Euclidean geometries* were discovered in which the other axioms hold but the parallel postulate does not.

parameter (in pure mathematics) A variable that is to take different values, thereby giving different values to certain other variables. For example, a parameter t could be used to write the solutions of the equation $5x_1 + 4x_2 = 7$ as $x_1 = 3 - 4t$, $x_2 = -2 + 5t$ $(t \in \mathbf{R})$. See also *parametric equations* (of a line in space) and *parametrization*.

parameter (in statistics) A **parameter** for a *population* is some quantity that relates to the population, such as its mean or median. A parameter for a population may be estimated from a sample by using an appropriate *statistic* as an *estimator*. For a *distribution*, a constant that appears in the *probability mass function* or *probability density function* of the distribution is called a parameter. In this sense, the *normal distribution* has two parameters and the *Poisson distribution* has one parameter, for example.

parametric equations (of a curve) See *parametrization*.

parametric equations (of a line in space) Given a line in 3-dimensional space, let (x_1, y_1, z_1) be coordinates of a point on the line, and l, m, n be *direction ratios* of a direction along the line. Then the line consists of all points P whose coordinates (x, y, z) are given by

$$x = x_1 + tl, \qquad y = y_1 + tm, \qquad z = z_1 + tn,$$

for some value of the *parameter* t. These are **parametric equations** for the line. They are most easily established by using the *vector equation* of the line and taking components. If none of l, m, n is zero, the equations can be written

$$\frac{x-x_1}{l}=\frac{y-y_1}{m}=\frac{z-z_1}{n} \quad (=t),$$

which can be considered to be another form of the parametric equations, or called the equations of the line in 'symmetric form'. If, say, $n=0$ and l and m are both non-zero, the equations are written

$$\frac{x-x_1}{l}=\frac{y-y_1}{m}, \qquad z=z_1.$$

If, say, $m=n=0$, they become $y=y_1, z=z_1$.

parametrization (of a curve) A method of associating, with every value of a *parameter* t in some interval I (or some other subset of $\mathbf{R}$), a point $P(t)$ on the curve such that every point of the curve corresponds to some value of t. Often this is done by giving the x- and y-coordinates of P as functions of t, so that the coordinates of P may be written $(x(t), y(t))$. The equations that give x and y as functions of t are **parametric equations** for the curve. For example, $x=at^2, y=2at$ $(t \in \mathbf{R})$ are parametric equations for the parabola $y^2=4ax$; and $x=a\cos\theta, y=b\sin\theta$ $(\theta \in [0,2\pi))$ are parametric equations for the ellipse

$$\frac{x^2}{a^2}+\frac{y^2}{b^2}=1.$$

The gradient dy/dx of the curve at any point can be found, if $x'(t) \neq 0$, from $dy/dx = y'(t)/x'(t)$.

parity The attribute of an integer of being even or odd. Thus, it can be said that 6 and 14 have the same parity (both are even), whereas 7 and 12 have opposite parity.

parity check See *check digit*.

partial derivative Suppose that f is a real function of n variables $x_1, x_2, \ldots, x_n$, so that its value at a typical point is $f(x_1, x_2, \ldots, x_n)$. If

$$\frac{f(x_1+h, x_2, \ldots, x_n)-f(x_1, x_2, \ldots, x_n)}{h}$$

tends to a limit as $h \to 0$, this limit is the **partial derivative** of f, at the point $(x_1, x_2, \ldots, x_n)$, with respect to x_1; it is denoted by $f_1(x_1, x_2, \ldots, x_n)$ or $\partial f/\partial x_1$ (read as 'partial df by dx_1'). For a particular function, this partial derivative may be found using the normal rules of differentiation, by differentiating as though the function were a function of x_1 only and treating $x_2, \ldots, x_n$ as constants. The other partial derivatives,

$$\frac{\partial f}{\partial x_2}, \ldots, \frac{\partial f}{\partial x_n},$$

are defined similarly. The partial derivatives may also be denoted by $f_{x_1}, f_{x_2}, \ldots, f_{x_n}$. For example, if $f(x,y)=xy^3$, then the partial derivatives are $\partial f/\partial x$, or f_x, and $\partial f/\partial y$, or f_y; and $f_x=y^3$, $f_y=3xy^2$. With the value of

f at (x, y) denoted by $f(x, y)$, the notation f_1 and f_2 is also used for the partial derivatives f_x and f_y, and this can be extended to functions of more variables. See also *higher-order partial derivative.*

partial differential equation See *differential equation.*

partial differentiation The process of obtaining one of the *partial derivatives* of a function of more than one variable. The partial derivative $\partial f/\partial x_i$ is said to be obtained from f by 'differentiating partially with respect to x_i'.

partial fractions Suppose that $f(x)/g(x)$ defines a *rational function*, so that $f(x)$ and $g(x)$ are polynomials, and suppose that the degree of $f(x)$ is less than the degree of $g(x)$. In general, $g(x)$ can be factorized into a product of some different linear factors, each to some index, and some different irreducible quadratic factors, each to some index. Then the original expression $f(x)/g(x)$ can be written as a sum of terms: corresponding to each $(x-\alpha)^n$ in $g(x)$, there are terms

$$\frac{A_1}{x-\alpha}+\frac{A_2}{(x-\alpha)^2}+\cdots+\frac{A_n}{(x-\alpha)^n},$$

and corresponding to each $(ax^2+bx+c)^n$ in $g(x)$, there are terms

$$\frac{B_1x+C_1}{ax^2+bx+c}+\frac{B_2x+C_2}{(ax^2+bx+c)^2}+\cdots+\frac{B_nx+C_n}{(ax^2+bx+c)^n},$$

where the real numbers denoted here by capital letters are uniquely determined. The expression $f(x)/g(x)$ is said to have been written in **partial fractions**. The method, which sounds complicated when stated in general, as above, is more easily understood from examples:

$$\frac{3}{(x-1)(x+2)}=\frac{A}{x-1}+\frac{B}{x+2},$$

$$\frac{3x^2+2x+1}{(x-1)^3}=\frac{A}{x-1}+\frac{B}{(x-1)^2}+\frac{C}{(x-1)^3},$$

$$\frac{3x+2}{(x-1)(x^2+x+1)^2}=\frac{A}{x-1}+\frac{Bx+C}{x^2+x+1}+\frac{Dx+E}{(x^2+x+1)^2},$$

$$\frac{3x+2}{(x-1)^2(x^2+x+1)}=\frac{A}{x-1}+\frac{B}{(x-1)^2}+\frac{Cx+D}{x^2+x+1}.$$

The values for the numbers $A, B, C, \ldots$ are found by first multiplying both sides of the equation by the denominator $g(x)$. In the last example, this gives

$$3x+2=A(x-1)(x^2+x+1)+B(x^2+x+1)+(Cx+D)(x-1)^2.$$

This has to hold for all values of x, so the coefficients of corresponding powers of x on the two sides can be equated, and this determines the unknowns. In some cases, setting x equal to particular values (in this example, $x=1$) may determine some of the unknowns more quickly. The method of partial fractions is used in the integration of rational functions.

partial product When an infinite product

$$\prod_{r=1}^{\infty} a_r$$

is formed from a sequence $a_1, a_2, a_3, \ldots$, the product $a_1 a_2 \ldots a_n$ of the first n terms is called the n-th **partial product**.

partial sum The n-th **partial sum** s_n of a series $a_1 + a_2 + \cdots$ is the sum of the first n terms; thus $s_n = a_1 + a_2 + \cdots + a_n$.

particle An object considered as having no size, but having mass, position, velocity, acceleration, and so on. It is used in *mathematical models* to represent an object in the real world of negligible size. The subject of particle dynamics is concerned with the study of the motion of one or more particles experiencing a system of forces.

particular integral See *linear differential equation with constant coefficients.*

particular solution See *differential equation.*

partition (of an interval) Let $[a, b]$ be a closed interval. A set of $n + 1$ points $x_0, x_1, \ldots, x_n$ such that

$$a = x_0 < x_1 < x_2 < \cdots < x_{n-1} < x_n = b$$

is a **partition** of the interval $[a, b]$. A partition divides the interval into n subintervals $[x_i, x_{i+1}]$. The **norm** of the partition P is equal to the length of the largest subinterval and is denoted by $\|P\|$. Such partitions are used in defining the Riemann integral of a function over $[a, b]$ (see *integral*).

partition (of a positive integer) A **partition** of the positive integer n is obtained by writing $n = n_1 + n_2 + \cdots + n_k$, where $n_1, n_2, \ldots, n_k$ are positive integers, and the order in which $n_1, n_2, \ldots, n_k$ appear is unimportant. The number of partitions of n is denoted by $p(n)$. For example, the partitions of 5 are:

$$5,\ 4+1,\ 3+2,\ 3+1+1,\ 2+2+1,\ 2+1+1+1,\ 1+1+1+1+1,$$

and hence $p(5) = 7$. The values of $p(n)$ for small values of n are as follows:

n	1	2	3	4	5	6	7	8	9	10
$p(n)$	1	2	3	5	7	11	15	22	30	42

partition (of a set) A **partition** of a set S is a collection of non-empty subsets of S such that every element of S belongs to exactly one of the subsets in the collection. Thus S is the union of these subsets, and any two distinct subsets are disjoint. Given a partition of a set S, an *equivalence relation* $\sim$ on S can be obtained by defining $a \sim b$ if a and b belong to the same subset in the partition. It is an important fact that, conversely, from any equivalence relation on S a partition of S can be obtained.

Pascal, Blaise (1623–1662) French mathematician and religious philosopher who in mathematics is noted for his work in geometry, hydrostatics and probability. *Pascal's triangle*, the arrangement of the binomial coefficients, was not invented by him but he did use it in his studies in probability, on which he corresponded with Fermat. He also joined in the work on finding the areas of curved figures, work which was soon to lead to the calculus. Here, his main contribution was to find the area of an arch in the shape of a *cycloid.*

pascal The SI unit of *pressure*, abbreviated to 'Pa'. One pascal is equal to one *newton* per square metre.

Pascal's triangle Below are shown the first seven rows of the arrangement of numbers known as **Pascal's triangle**. In general, the n-th row consists of the *binomial coefficients* $\binom{n}{r}$, or ${}^nC_r$, with $r = 0, 1, \ldots, n$. With the numbers set out in this fashion, it can be seen how the number $\binom{n+1}{r}$ is equal to the sum of the two numbers $\binom{n}{r-1}$ and $\binom{n}{r}$, which are situated above it to the left and right. For example, $\binom{7}{3}$ equals 35, and is the sum of $\binom{6}{2}$, which equals 15, and $\binom{6}{3}$, which equals 20.

$$
\begin{array}{ccccccccccccccc}
 & & & & & & 1 & & 1 & & & & & & \\
 & & & & & 1 & & 2 & & 1 & & & & & \\
 & & & & 1 & & 3 & & 3 & & 1 & & & & \\
 & & & 1 & & 4 & & 6 & & 4 & & 1 & & & \\
 & & 1 & & 5 & & 10 & & 10 & & 5 & & 1 & & \\
 & 1 & & 6 & & 15 & & 20 & & 15 & & 6 & & 1 & \\
1 & & 7 & & 21 & & 35 & & 35 & & 21 & & 7 & & 1
\end{array}
$$

path (in a graph) Let u and v be vertices of a *graph*. A path from u to v is traced out by travelling from u to v along edges. In a precise definition, it is normal to insist that, in a path, no edge is used more than once and no vertex is visited more than once. Thus a **path** may be defined as a sequence $v_0, e_1, v_1, \ldots, e_k, v_k$ of alternately vertices and edges (where e_i is an edge joining v_{i-1} and v_i), with all the edges different and all the vertices different; this is a path from v_0 to v_k.

p.d.f. = *probability density function.*

Peano, Giuseppe (1858–1932) Italian mathematician whose work towards the axiomatization of mathematics was very influential. He was responsible for the development of symbolic logic, for which he introduced important notation. His axioms for the integers were an important development in the formal analysis of arithmetic. He was also the first person to produce examples of so-called space-filling curves.

Peano curve Take the diagonal of a square, as shown on the left in the diagram. Replace this by the nine diagonals of smaller squares (drawn here in a way that indicates the order in which the diagonals are to be traced out). Now replace each straight section of this by the nine diagonals of even smaller squares, to obtain the result shown on the right. The **Peano curve** is the curve obtained when this process is continued indefinitely. It has the remarkable property that it passes through every point of the square, and it is therefore described as **space-filling**. Any similar space-filling curve constructed in the same kind of way may also be called a Peano curve.

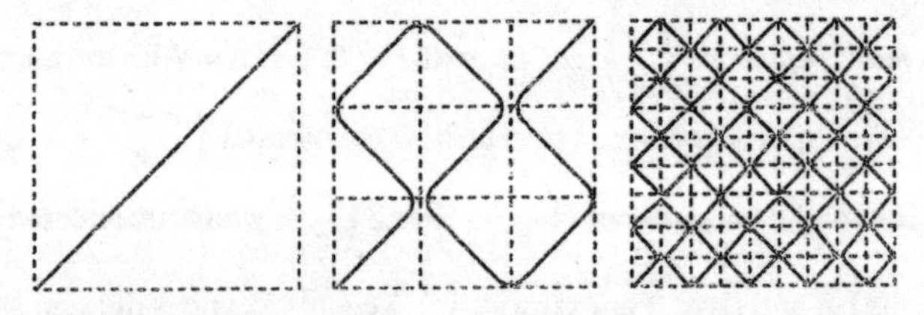

Pearson, Karl (1857–1936) British statistician influential in the development of statistics for application to the biological and social sciences. Stimulated by problems in evolution and heredity, he developed fundamental concepts such as the *standard deviation*, the *coefficient of variation* and, in 1900, the *chi-squared test*. This work was contained in a series of important papers written while he was professor of applied mathematics, and then eugenics, at University College, London.

pedal triangle In the triangle with vertices A, B, and C, let D, E and F be the feet of the perpendiculars from A, B and C, respectively, to the opposite sides, so that AD, BE and CF are the *altitudes*. The triangle DEF is called the **pedal triangle**. The altitudes of the original triangle bisect the angles of the pedal triangle.

Pell, John (1610–1685) English mathematician remembered in the name of *Pell's equation*, which was wrongly attributed to him.

Pell's equation The *Diophantine equation* $x^2 = ny^2 + 1$, where n is a positive integer that is not a perfect square. Methods of solving such an equation have been sought from as long ago as the time of Archimedes. Bhāskara solved particular cases. Fermat apparently understood that infinitely many solutions always exist, a fact that was proved by Lagrange. Pell's name was incorrectly given to the equation by Euler.

pendulum See *simple pendulum*, *conical pendulum*, *compound pendulum* and *Foucault pendulum*.

pentagon A five-sided polygon.

pentagram The plane figure shown below, formed by joining alternate vertices of a regular *pentagon*.

percentage error The *relative error* expressed as a percentage. When 1.9 is used as an approximation for 1.875, the relative error equals $0.025/1.875$ (or $0.025/1.9) = 0.013$, to 2 significant figures. So the percentage error is 1.3%, to 2 significant figures.

percentile See *quantile*.

perfect number An integer that is equal to the sum of its positive divisors (not including itself). Thus, 6 is a perfect number, since its positive divisors (not including itself) are 1, 2 and 3, and $1 + 2 + 3 = 6$; so too are 28 and 496, for example. At present there are over 30 known perfect numbers, all even. If $2^p - 1$ is prime (so that it is a *Mersenne prime*), then $2^{p-1}(2^p - 1)$ is perfect; moreover, these are the only even perfect numbers. It is not known if there are any odd perfect numbers; none has been found, but it has not been proved that one cannot exist.

perfect square An integer of the form n^2, where n is a positive integer. The perfect squares are $1, 4, 9, 16, 25, \ldots$.

perigee See *apse*.

perihelion See *apse*.

perimeter The **perimeter** of a plane figure is the length of its boundary. Thus, the perimeter of a rectangle of length L and width W is $2L + 2W$. The perimeter of a circle is the length of its circumference.

period, periodic If, for some value p, $f(x + p) = f(x)$ for all x, the real function f is **periodic** and has **period** p. For example, $\cos x$ is periodic with period 2π, since $\cos(x + 2\pi) = \cos x$ for all x; or, using degrees, $\cos x°$ is periodic with period 360, since $\cos(x + 360)° = \cos x°$ for all x. Some authors restrict the use of the term 'period' to the smallest positive value of p with this property.

In mechanics, any phenomenon that repeats regularly may be called periodic, and the time taken before the phenomenon repeats itself is then called the period. The motion may be said to consist of repeated **cycles** the period being the time taken for the execution of one cycle. Suppose that $x = A \sin(\omega t + \alpha)$, where $A\ (> 0)$, ω and α are constants.

This may, for example, give the displacement x of a particle, moving in a straight line, at time t. The particle is thus oscillating about the origin. The **period** is the time taken for one complete oscillation and is equal to $2\pi/\omega$.

permutation At an elementary level, a **permutation** of n objects can be thought of as an arrangement or a rearrangement of the n objects. The number of permutations of n objects is equal to $n!$. The number of 'permutations of n objects taken r at a time' is denoted by ^nP_r, and equals $n(n-1)\dots(n-r+1)$, which equals $n!/(n-r)!$. For example, there are 12 permutations of A, B, C, D taken two at a time: $AB, AC, AD, BA, BC, BD, CA, CB, CD, DA, DB, DC$.

Suppose that n objects are of k different kinds, with r_1 alike of one kind, r_2 alike of a second kind, and so on. Then the number of distinct permutations of the n objects equals

$$\frac{n!}{r_1!\, r_2! \dots r_k!},$$

where $r_1 + r_2 + \dots + r_k = n$. For example, the number of different anagrams of the word '*CHEESES*' is $7!/3!\,2!$, which equals 420. At a more advanced level, a **permutation** of a set X is defined as a *one-to-one onto mapping* from X to X.

Perpendicular Axis Theorem The following theorem about *moments of inertia* of a *lamina*:

THEOREM: Suppose that a rigid body is in the form of a lamina lying in the plane $z = 0$. Let I_{Ox}, I_{Oy} and I_{Oz} be the moments of inertia of the body about the three coordinate axes. Then $I_{Oz} = I_{Ox} + I_{Oy}$.

perpendicular bisector The **perpendicular bisector** of a line segment AB is the straight line perpendicular to AB through the midpoint of AB.

perpendicular lines In coordinate geometry of the plane, a useful necessary and sufficient condition that two lines, with gradients m_1 and m_2, are **perpendicular** is that $m_1m_2 = -1$. (This is taken to include the cases when $m_1 = 0$ and m_2 is infinite, and vice versa.)

perpendicular planes Two planes in 3-dimensional space are **perpendicular** if *normals* to the two planes are perpendicular. If $\mathbf{n}_1$ and $\mathbf{n}_2$ are vectors normal to the planes, this is so if $\mathbf{n}_1 . \mathbf{n}_2 = 0$.

peta- Prefix used with *SI units* to denote multiplication by 10^{15}.

phase Suppose that $x = A\sin(\omega t + \alpha)$, where $A\ (> 0)$, ω and α are constants. This may, for example, give the displacement x of a particle, moving in a straight line, at time t. The particle is thus oscillating about the origin. The constant α is the **phase**. Two particles oscillating like this with the same *amplitude* and *period* (of oscillation) but with different phases are executing the same motion apart from a shift in time.

pi For circles of all sizes, the length of the circumference divided by the length of the diameter is the same, and this number is the value of π. It is equal to 3.141 592 65 to 8 decimal places. Sometimes π is taken to be equal to $\frac{22}{7}$, but it must be emphasized that this is only an approximation. The decimal expansion of π is neither finite nor recurring, for it was shown in 1761 by Lambert, using continued fractions, that π is an *irrational number*. In 1882, Lindemann proved that π is also a *transcendental number*. The number appears in some contexts that seem to have no connection with the definition relating to a circle. For example:

$$\frac{\pi}{4} = 1 - \frac{1}{3} + \frac{1}{5} - \frac{1}{7} + \cdots, \qquad \frac{\pi^2}{6} = 1 + \frac{1}{2^2} + \frac{1}{3^2} + \frac{1}{4^2} + \cdots,$$

$$\frac{\pi}{2} = \frac{2}{1} \times \frac{2}{3} \times \frac{4}{3} \times \frac{4}{5} \times \frac{6}{5} \times \frac{6}{7} \times \frac{8}{7} \times \frac{8}{9} \times \cdots$$

$$= \prod_{n=1}^{\infty} \frac{4n^2}{4n^2 - 1} \quad \textbf{(Wallis's Product)}.$$

pico- Prefix used with *SI units* to denote multiplication by 10^{-12}.

pie chart Suppose that some finite set is partitioned into subsets, and that the proportion of elements in each subset is expressed as a percentage. A **pie chart** is a diagram consisting of a circle divided into sectors whose areas are proportional to these percentages. In a similar way, when nominal data have been collected the data can be presented as a pie chart in which the circle is divided into sectors whose areas are proportional to the frequencies. The figure shows the kinds of vehicles recorded in a small traffic survey.

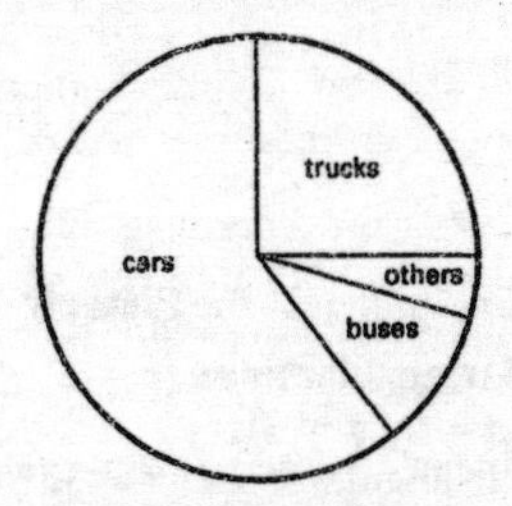

Pigeonhole Principle The observation that, if m objects are distributed into n boxes, with $m > n$, then at least one box receives at least two objects. This can be applied in some obvious ways; for example, if you take any 13 people, then at least two of them have birthdays that fall in the same month. It also has less trivial applications.

pivot, pivot operation Each step in *Gaussian elimination* or *Gauss–Jordan elimination* consists of making one of the entries a_{ij} of the matrix equal to 1 and using this to produce zeros elsewhere in the j-th column. This process is known as a **pivot operation**, with a_{ij} as **pivot**. The pivot, of course, must be non-zero. This corresponds to solving one of the equations for one of the unknowns x_j and substituting this into the other equations.

planar graph A *graph* that can be drawn in such a way that no two edges cross. The *complete graph* K_4, for example, is planar as either drawing of it in the figure shows. Neither the complete graph K_5 nor the complete bipartite graph $K_{3,3}$, for example, is planar.

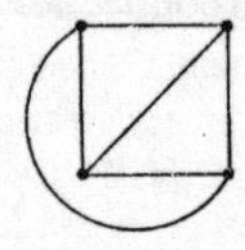

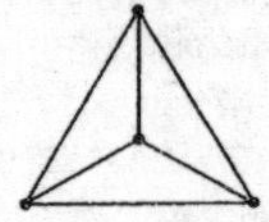

plane (in Cartesian coordinates) A **plane** is represented by a linear equation, in other words by an equation of the form $ax + by + cz + d = 0$, where the constants a, b and c are not all zero. Here a, b and c are *direction ratios* of a direction *normal* to the plane. See also *vector equation* (of a plane).

Platonic solid A convex *polyhedron* is **regular** if all its faces are alike and all its vertices are alike. More precisely, this means that

(i) all the faces are regular polygons having the same number p of edges, and
(ii) the same number q of edges meet at each vertex.

Notice that the polyhedron shown here, with 6 triangular faces, satisfies (i), but is not regular because it does not satisfy (ii).

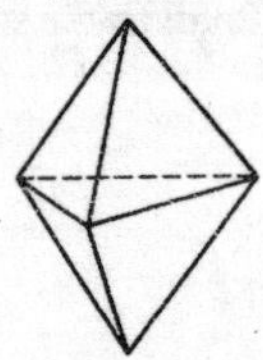

There are five regular convex polyhedra, known as the **Platonic solids**:

(i) the regular *tetrahedron*, with 4 triangular faces ($p = 3,\ q = 3$),
(ii) the *cube*, with 6 square faces ($p = 4,\ q = 3$),
(iii) the regular *octahedron*, with 8 triangular faces ($p = 3,\ q = 4$),
(iv) the regular *dodecahedron*, with 12 pentagonal faces ($p = 5$, $q = 3$),
(v) the regular *icosahedron*, with 20 triangular faces ($p = 3,\ q = 5$).

Playfair's axiom The axiom which says that, given a point not on a given line, there is precisely one line through the point parallel to the line. It is equivalent to the axiom of Euclid's known as the *parallel postulate*. The name arises from its occurrence in a book by John Playfair (1748–1819), a British geologist and mathematician.

p.m.f. = *probability mass function.*

Poincaré, (Jules) Henri (1854–1912) French mathematician, who can be regarded as the last mathematician to be active over the entire field of mathematics. He moved from one area of mathematics to another, making major contributions to most of them—'a conqueror rather than a colonist', it was said. Within pure mathematics, he is seen as one of the main founders of topology and the discoverer of what are called automorphic functions. In applied mathematics, he is remembered for his theoretical work on the qualitative aspects of celestial mechanics, which was probably the most important work in that area since Laplace and Lagrange. Connecting and motivating both was his qualitative theory of differential equations. Not the least of his achievements was a stream of books, popular in both senses, still strongly recommended to any aspiring young mathematician.

point estimate See *estimate*.

point of application In the real world, a force acting on a body is usually applied over a region, such as a part of the surface of the body. When someone is pushing a trolley by hand, there is a region of the handle of the trolley where the hand pushes. When a tugboat is towing a ship, the rope that is being used is attached to a bollard on the deck of the ship. Some forces, such as gravitational forces, act throughout a body. In contrast, in a *mathematical model*, it is assumed that every force acts at a particular point, the **point of application** of the force. In some mathematical models, each point of the surface of a body may experience a force (when the surface is experiencing a *pressure*), or every point throughout the volume occupied by a body may experience a force (when the body is experiencing a *field of force*).

point of inflexion A point on a graph $y = f(x)$ at which the *concavity* changes. Thus, there is a point of inflexion at a if, for some $\delta > 0, f''(x)$ exists and is positive in the open interval $(a - \delta, a)$ and $f''(x)$ exists and is negative in the open interval $(a, a + \delta)$, or vice versa.

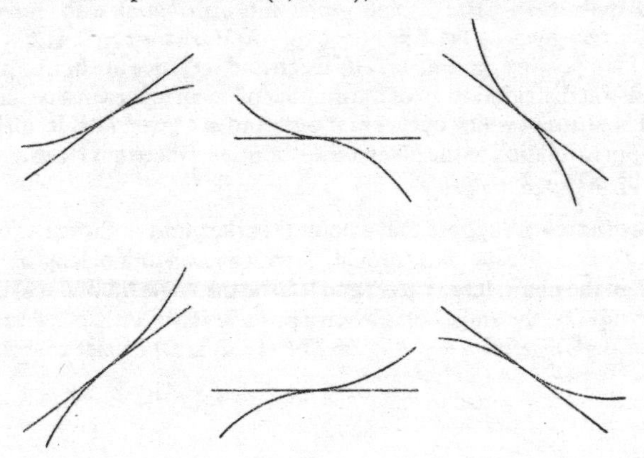

If the change (as x increases) is from concave up to concave down, the graph and its tangent at the point of inflexion look like one of the cases shown in the first row of the figure. If the change is from concave down to concave up, the graph and its tangent look like one of the cases shown in the second row. The middle diagram in each row shows a point of inflexion that is also a *stationary point*, since the tangent is horizontal.

If f'' is continuous at a, then for $y = f(x)$ to have a point of inflexion at a it is necessary that $f''(a) = 0$, and so this is the usual method of finding possible points of inflexion. However, the condition $f''(a) = 0$ is not sufficient to ensure that there is a point of inflexion at a: it must be shown that $f''(x)$ is positive to one side of a and negative to the other side. Thus, if $f(x) = x^4$ then $f''(0) = 0$; but $y = x^4$ does not have a point of inflexion at 0 since $f''(x)$ is positive to both sides of 0. Finally, there may be a point of inflexion at a point where $f''(x)$ does not exist, for example at the origin on the curve $y = x^{1/3}$, as shown in the figure below.

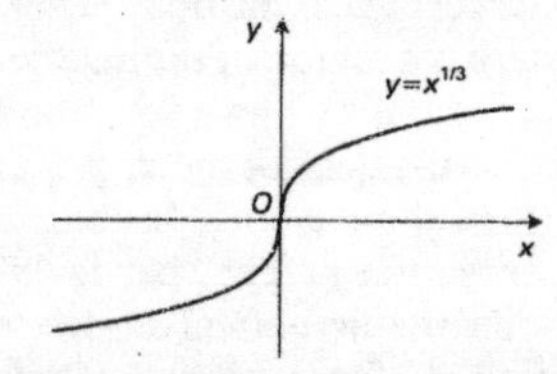

Poisson, Siméon-Denis (1781–1840) French mathematician who was a pupil and friend of Laplace and Lagrange. He extended their work in celestial mechanics and made outstanding contributions in the fields of electricity and magnetism. In 1837, in an important paper on probability, he introduced the distribution that is named after him and formulated the law of large numbers.

Poisson distribution The discrete probability *distribution* with *probability mass function* given by $\Pr(X = r) = \exp(-\lambda)\lambda^r/r!$, for $r = 0, 1, 2, \ldots$, where λ is a positive parameter. The mean and variance are both equal to λ. The Poisson distribution gives the number of occurrences in a certain time period of an event which occurs randomly but at a given rate. It can be used as an approximation to the *binomial distribution*, where n is large and p is small, by taking $\lambda = np$.

polar coordinates Suppose that a point O in the plane is chosen as origin, and let Ox be a directed line through O, with a given unit of length. For any point P in the plane, let $r = |OP|$ and let θ be the angle (in radians) that OP makes with Ox, the angle being given a positive sense anticlockwise from Ox. The angle θ satisfies $0 \leq \theta < 2\pi$. Then (r, θ) are the **polar coordinates**

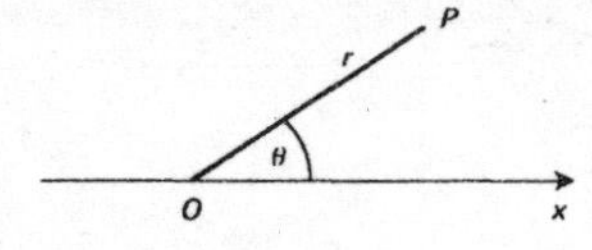

of P. (The point O gives no corresponding value for θ, but is simply said to correspond to $r = 0$.) If P has polar coordinates (r, θ) then $(r, \theta + 2k\pi)$, where k is an integer, may also be permitted as polar coordinates of P.

Suppose that Cartesian coordinates are taken with the same origin and the same unit of length, with positive x-axis along the directed line Ox. Then the Cartesian coordinates (x, y) of a point P can be found from (r, θ), by $x = r\cos\theta$, $y = r\sin\theta$. Conversely, the polar coordinates can be found from the Cartesian coordinates as follows: $r = \sqrt{x^2 + y^2}$, and θ is such that

$$\cos\theta = \frac{x}{\sqrt{x^2 + y^2}}, \qquad \sin\theta = \frac{y}{\sqrt{x^2 + y^2}}.$$

(It is true that, when $x \neq 0$, $\theta = \tan^{-1}(y/x)$, but this is not sufficient to determine θ.) In certain circumstances, authors may allow r to be negative, in which case the polar coordinates $(-r, \theta)$ give the same point as $(r, \theta + \pi)$.

In mechanics, it is useful, when a point P has polar coordinates (r, θ), to define vectors $\mathbf{e}_r$ and $\mathbf{e}_\theta$ by $\mathbf{e}_r = \mathbf{i}\cos\theta + \mathbf{j}\sin\theta$ and $\mathbf{e}_\theta = -\mathbf{i}\sin\theta + \mathbf{j}\cos\theta$, where $\mathbf{i}$ and $\mathbf{j}$ are unit vectors in the directions of the positive x- and y-axes. Then $\mathbf{e}_r$ is a unit vector along OP in the direction of increasing r, and $\mathbf{e}_\theta$ is a unit vector perpendicular to this in the direction of increasing θ. These vectors satisfy $\mathbf{e}_r \,.\, \mathbf{e}_\theta = 0$ and $\mathbf{e}_r \times \mathbf{e}_\theta = \mathbf{k}$, where $\mathbf{k} = \mathbf{i} \times \mathbf{j}$.

polar equation An equation of a curve in *polar coordinates* is usually written in the form $r = f(\theta)$. **Polar equations** of some common curves are given in the table.

Curve		Polar equation	
Circle	$x^2 + y^2 = 1$	$r = 1$	
Half-line	$y = x\ (x > 0)$	$\theta = \pi/4$	
Line	$x = 1$	$r = \sec\theta$	$(-\frac{1}{2}\pi < \theta < \frac{1}{2}\pi)$
Line	$y = 1$	$r = \operatorname{cosec}\theta$	$(0 < \theta < \pi)$
Circle	$x^2 + y^2 - 2ax = 0\ (a > 0)$	$r = 2a\cos\theta$	$(-\frac{1}{2}\pi < \theta \le \frac{1}{2}\pi)$
Circle	$x^2 + y^2 - 2by = 0\ (b > 0)$	$r = 2b\sin\theta$	$(0 \le \theta < \pi)$
Cardioid	See *cardioid*	$r = 2a(1 + \cos\theta)$	$(-\pi < \theta \le \pi)$
Conic	See *conic*	$l/r = 1 + e\cos\theta$	$(\cos\theta \neq -1/e)$

polar form of a complex number For a complex number z, let $r = |z|$ and $\theta = \arg z$. Then $z = r(\cos\theta + i\sin\theta)$, and this is the **polar form** of z. It may also be written $z = re^{i\theta}$.

polygon A plane figure bounded by some number of straight sides. This definition would include polygons like the one shown below, but often

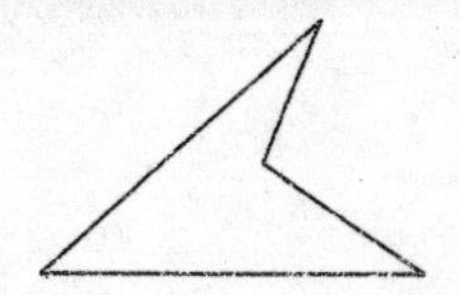

figures of this sort are intended to be excluded and it is implicitly assumed that a polygon is to be *convex*. Thus a (convex) polygon is a plane figure consisting of a finite region bounded by some finite number of straight lines, in the sense that the region lies entirely on one side of each line. The exterior angles of the polygon shown below are those marked, and the sum

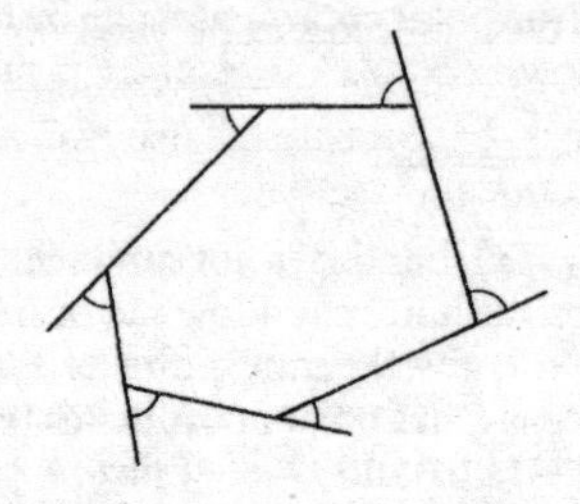

of the exterior angles is always equal to 360°. The sum of the interior angles of an n-sided (convex) polygon equals $2n - 4$ right angles. In a **regular** (convex) **polygon**, all the sides have equal length and the interior angles have equal size; and the vertices lie on a circle.

Polygons with the following numbers of edges have the names given:

5	pentagon	9	enneagon (or nonagon)
6	hexagon	10	decagon
7	heptagon	11	hendecagon
8	octagon	12	dodecagon

polygon of forces Suppose that the forces $\mathbf{F}_1, \mathbf{F}_2, \ldots, \mathbf{F}_n$ act on a particle. Make a construction of directed line-segments as follows. Let $\overrightarrow{A_1A_2}$ represent $\mathbf{F}_1$, $\overrightarrow{A_2A_3}$ represent $\mathbf{F}_2$, and so on, with $\overrightarrow{A_nA_{n+1}}$ representing $\mathbf{F}_n$. (These line-segments are not necessarily coplanar.) Then the total force $\mathbf{F}$ acting on the particle, given by $\mathbf{F} = \sum \mathbf{F}_i$, is represented by the directed line-segment $\overrightarrow{A_1A_{n+1}}$. In particular, if the total force is zero, the point A_{n+1} coincides with A_1 and the directed line-segments form the n sides of a polygon, called a **polygon of forces.**

polyhedron (plural: polyhedra) A solid figure bounded by some number of plane polygonal faces. This definition would include the one shown here,

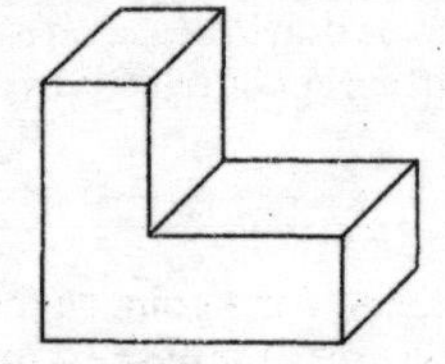

and it may well be that figures such as this are to be excluded from consideration. So it is often assumed that a polyhedron is *convex*. Thus a (convex) polyhedron is a finite region bounded by some number of planes, in the sense that the region lies entirely on one side of each plane. Each edge of the polyhedron joins two vertices, and each edge is the common edge of two faces.

The numbers of vertices, edges and faces of a convex polyhedron are connected by an equation given in *Euler's Theorem* (for polyhedra).

Certain polyhedra are called regular, and these are the five *Platonic solids*; the thirteen *Archimedean solids* are semi-regular.

polynomial Let $a_0, a_1, \ldots, a_n$ be real numbers. Then

$$a_n x^n + a_{n-1} x^{n-1} + \cdots + a_1 x + a_0$$

is a **polynomial** in x (with real coefficients). When $a_0, a_1, \ldots, a_n$ are not all zero, it can be assumed that $a_n \neq 0$ and the polynomial has **degree** n. For example, $x^2 - \sqrt{2}x + 5$ and $3x^4 + \frac{7}{2}x^2 - x$ are polynomials of degrees 2 and 4 respectively. The number a_r is the **coefficient** of x^r (for $r = 1, 2, \ldots, n$) and a_0 is the **constant term**. A polynomial can be denoted by $f(x)$ (so that f is a *polynomial function*), and then $f(-1)$, for example, denotes the value of the polynomial when x is replaced by -1. In the same way, it is possible to consider polynomials in x, or more commonly in z, with complex coefficients, such as $z^2 + 2(1 - i)z + (15 + 6i)$, and the same terminology is used.

polynomial equation An equation $f(x) = 0$, where $f(x)$ is a *polynomial*.

polynomial function In real analysis, a **polynomial function** is a function f defined by a formula $f(x) = a_n x^n + a_{n-1} x^{n-1} + \cdots + a_1 x + a_0$ for all x in $\mathbf{R}$, where $a_0, a_1, \ldots, a_n$ are real numbers. See also *polynomial*; terminology applying to a polynomial is also used to apply to the corresponding polynomial function.

Poncelet, Jean-Victor (1788–1867) French military engineer and mathematician, recognized in mathematics as the founder of projective geometry.

pons asinorum The Latin name, meaning 'bridge of asses', familiarly given to the fifth proposition in the first book of Euclid: that the base angles of an isosceles triangle are equal. It is said to be the first proof found difficult by some readers.

population A set about which *inferences* are to be drawn based on a sample taken from the set. The sample may be used to accept or reject a hypothesis about the population.

population mean See *mean*.

position ratio If A and B are two given points, and P is a point on the line through A and B, then the ratio $AP : PB$ may be called the **position ratio** of P.

position vector Suppose that a point O is chosen as origin (in the plane, or in 3-dimensional space). Given any point P, the **position vector p** of the point P is the *vector* represented by the directed line-segment $\overrightarrow{OP}$. Authors who refer to $\overrightarrow{OP}$ as a vector call $\overrightarrow{OP}$ the position vector of P.

In mechanics, the position vector of a particle is often denoted by **r**, which is a function of the time t, giving the position of the particle as it moves. If Cartesian coordinates are used, then $\mathbf{r} = x\mathbf{i} + y\mathbf{j} + z\mathbf{k}$, where x, y and z are functions of t.

positive direction See *directed line*.

posterior distribution See *prior distribution*.

posterior probability See *prior probability*.

post-multiplication When the product **AB** of matrices **A** and **B** is formed (see *multiplication* (of matrices)), **A** is said to be **post-multiplied** by **B**.

postulate = *axiom*.

potential energy With every *conservative force* there is associated a **potential energy**, denoted by E_p or V, defined up to an additive constant as follows. The change in potential energy during a time interval is defined to be the negative of the *work done* by the force during that interval. In detail, the change in E_p during the time interval from $t = t_1$ to $t = t_2$ is equal to

$$-\int_{t_1}^{t_2} \mathbf{F}\,.\,\mathbf{v}\,dt,$$

where **v** is the velocity of the point of application of the conservative force **F**. For example, if $\mathbf{F} = -k(x - l)\mathbf{i}$, as in *Hooke's law*, then $E_p = \frac{1}{2}k(x - l)^2$, taking the potential energy to be zero when $x = l$. In problems involving *gravity* near the Earth's surface it may be assumed that the gravitational force is given by $\mathbf{F} = -mg\mathbf{k}$, and then, taking the potential energy to be zero when $z = 0$, $E_p = mgz$. See also *gravitational potential energy*.

Potential energy has the dimensions $\mathrm{ML}^2\mathrm{T}^{-2}$, and the SI unit of measurement is the *joule*.

power (of a matrix) If **A** is a square matrix, then **AA**, **AAA**, **AAAA**, ... are defined, and these **powers** of **A** are denoted by $\mathbf{A}^2, \mathbf{A}^3, \mathbf{A}^4, \ldots$. For all positive integers p and q, (i) $\mathbf{A}^p\mathbf{A}^q = \mathbf{A}^{p+q} = \mathbf{A}^q\mathbf{A}^p$, and (ii) $(\mathbf{A}^p)^q = \mathbf{A}^{pq}$. By definition, $\mathbf{A}^0 = \mathbf{I}$. Moreover, if **A** is *invertible*, then $\mathbf{A}^2, \mathbf{A}^3, \ldots$ are invertible, and it can be shown that $(\mathbf{A}^{-1})^p$ is the inverse of $\mathbf{A}^p$, that is, that $(\mathbf{A}^{-1})^p = (\mathbf{A}^p)^{-1}$. So either of these is denoted by $\mathbf{A}^{-p}$. Thus, when **A** is invertible, $\mathbf{A}^p$ has been defined for all integers (positive, zero and negative) and properties (i) and (ii) above hold for *all* integers p and q.

power (in mechanics) The **power** associated with a force is the rate at which the force does work. The power P is given by $P = \mathbf{F}\,.\,\mathbf{v}$, where **v** is the velocity of the point of application of the force **F**. The *work done* by the force during a certain time interval is equal to the definite integral, with respect to t, of the power, between the appropriate limits.

In everyday usage, the power of an engine can be thought of as the rate at which the engine can do work, or as the rate at which it can produce energy.

Power has the dimensions ML^2T^{-3}, and the SI unit of measurement is the *watt*.

power (of a real number) When a real number a is raised to the *index* p to give a^p, the result is a **power** of a. The same index notation is used in other contexts, for example to give a power $\mathbf{A}^p$ of a square matrix (see *power* (of a matrix)) or a power g^p of an element g in a multiplicative group. When a^p is formed, p is sometimes called the power, but it is more correctly called the index to which a is raised.

power (of a test) In *hypothesis testing*, the probability that a test rejects the null hypothesis when it is indeed false. It is equal to $1 - \beta$, where β is the probability of a Type II error.

power series A series $a_0 + a_1x + a_2x^2 + a_3x^3 + \cdots + a_nx^n + \cdots$, in ascending powers of x, with coefficients $a_0, a_1, a_2, \ldots$, is a **power series** in x. For example, the *geometric series* $1 + x + x^2 + \cdots + x^n + \cdots$ is a power series; this has a sum to infinity (see *series*) only if $-1 < x < 1$. Further examples of power series are given in Appendix 4. Notice that it is necessary to say for what values of x each series has the given sum; this is the case with any power series.

power set The set of all subsets of a set S is the **power set** of S, denoted by $\mathcal{P}(S)$. Suppose that S has n elements $a_1, a_2, \ldots, a_n$, and let A be a subset of S. For each element a_i of S, there are two possibilities: either $a_i \in A$ or not. Considering all n elements leads to 2^n possibilities in all. Hence, S has 2^n subsets; that is, $\mathcal{P}(S)$ has 2^n elements. If $S = \{a, b, c\}$, the $8(= 2^3)$ elements of $\mathcal{P}(S)$ are $\emptyset$, $\{a\}$, $\{b\}$, $\{c\}$, $\{a, b\}$, $\{a, c\}$, $\{b, c\}$ and $\{a, b, c\}$.

predictor variable = *explanatory variable*.

pre-multiplication When the product **AB** of matrices **A** and **B** is formed (see *multiplication* (of matrices)), **B** is said to be **pre-multiplied** by **A**.

pressure Force per unit area. The simplest situation to consider is when a force is uniformly distributed over a plane surface, the direction of the force being perpendicular to the surface. Suppose that a rectangular box, of length a and width b, has weight mg. Then the area of the base is ab, and the gravitational force creates a pressure of mg/ab acting at each point of the ground on which the box rests.

Now consider a liquid of density ρ at rest in an open container. The weight of liquid supported by a horizontal surface of area ΔA situated at a depth h is $\rho(\Delta A)hg$. The resulting pressure (in excess of *atmospheric pressure*) at each point of this horizontal surface at a depth h is ρgh. Further investigations involving elemental regions show that pressure can be defined at a point in a liquid, and that pressure at a point is independent of direction. In general, pressure in a fluid (a liquid or a gas) is a scalar function of position and time.

Pressure has dimensions $ML^{-1}T^{-2}$, and the SI unit of measurement is the *pascal.*

prime A positive integer p is a **prime**, or a **prime number**, if $p \neq 1$ and its only positive divisors are 1 and itself.

It is known that there are infinitely many primes. Euclid's proof argues by contradiction as follows. Suppose that there are finitely many primes p_1, $p_2, \ldots, p_n$. Consider the number $p_1 p_2 \ldots p_n + 1$. This is not divisible by any of $p_1, p_2, \ldots, p_n$, so it is either another prime itself or is divisible by primes not so far considered. It follows that the number of primes is not finite.

Large prime numbers can be discovered by using computers. At any time, the largest known prime is usually the largest known *Mersenne prime.*

prime decomposition See *Unique Factorization Theorem.*

prime factorization, prime representation = *prime decomposition.*

prime meridian The *meridian* through Greenwich from which *longitude* is measured.

Prime Number Theorem For a positive real number x, let $\pi(x)$ be the number of *primes* less than or equal to x. The **Prime Number Theorem** says that, as $x \to \infty$,

$$\frac{\pi(x)}{x/\ln x} \to 1.$$

In other words, for large values of x, $\pi(x)$ is approximately equal to $x/\ln x$. This gives, in a sense, an idea of what proportion of integers are prime. Proved first in 1896 by Jacques Hadamard and Charles De La Vallée-Poussin independently, all proofs are either extremely complicated or based on advanced mathematical ideas.

prime to each other = *relatively prime.*

primitive (*n*-th root of unity) An *n-th root of unity* z is **primitive** if every n-th root of unity is a power of z. For example, i is a primitive fourth root of unity, but -1 is a fourth root of unity that is not primitive.

principal axes (in mechanics) See *products of inertia.*

principal axes (of a quadric) A set of **principal axes** of a *quadric* is a set of axes of a coordinate system in which the quadric has an equation in *canonical form.*

principal moments of inertia See *products of inertia.*

principal value See *argument.*

principle of conservation of energy See *conservation of energy.*

principle of conservation of linear momentum See *conservation of linear momentum.*

principle of mathematical induction See *mathematical induction.*

principle of moments The principle that, when a system of coplanar forces acts on a body and produces a state of equilibrium, the sum of the *moments* of the forces about any point in the plane is zero.

Suppose, for example, that a light rod is supported by a pivot and that a particle of mass m_1 is suspended from the rod at a distance d_1 to the right of the pivot, and another of mass m_2 is suspended from the rod at a distance d_2 to the left of the pivot. The force due to gravity on the first particle has magnitude m_1g, and its moment about the pivot is equal to $(m_1g)d_1$ clockwise. Corresponding to the second particle, there is a moment $(m_2g)d_2$ anticlockwise. By the principle of moments, there is equilibrium if $(m_1g)d_1 = (m_2g)d_2$.

To use the vector definition of moment in this example, take the pivot as the origin O, the x-axis along the rod with the positive direction to the right and the y-axis vertical with the positive direction upwards. The force due to gravity on the first particle is $-m_1g\mathbf{j}$, and the moment of the force about O equals $(d_1\mathbf{i}) \times (-m_1g\mathbf{j}) = -(m_1g)d_1\mathbf{k}$. Corresponding to the second particle, the moment equals $(-d_2\mathbf{i}) \times (-m_2g\mathbf{j}) = (m_2g)d_2\mathbf{k}$. The principle of moments gives the same condition for equilibrium as before.

principle of the excluded middle The statement that every proposition is either true or false.

prior distribution The *distribution* attached to a *parameter* before certain data are obtained. The **posterior distribution** is the distribution attached to it after the data are obtained. The posterior distribution may then be considered as the prior distribution before further data are obtained.

prior probability The probability attached to an event before certain data are obtained. The **posterior probability** is the probability attached to it after the data are obtained. The posterior probability may be calculated by using *Bayes' Theorem*. It may then be considered as the prior probability before further data are obtained.

prism A convex *polyhedron* with two 'end' faces that are congruent convex polygons lying in parallel planes in such a way that, with edges joining corresponding vertices, the remaining faces are parallelograms. A **right-regular prism** is one in which the two end faces are regular polygons and the

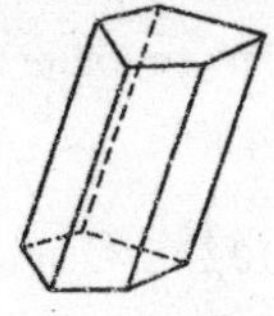

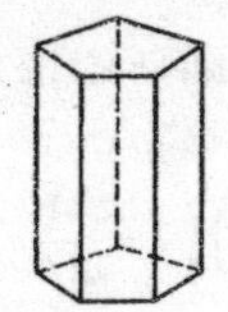

remaining faces are rectangular. A right-regular prism in which the rectangular faces are square is semi-regular (see *Archimedean solid*).

probability The **probability** of an *event* A, denoted by $\Pr(A)$, is a measure of the possibility of the event occurring as the result of an experiment. For any event A, $0 \leq \Pr(A) \leq 1$. If A never occurs, then $\Pr(A) = 0$; if A always occurs, then $\Pr(A) = 1$. If an experiment could be repeated n times and the event A occurs m times, then the limit of m/n as $n \to \infty$ is equal to $\Pr(A)$.

If the *sample space* S is finite and the possible outcomes are all equally likely, then the probability of the event A is equal to $n(A)/n(S)$, where $n(A)$ and $n(S)$ denote the number of elements in A and S. The probability that a randomly selected element from a finite population belongs to a certain category is equal to the proportion of the population belonging to that category.

The probability that a discrete *random variable* X takes the value x_i is denoted by $\Pr(X = x_i)$. The probability that a continuous random variable X takes a value less than or equal to x is denoted by $\Pr(X \leq x)$. This notation may be extended in a natural way.

See also *conditional probability*, *prior probability* and *posterior probability*.

probability density function For a continuous *random variable* X, the **probability density function** (or **p.d.f.**) of X is the function f such that

$$\Pr(a \leq X \leq b) = \int_a^b f(x)\,dx.$$

Suppose that, for the random variable X, a sample is taken and a corresponding *histogram* is drawn with class intervals of a certain width. Then, loosely speaking, as the number of observations increases and the width of the class intervals decreases, the histogram assumes more closely the shape of the graph of the probability density function.

probability mass function For a discrete *random variable* X, the **probability mass function** (or **p.m.f.**) of X is the function p such that $p(x_i) = \Pr(X = x_i)$, for all i.

product (of complex numbers) See *multiplication* (of complex numbers).

product (of matrices) See *multiplication* (of matrices).

product notation For a finite sequence $a_1, a_2, \ldots, a_n$, the product $a_1 a_2 \ldots a_n$ may be denoted, using the capital Greek letter pi, by

$$\prod_{r=1}^{n} a_r.$$

(The letter r used here could equally well be replaced by any other letter.) For example,

$$\prod_{r=1}^{9}\left(1 - \frac{1}{r+1}\right) = \left(1 - \frac{1}{2}\right)\left(1 - \frac{1}{3}\right)\ldots\left(1 - \frac{1}{10}\right).$$

Similarly, from an infinite sequence $a_1, a_2, a_3, \ldots$, an *infinite product* $a_1 a_2 a_3 \ldots$ can be formed, and it is denoted by

$$\prod_{r=1}^{\infty} a_r.$$

This notation is also used for the value of the infinite product, if it exists. For example:

$$\prod_{r=2}^{\infty}\left(1-\frac{1}{r^2}\right)=\frac{1}{2}.$$

product of inertia A quantity similar to a *moment of inertia*, but relating to a rigid body and a pair of given perpendicular axes. The products of inertia occur in the *inertia matrix* which arises in connection with the *angular momentum* and the *kinetic energy* of the rigid body. With Cartesian coordinates, let I_{yz}, I_{zx} and I_{xy} denote the products of inertia of the rigid body associated with the y- and z-axes, the z- and x-axes, and the x- and y-axes. Suppose that a rigid body consists of a system of n particles $P_1, P_2, \ldots, P_n$, where P_i has mass m_i and position vector $\mathbf{r}_i$, with $\mathbf{r}_i = x_i\mathbf{i} + y_i\mathbf{j} + z_i\mathbf{k}$. Then

$$I_{yz}=\sum_{i=1}^{n} m_i y_i z_i, \qquad I_{zx}=\sum_{i=1}^{n} m_i z_i x_i, \qquad I_{xy}=\sum_{i=1}^{n} m_i x_i y_i.$$

For solid bodies, the products of inertia are defined by corresponding integrals.

When the coordinate planes are planes of symmetry of the rigid body, the products of inertia are zero. Furthermore, it can be shown that, for any point fixed in a rigid body, such as the centre of mass, there is a set of three perpendicular axes, with origin at the point, such that the corresponding products of inertia are all zero. These are called the **principal axes**, and the corresponding moments of inertia are the **principal moments of inertia**. The use of the principal axes produces a considerable simplification in the expressions for the angular momentum and the kinetic energy of the rigid body.

product rule (for differentiation) See *differentiation.*

product set = *Cartesian product.*

progression A sequence in which each term is obtained from the previous one by some rule. The most common progressions are *arithmetic sequences*, *geometric sequences* and *harmonic sequences.*

projectile When a body is travelling near the Earth's surface, subject to no forces except the uniform gravitational force and possibly air resistance, it may be called a **projectile**. The standard *mathematical model* uses a particle to represent the body and a horizontal plane to represent the Earth's surface.

When there is no air resistance, the trajectory, the path traced out by the projectile, is a parabola whose vertex corresponds to the point at which the projectile attains its maximum height. Take the origin at the point of projection, the x-axis horizontal and the y-axis vertical with the positive direction upwards. Then the equation of motion is $m\ddot{\mathbf{r}} = -mg\mathbf{j}$, subject to $\mathbf{r} = \mathbf{0}$ and $\dot{\mathbf{r}} = (v\cos\theta)\mathbf{i} + (v\sin\theta)\mathbf{j}$ at $t = 0$, where v is the speed of projection and θ is the *angle of projection*. This gives $x = (v\cos\theta)t$ and $y = (v\sin\theta)t - \frac{1}{2}gt^2$, for $t \geq 0$. From these, it is found that $y = 0$ when $t = \dfrac{2v\sin\theta}{g}$ and then $x = \dfrac{v^2}{g}\sin 2\theta$. Hence the *range* on the horizontal plane through the point of projection is equal to $\dfrac{v^2}{g}\sin 2\theta$, and the time of flight is $\dfrac{2v\sin\theta}{g}$. The maximum height is attained at $t = \dfrac{v\sin\theta}{g}$, halfway through the flight. The maximum range, for any given value of v, is obtained when $\theta = \frac{1}{4}\pi$. Similar investigations can be carried out for a projectile projected from a point a given height above a horizontal plane, or projected up or down an inclined plane.

For a body projected vertically upwards with initial speed v, the equation $y = vt - \frac{1}{2}gt^2$ is obtained. The time of flight is $2v/g$, and the maximum height attained is $v^2/2g$.

A more sophisticated model, based on a sphere, is necessary if the motion of a long-range projectile such as an intercontinental missile is to be investigated. The rotation of the sphere needs to be considered if the effect of the *Coriolis force* is to be included.

projection (of a line on a plane) Given a line l and a plane p, the locus of all points N in the plane p such that N is the projection on p of some point on l is a straight line, the **projection** of l on p.

projection (of a point on a line) Given a line l and a point P not on l, the **projection** of P on l is the point N on l such that PN is perpendicular to l. The length $|PN|$ is the distance from P to l. The point N is called the **foot of the perpendicular** from P to l.

projection (of a point on a plane) Given a plane p and a point P not in p, the **projection** of P on p is the point N in p such that PN is perpendicular to p. The length $|PN|$ is the distance from P to p. The point N is called the **foot of the perpendicular** from P to p.

projection (of a vector on a vector) See *vector projection* (of a vector on a vector).

proof by contradiction A *direct proof* of a statement q is a logically correct argument establishing the truth of q. A **proof by contradiction** assumes that q is false and derives the truth of some statement r and of its negation $\neg r$. This contradiction shows that the initial assumption cannot hold, hence establishing the truth of q. A more complicated example is a proof that 'if p

then q'. A proof by contradiction assumes that p is true and that q is false, and derives the truth of some statement r and of its negation $\neg r$. This contradiction shows that the initial assumptions cannot both hold, and so a valid proof has been given that, if p is true, then q is true.

proper fraction See *fraction*.

properly included See *proper subset*.

proper subset Let A be a subset of B. Then A is a **proper subset** of B if A is not equal to B itself. Thus there is some element of B not in A. The subset A is then said to be **properly** (or **strictly**) **included** in B, and this is written $A \subset B$. Some authors use $A \subset B$ to mean $A \subseteq B$ (see *subset*), but they then have no easy means of indicating proper inclusion.

proportion If two quantities x and y are related by an equation $y = kx$, where k is a constant, then y is said to be **(directly) proportional** to x, which may be written $y \propto x$. The constant k is the **constant of proportionality**. It is also said that y **varies (directly)** as x. When y is plotted against x, the graph is a straight line through the origin.

If $y = k/x$, then y is **inversely proportional** to x. This is written $y \propto 1/x$, and it is said that y **varies inversely** as x.

proposition A mathematical statement for which a proof is either required or provided.

proposition (in mathematical logic) = *statement*.

pseudo-prime The positive integer n is a **pseudo-prime** if $a^n \equiv a \pmod{n}$, for all integers a. According to *Fermat's Little Theorem*, all primes are pseudo-primes. There are comparatively few pseudo-primes that are not primes; the first is 561. To determine whether an integer is prime or composite, it may be useful to test first whether or not it is a pseudo-prime. For most composite numbers, this will establish that they are composite.

pseudo-random numbers See *random numbers*.

Ptolemy (Claudius Ptolemaeus) (2nd century AD) Greek astronomer and mathematician, responsible for the most significant work of trigonometry of ancient times. Usually known by its Arabic name *Almagest* ('The Greatest'), it contains, amongst other things, tables of chords, equivalent to a modern table of sines, and an account of how they were obtained. Use is made of *Ptolemy's Theorem*, from which the familiar addition formulae of trigonometry can be shown to follow.

Ptolemy's Theorem The following theorem of Euclidean geometry:

THEOREM: Suppose that a convex quadrilateral has vertices A, B, C and D, in that order. Then the quadrilateral is cyclic if and only if

$$AB \,.\, CD + AD \,.\, BC = AC \,.\, BD.$$

pulley A grooved wheel around which a rope can pass. When supported on an axle in some way, the device can be used to change the direction of a force: for example, pulling down on a rope may enable a load to be lifted. If the contact between the pulley and the axle is smooth, the magnitude of the force is not changed. A system of pulleys can be constructed that enables a large load to be raised a small distance by a small effort moving through a large distance. In such a *machine*, the velocity ratio is equal to the number of ropes supporting the load or, equivalently, the number of pulleys in the system.

pure imaginary A *complex number* is **pure imaginary** if its real part is zero.

pure strategy In a *matrix game*, if a player always chooses one particular row (or column) this is a **pure strategy**. Compare this with *mixed strategy*.

pyramid A convex *polyhedron* with one face (the **base**) a convex polygon and all the vertices of the base joined by edges to one other vertex (the **apex**); thus the remaining faces are all triangular. A **right-regular pyramid** is one in which the base is a regular polygon and the remaining faces are isosceles triangles.

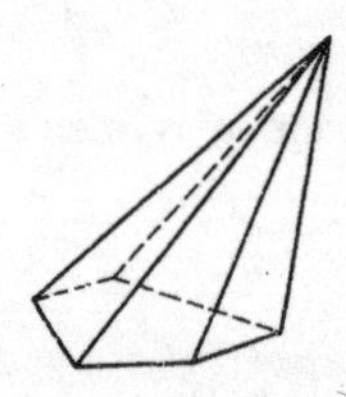

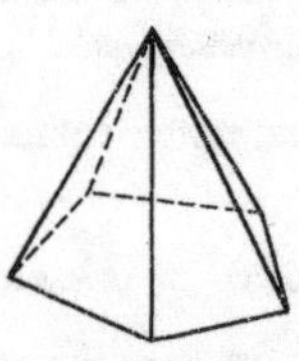

Pythagoras (died about 500 BC) Greek philosopher and mystic who, with his followers, seems to have been the first to take mathematics seriously as a study in its own right as opposed to being a collection of formulae for practical calculation. The Pythagoreans are credited with the discovery of the well-known *Pythagoras' Theorem* on right-angled triangles. They were also much concerned with *figurate numbers*, for semi-philosophical reasons. It is said that they regarded whole numbers as the fundamental constituents of reality, a view that was shattered by the discovery of *irrational numbers*.

Pythagoras' Theorem Probably the best-known theorem in geometry, which gives the relationship between the lengths of the sides of a right-angled triangle:

THEOREM: In a right-angled triangle, the square on the hypotenuse is equal to the sum of the squares on the other two sides.

Thus, if the hypotenuse, the side opposite the right angle, has length c and the other two sides have lengths a and b, then $a^2 + b^2 = c^2$. One elegant proof is obtained by dividing up a square of side $a + b$ in two different ways as shown in the figure, and equating areas.

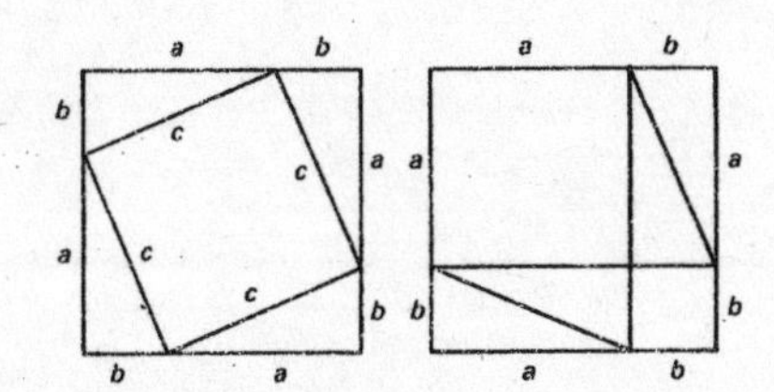

Pythagorean triple A set of three positive integers a, b and c such that $a^2 + b^2 = c^2$ (see *Pythagoras' Theorem*). If $\{a, b, c\}$ is a Pythagorean triple, then so is $\{ka, kb, kc\}$ for any positive integer k. Pythagorean triples that have *greatest common divisor* equal to 1 include the following: $\{3, 4, 5\}$, $\{5, 12, 13\}$, $\{8, 15, 17\}$, $\{7, 24, 25\}$ and $\{20, 21, 29\}$.

Q

Q See *rational number.*

quadrangle See *complete quadrangle.*

quadrant In a Cartesian coordinate system in the plane, the axes divide the rest of the plane into four regions called **quadrants**. By convention, they are usually numbered as follows: the first quadrant is $\{(x, y) \mid x > 0, y > 0\}$, the second is $\{(x, y) \mid x < 0, y > 0\}$, the third is $\{(x, y) \mid x < 0, y < 0\}$, the fourth is $\{(x, y) \mid x > 0, y < 0\}$.

quadratic equation A **quadratic equation** in the unknown x is an equation of the form $ax^2 + bx + c = 0$, where a, b and c are given real numbers, with $a \neq 0$. This may be solved by *completing the square* or by using the formula

$$x = \frac{-b \pm \sqrt{b^2 - 4ac}}{2a},$$

which is established by completing the square. If $b^2 > 4ac$, there are two distinct real roots; if $b^2 = 4ac$, there is a single real root (which it may be convenient to treat as two equal or coincident roots); and, if $b^2 < 4ac$, the equation has no real roots, but there are two complex roots:

$$x = -\frac{b}{2a} \pm i\frac{\sqrt{4ac - b^2}}{2a}.$$

If α and β are the roots of the quadratic equation $ax^2 + bx + c = 0$, then $\alpha + \beta = -b/a$ and $\alpha\beta = c/a$. Thus a quadratic equation with given numbers α and β as its roots is $x^2 - (\alpha + \beta)x + \alpha\beta = 0$.

quadratic function In real analysis, a **quadratic function** is a *real function* f such that $f(x) = ax^2 + bx + c$ for all x in $\mathbf{R}$, where a, b and c are real numbers, with $a \neq 0$. (In some situations, $a = 0$ may be permitted.) The graph $y = f(x)$ of such a function is a *parabola* with its axis parallel to the y-axis, and with its vertex downwards if $a > 0$ and upwards if $a < 0$. The graph cuts the x-axis where $ax^2 + bx + c = 0$, so the points (if any) are given by the roots (if real) of this *quadratic equation*. The position of the vertex can be determined by completing the square or by finding the *stationary point* of the function using differentiation. If the graph cuts the x-axis in two points, the x-coordinate of the vertex is midway between these two points. In this way the graph of the quadratic function can be sketched, and information can be deduced.

quadratic polynomial A polynomial of degree two.

quadrature A method of **quadrature** is a numerical method that finds an approximation to the area of a region with a curved boundary; the area of the region may then be found by some kind of limiting process.

quadric or **quadric surface** A locus in 3-dimensional space that can be represented in a Cartesian coordinate system by a polynomial equation in x, y and z of the second degree; that is, an equation of the form

$$ax^2 + by^2 + cz^2 + 2fyz + 2gzx + 2hxy + 2ux + 2vy + 2wz + d = 0,$$

where the constants a, b, c, f, g and h are not all zero. When the equation represents a non-empty locus, it can be reduced by translation and rotation of axes to one of the following **canonical forms**, and hence classified:

(i) *Ellipsoid*: $\dfrac{x^2}{a^2} + \dfrac{y^2}{b^2} + \dfrac{z^2}{c^2} = 1.$

(ii) *Hyperboloid of one sheet*: $\dfrac{x^2}{a^2} + \dfrac{y^2}{b^2} - \dfrac{z^2}{c^2} = 1.$

(iii) *Hyperboloid of two sheets*: $\dfrac{x^2}{a^2} + \dfrac{y^2}{b^2} - \dfrac{z^2}{c^2} = -1.$

(iv) *Elliptic paraboloid*: $\dfrac{x^2}{a^2} + \dfrac{y^2}{b^2} = \dfrac{2z}{c}.$

(v) *Hyperbolic paraboloid*: $\dfrac{x^2}{a^2} - \dfrac{y^2}{b^2} = \dfrac{2z}{c}.$

(vi) *Quadric cone*: $\dfrac{x^2}{a^2} + \dfrac{y^2}{b^2} = \dfrac{z^2}{c^2}.$

(vii) *Elliptic cylinder*: $\dfrac{x^2}{a^2} + \dfrac{y^2}{b^2} = 1.$

(viii) *Hyperbolic cylinder*: $\dfrac{x^2}{a^2} - \dfrac{y^2}{b^2} = 1.$

(ix) *Parabolic cylinder*: $\dfrac{x^2}{a^2} = \dfrac{2y}{b}.$

(x) Pair of non-parallel planes: $\dfrac{x^2}{a^2} = \dfrac{y^2}{b^2}$ $\left(\text{that is, } y = \pm\dfrac{b}{a}x\right).$

(xi) Pair of parallel planes: $\dfrac{x^2}{a^2} = 1$ (that is, $x = \pm a$).

(xii) Plane: $\dfrac{x^2}{a^2} = 0$ (that is, $x = 0$).

(xiii) Line: $\dfrac{x^2}{a^2} + \dfrac{y^2}{b^2} = 0$ (that is, $x = y = 0$).

(xiv) Point: $\frac{x^2}{a^2}+\frac{y^2}{b^2}+\frac{z^2}{c^2}=0$ (that is, $x=y=z=0$).

Forms (i), (ii), (iii), (iv) and (v) are the non-*degenerate quadrics*.

quadric cone A *quadric* whose equation in a suitable coordinate system is

$$\frac{x^2}{a^2}+\frac{y^2}{b^2}=\frac{z^2}{c^2}.$$

Sections by planes parallel to the *xy*-plane are ellipses (circles if $a=b$), and sections by planes parallel to the other axial planes are hyperbolas.

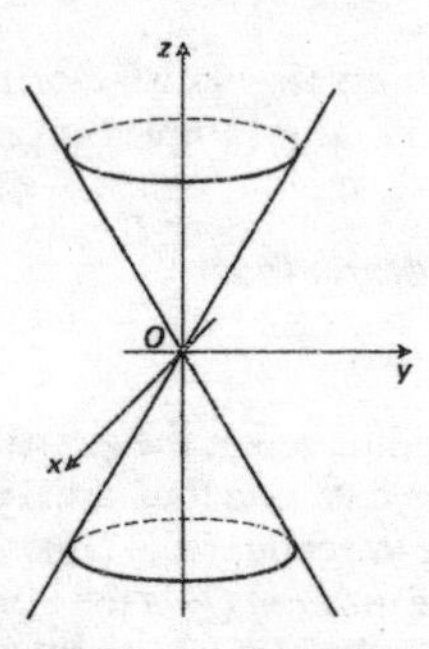

quadrilateral A polygon with four sides. See also *complete quadrilateral.*

quadrillion In Britain, a million to the fourth power (10^{24}); in the United States, the number 10^{15}. Use of the word is rare and, as with *billion*, can lead to ambiguity.

quadruple Four objects normally taken in a particular order, denoted, for example, by (x_1, x_2, x_3, x_4).

quantifier The two expressions 'for all ...' and 'there exists ...' are called **quantifiers.** A phrase such as 'for all *x*' or 'there exists *x*' may stand in front of a sentence involving a symbol *x* and thereby create a statement that makes sense and is either true or false. There are different ways in English of expressing the same sense as 'for all *x*', but it is sometimes useful to standardize the language to this particular form. This is known as a **universal quantifier** and is written in symbols as '$\forall x$'. Similarly, 'there exists *x*' may be used as the standard form to replace any phrase with this meaning, and is an **existential quantifier**, written in symbols as '$\exists x$'.

For example, the statements, 'if *x* is any number greater than 3 then *x* is positive' and 'there is a real number satisfying $x^2=2$', can be written in more standard form: 'for all *x*, if *x* is greater than 3 then *x* is positive', and 'there exists *x* such that *x* is real and $x^2=2$'. These can be written, using the symbols of mathematical logic, as: $(\forall x)(x>3 \Rightarrow x>0)$, and $(\exists x)\,(x \in \mathbf{R} \wedge x^2=2)$.

quantile Let X be a continuous *random variable*. For $0 < p < 1$, the p-th **quantile** is the value x_p such that $\Pr(X \leq x_p) = p$. In other words, the fraction of the population less than or equal to x_p is p. For example, $x_{0.5}$ is the (population) median.

Often percentages are used. The n-th **percentile** is the value $x_{n/100}$ such that n per cent of the population is less than or equal to $x_{n/100}$. For example, 30% of the population is less than or equal to the 30th percentile. The 25th, 50th and 75th percentiles are the *quartiles*.

Alternatively, the population may be divided into tenths. The n-th **decile** is the value $x_{n/10}$ such that n tenths of the population is less than or equal to $x_{n/10}$. For example, three-tenths of the population is less than or equal to the 3rd decile.

The terms can be modified, though not always very satisfactorily, to be applicable to a discrete random variable or to a large sample ranked in ascending order.

quartic equation A polynomial equation of degree four.

quartic polynomial A polynomial of degree four.

quartiles For numerical data ranked in ascending order, the **quartiles** are values derived from the data which divide the data into four equal parts. If there are n observations, the first quartile (or lower quartile) Q_1 is the $\frac{1}{4}(n+1)$-th, the second quartile (which is the *median*) Q_2 is the $\frac{1}{2}(n+1)$-th and the third quartile (or upper quartile) Q_3 is the $\frac{3}{4}(n+1)$-th in ascending order. When $\frac{1}{4}(n+1)$ is not a whole number, it is sometimes thought necessary to take the (weighted) average of two observations, as is done for the median. However, unless n is very small, an observation that is nearest will normally suffice. For example, for the sample 15, 37, 43, 47, 54, 55, 57, 64, 76, 98, we may take $Q_1 = 43$, $Q_2 = 54.5$ and $Q_3 = 64$.

For a random variable, the quartiles are the *quantiles* $x_{0.25}$, $x_{0.5}$ and $x_{0.75}$; that is, the 25th, 50th and 75th percentiles.

quartile deviation = *semi-interquartile range*.

quaternion The complex number system can be obtained by taking a complex number to be an expression $a + bi$, where a and b are real numbers, and defining addition and multiplication in the natural way with the understanding that $i^2 = -1$. In an extension of this idea, Hamilton introduced the following notion, originally for use in mechanics. Define a **quaternion** to be an expression $a + bi + cj + dk$, where a, b, c and d are real numbers, and define addition and multiplication in the natural way, with

$$i^2 = j^2 = k^2 = -1, \quad ij = -ji = k, \quad jk = -kj = 1, \quad ki = -ik = j.$$

All the normal laws of algebra hold except that multiplication is not *commutative*. That is to say, the quaternions form a *ring* which is not commutative but in which every non-zero element has an *inverse*.

quintic equation A polynomial equation of degree five.

quintic polynomial A polynomial of degree five.

quota sample See *sample.*

quotient See *Division Algorithm.*

quotient rule (for differentiation) See ***differentiation.***

R See *real number*.

radial and **transverse components** When a point P has polar coordinates (r, θ), the vectors $\mathbf{e}_r$ and $\mathbf{e}_\theta$ are defined by $\mathbf{e}_r = \mathbf{i}\cos\theta + \mathbf{j}\sin\theta$ and $\mathbf{e}_\theta = -\mathbf{i}\sin\theta + \mathbf{j}\cos\theta$, where **i** and **j** are unit vectors in the directions of the positive x- and y-axes. Then $\mathbf{e}_r$ is a unit vector along OP in the direction of increasing r, and $\mathbf{e}_\theta$ is a unit vector perpendicular to this in the direction of increasing θ. Any vector **v** can be written in terms of its components in the directions of $\mathbf{e}_r$ and $\mathbf{e}_\theta$. Thus $\mathbf{v} = v_1\mathbf{e}_r + v_2\mathbf{e}_\theta$, where $v_1 = \mathbf{v} \,.\, \mathbf{e}_r$ and $v_2 = \mathbf{v} \,.\, \mathbf{e}_\theta$. The component v_1 is the **radial component**, and the component v_2 is the **transverse component**.

radian In elementary work, angles are measured in degrees, where one revolution measures 360°. In more advanced work, it is essential that angles are measured differently. Suppose that a circle centre O meets two lines through O at A and B. Take the length of arc AB and divide it by the length of OA. This value is independent of the radius of the circle, and depends only upon the size of $\angle AOB$. So the value is called the size of $\angle AOB$, measured in **radians**.

The angle measures 1 radian when the length of the arc AB equals the length of OA. This happens when $\angle AOB$ is about 57°. More accurately, 1 radian $= 57.296° = 57°17'45''$, approximately. Since the length of the circumference of a circle of radius r is $2\pi r$, one revolution measures 2π radians. Consequently, $x° = \pi x/180$ radians. In much theoretical work, particularly involving calculus, radian measure is essential. When *trigonometric functions* are evaluated with a calculator, it is essential to be sure that the correct measure is being used.

The radian is the SI unit for measuring angle, and is abbreviated to 'rad'.

radical axis The **radical axis** of two circles is the straight line containing all points P such that the lengths of the tangents from P to the two circles are equal. Each figure below shows a point P on the radical axis and tangents touching the two circles at T_1 and T_2, with $PT_1 = PT_2$. If the circles intersect in two points, as in the figure on the right, the radical axis is the straight line through the two points of intersection. In this case, there are

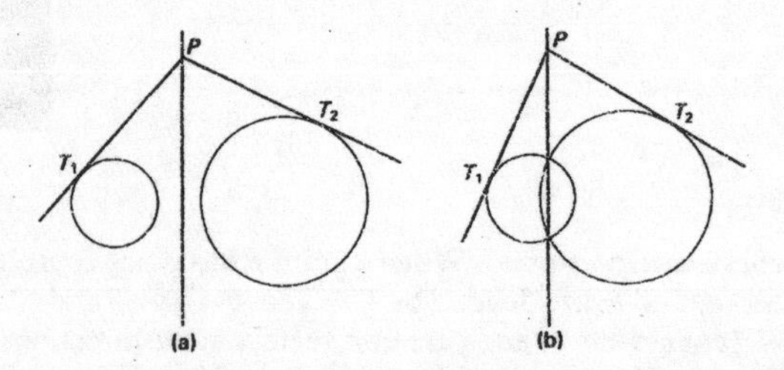

some points on the radical axis inside the circles, from which tangents to the two circles cannot be drawn. For the circles with equations $x^2+y^2+2g_1x+2f_1y+c_1=0$ and $x^2+y^2+2g_2x+2f_2y+c_2=0$, the radical axis has equation $2(g_1-g_2)x+2(f_1-f_2)y+(c_1-c_2)=0$.

radical sign The sign $\sqrt{}$ used in connection with square roots, cube roots and *n*-th roots, for larger values of *n*. The notation $\sqrt{a}$ indicates square root, $\sqrt[3]{a}$ indicates cube root and $\sqrt[n]{a}$ indicates *n*-th root of *a*. See *square root* and *n-th root* for detailed explanation of the correct usage of the notation.

radius (plural: radii) A **radius** of a circle is a line segment joining the centre of the circle to a point on the circle. All such line segments have the same length, and this length is also called the radius of the circle. The term also applies in both senses to a sphere. See also *circle* and *sphere.*

radius of gyration The square root of the ratio of a *moment of inertia* to the mass of a rigid body. Thus if I is the moment of inertia of a rigid body of mass m about a specified axis, and k is the radius of gyration, then $I=mk^2$. It means that, when rotating about this axis, the body has the same moment of inertia as a ring of the same mass and of radius k.

radius vector Suppose that a point O is taken as origin in the plane. If a point P in the plane has *position vector* $\mathbf{p}$, then $\mathbf{p}$ may also be called the **radius vector**, particularly when P is a typical point on a certain curve or when P is thought of as a point or particle moving in the plane.

Ramanujan, Srinivasa (1887–1920) The outstanding Indian mathematician of modern times. Originally a clerk in Madras, he studied and worked on mathematics totally unaided. Following correspondence with G. H. Hardy, he accepted an invitation to visit Britain in 1914. He studied and collaborated with Hardy on the subject of *partitions* and other topics, mainly in number theory. He was considered a genius for his inexplicable ability in, for example, the handling of series and *continued fractions*. Because of ill-health he returned to India a year before he died.

random numbers Tables of **random numbers** give lists of the digits $0,1,2,\ldots,9$ in which each digit is equally likely to occur at any stage. There is no way of predicting the next digit. Such tables can be used to select items at random from a *population*. Numbers generated by a deterministic algorithm that appear to pass statistical tests for randomness are called **pseudo-random numbers**. Such an algorithm may be called a **random number generator**.

random sample See *sample*.

random variable A quantity that takes different numerical values according to the result of a particular experiment. A random variable is **discrete** if the set of possible values is finite or countably infinite. For a discrete random variable, the probability of its taking any particular value is given by the *probability mass function*. A random variable is **continuous** if the set of possible values forms an interval, finite or infinite. For a continuous random variable, the probability of its taking a value in a particular sub-interval may be calculated from the *probability density function*.

random walk Consider a *Markov chain* $X_1, X_2, X_3, \ldots$ in which the state space is $\{\ldots, -2, -1, 0, 1, 2, 3, \ldots\}$. With the integers positioned as they occur on the real line, imagine an object moving from integer to integer a step at a time. That is to say, from position i the object either moves to $i - 1$ or $i + 1$, or possibly stays where it is. Such a Markov chain is called a one-dimensional **random walk**: if $X_n = i$, then $X_{n+1} = i - 1, i$ or $i + 1$. The state i is an **absorbing state** if, whenever $X_n = i$, then $X_{n+1} = i$; in other words, when the object reaches this position it stays there. When a gambler, playing a sequence of games, either wins or loses a fixed amount in each game, his winnings give an example of a random walk.

Random walks in two or more dimensions can be defined similarly.

range (of a function or mapping) See *function* and *mapping*.

range (in mechanics) The **range** of a *projectile* on a horizontal or inclined plane passing through the point of projection is the distance from the point of projection to the point at which the projectile lands on the plane.

range (in statistics) The difference between the maximum and minimum observations in a set of numerical data. It is a possible measure of *dispersion* of a sample.

rank (of a matrix) Let **A** be an $m \times n$ matrix. The **column rank** of **A** is the largest number of elements in a *linearly independent* set of columns of **A**. The **row rank** of **A** is the largest number of elements in a linearly independent set of rows of **A**. It can be shown that elementary row operations on a matrix do not change the column rank or the row rank. Consequently, the column rank and row rank of **A** are equal, both being equal to the number of non-zero rows in the matrix in *reduced echelon form* to which **A** can be transformed. This common value is the **rank** of **A**. It can also be shown that the rank of **A** is equal to the number of rows and columns in the largest square submatrix of **A** that has non-zero determinant. An $n \times n$ matrix is *invertible* if and only if it has rank n.

rank (in statistics) The observations in a sample are said to be **ranked** when they are arranged in order according to some criterion. For example, numbers can be ranked in ascending or descending order, people can be ranked according to height or age, and products can be ranked according to their sales. The **rank** of an observation is its position in the list when the sample has been ranked. *Non-parametric methods* frequently make use of ranking rather than the exact values of the observations in the sample.

rate of change Suppose that the quantity y is a function of the quantity x, so that $y = f(x)$. If f is *differentiable*, the **rate of change** of y with respect to x is the derivative dy/dx or $f'(x)$. The rate of change is often with respect to time t. Suppose, now, that x denotes the *displacement* of a particle, at time t, on a directed line with origin O. Then the velocity is dx/dt or $\dot{x}$, the rate of change of x with respect to t, and the acceleration is d^2x/dt^2 or $\ddot{x}$, the rate of change of the velocity with respect to t.

In the reference to velocity and acceleration in the preceding paragraph, a common convention has been followed, in which the unit vector $\mathbf{i}$ in the positive direction along the line has been suppressed. Velocity and acceleration are in fact vector quantities, and in the one-dimensional case above are equal to $\dot{x}\mathbf{i}$ and $\ddot{x}\mathbf{i}$. When the motion is in 2 or 3 dimensions, vectors are used explicitly. If $\mathbf{r}$ is the *position vector* of a particle, the velocity of the particle is the vector $\dot{\mathbf{r}}$ and the acceleration is $\ddot{\mathbf{r}}$.

rational function In real analysis, a **rational function** is a *real function* f such that, for x in the domain, $f(x) = g(x)/h(x)$, where $g(x)$ and $h(x)$ are polynomials, which may be assumed to have no common factor of degree greater than or equal to 1. The domain is usually taken to be the whole of $\mathbf{R}$, with any zeros of the denominator $h(x)$ omitted.

rationalize To remove *radicals* from an expression or part of it without changing the value of the whole expression. For example, in the expression $1/(2-\sqrt{x})$ the denominator can be rationalized by multiplying the numerator and the denominator by $2+\sqrt{x}$ to give

$$\frac{2+\sqrt{x}}{4-x}.$$

rational number A number that can be written in the form a/b, where a and b are integers, with $b \neq 0$. The set of all rational numbers is usually denoted by $\mathbf{Q}$. A real number is rational if and only if, when expressed as a decimal, it has a finite or recurring expansion (see *decimal representation*). For example,

$$\frac{5}{4} = 1.25, \qquad \frac{2}{3} = 0.\dot{6}, \qquad \frac{20}{7} = 2.\dot{8}5714\dot{2}.$$

A famous proof, attributed to Pythagoras, shows that $\sqrt{2}$ is not rational, and e and π are also known to be irrational.

The same rational number can be expressed as a/b in different ways; for example, $\frac{2}{3} = \frac{6}{9} = \frac{-4}{-6}$. In fact, $a/b = c/d$ if and only if $ad = bc$. But a rational number can be expressed uniquely as a/b if it is insisted that a and b have *greatest common divisor* 1 and that $b > 0$. Accepting the different forms for the same rational number, the explicit rules for addition and multiplication are that

$$\frac{a}{b} + \frac{c}{d} = \frac{ad + bc}{bd} \quad \text{and} \quad \frac{a}{b} \cdot \frac{c}{d} = \frac{ac}{bd}.$$

The set of rational numbers is closed under addition, subtraction, multiplication and division (not allowing division by zero). Indeed, all the axioms for a *field* can be seen to hold.

A more rigorous approach sets up the field **Q** of rational numbers as follows. Consider the set of all *ordered pairs* (a, b), where a and b are integers, with $b \neq 0$. Introduce an *equivalence relation* $\sim$ on this set, by defining $(a, b) \sim (c, d)$ if $ad = bc$, and let $[(a, b)]$ be the *equivalence class* containing (a, b). The intuitive approach above suggests that addition and multiplication between equivalence classes should be defined by

$$[(a, b)] + [(c, d)] = [(ad + bc, bd)] \quad \text{and} \quad [(a, b)][(c, d)] = [(ac, bd)],$$

where it is necessary, in each case, to verify that the class on the right-hand side is independent of the choice of elements (a, b) and (c, d) taken as *representatives* of the equivalence classes on the left-hand side. It can then be shown that the set of equivalence classes, with this addition and multiplication, form a field **Q**, whose elements, according to this approach, are called rational numbers.

ray = *half-line.*

real axis In the *complex plane*, the *x*-axis is called the **real axis**; its points represent the real numbers.

real function A function from the set **R** of real numbers (or a subset of **R**) to **R**. Thus, if f is a real function, then, for every real number x in the domain, a corresponding real number $f(x)$ is defined. In analysis, a function $f: S \to \mathbf{R}$ is often defined by giving a formula for $f(x)$, without specifying the domain S (see *function*). In that case, it is usual to assume that the domain is the largest possible subset S of **R**. For example, if $f(x) = 1/(x-2)$, the domain would be taken to be $\mathbf{R}\backslash\{2\}$; that is, the set of all real numbers not equal to 2. If $f(x) = \sqrt{9 - x^2}$, the domain would be the closed interval $[-3, 3]$.

real line On a horizontal straight line, choose a point O as origin, and a point A, to the right of O, such that $|OA|$ is equal to 1 unit. Each positive real number x can be represented by a point on the line to the right of O, whose distance from O equals x units, and each negative number by a point on the line to the left of O. The origin represents zero. The line is called the **real line** when its points are taken in this way to represent the real numbers.

real number The numbers generally used in mathematics, in scientific work and in everyday life are the **real numbers.** They can be pictured as points of a line, with the integers equally spaced along the line and a real number b to the right of the real number a if $a < b$. The set of real numbers is usually denoted by **R**. It contains such numbers as $0, \frac{1}{2}, -2, 4.75, \sqrt{2}$ and π. Indeed, **R** contains all the rational numbers but also numbers such as $\sqrt{2}$ and π that are irrational. Every real number has an expression as an infinite *decimal fraction.*

The set of real numbers, with the familiar addition and multiplication, form a *field* and, since there is a notion of 'less than' that satisfies certain basic axioms, **R** is called an 'ordered' field. However, a statement of a complete set of axioms that characterize **R** will not be attempted here. There are too a number of rigorous approaches that, assuming the existence of the field **Q** of *rational numbers,* construct a system of real numbers with the required properties.

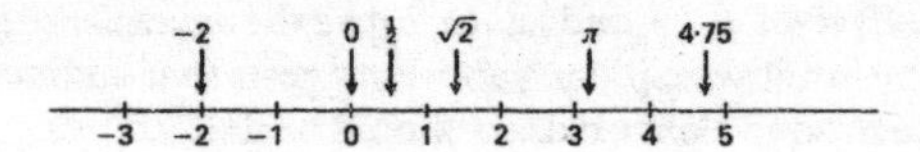

real part A *complex number* z may be written $x + yi$, where x and y are real, and then x is the **real part** of z. It is denoted by Re z or $\Re z$.

real world See *mathematical model.*

reciprocal The *multiplicative inverse* of a quantity may, when the operation of multiplication is *commutative,* be called its **reciprocal.** Thus the reciprocal of 2 is $\frac{1}{2}$, the reciprocal of $3x + 4$ is $1/(3x + 4)$ and the reciprocal of $\sin x$ is $1/\sin x$.

reciprocal rule (for differentiation) See *differentiation.*

rectangular coordinate system See *coordinates* (in the plane).

rectangular distribution = *uniform distribution.*

rectangular hyperbola A *hyperbola* whose asymptotes are perpendicular. With the origin at the centre and the coordinate x-axis along the transverse

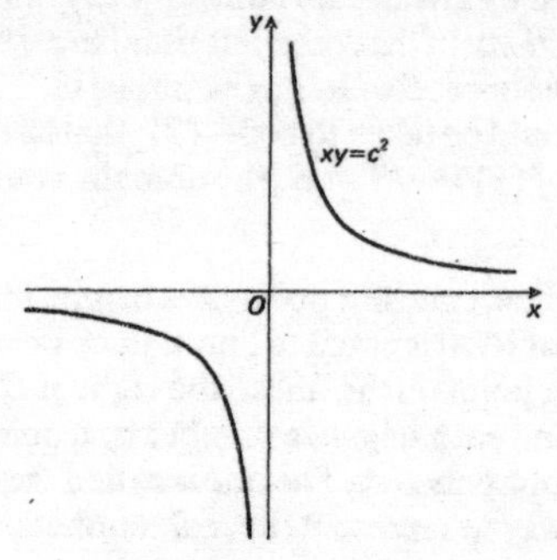

axis, it has equation $x^2 - y^2 = a^2$. Instead, the coordinate axes can be taken along the asymptotes in such a way that the two branches of the hyperbola are in the first and third quadrants. This coordinate system can be obtained from the other by a rotation of axes. The rectangular hyperbola then has equation of the form $xy = c^2$. For example, $y = 1/x$ is a rectangular hyperbola. For $xy = c^2$, it is customary to take $c > 0$ and to use, as *parametric equations*, $x = ct, y = c/t\ (t \neq 0)$.

recurrence relation = *difference equation.*

recurring decimal See *decimal representation.*

reduced echelon form Suppose that a row of a matrix is called zero if all its entries are zero. Then a matrix is in **reduced echelon form** if

(i) all the zero rows come below the non-zero rows,
(ii) the first non-zero entry in each non-zero row is 1 and occurs in a column to the right of the leading 1 in the row above,
(iii) the leading 1 in each non-zero row is the only non-zero entry in the column that it is in.

(If (i) and (ii) hold, the matrix is in *echelon form.*) For example, these two matrices are in reduced echelon form:

$$\begin{bmatrix} 1 & 6 & 0 & 0 & 2 \\ 0 & 0 & 1 & 0 & -3 \\ 0 & 0 & 0 & 1 & 5 \end{bmatrix}, \qquad \begin{bmatrix} 1 & 0 & -1 & 4 & 2 \\ 0 & 1 & 2 & -3 & 5 \\ 0 & 0 & 0 & 0 & 0 \end{bmatrix}.$$

Any matrix can be transformed to a matrix in reduced echelon form using elementary row operations, by a method known as *Gauss–Jordan elimination.* For any matrix, the reduced echelon form to which it can be transformed is unique. The solutions of a set of linear equations can be immediately obtained from the reduced echelon form to which the augmented matrix has been transformed. A set of linear equations is said to be in reduced echelon form if its augmented matrix is in reduced echelon form.

reduced set of residues For a positive integer n, the number of positive integers, less than n, *relatively prime* to n, is denoted by $\phi(n)$ (see *Euler's function*). A **reduced set of residues** modulo n is a set of $\phi(n)$ integers, one *congruent* (modulo n) to each of the positive integers less than n, relatively prime to n. Thus $\{1, 5, 7, 11\}$ is a reduced set of residues modulo 12, and so is $\{1, -1, 5, -5\}$.

reductio ad absurdum The Latin phrase meaning 'reduction to absurdity' used to describe the method of *proof by contradiction.*

reduction formula Let I_n be some quantity that is dependent upon the integer $n\ (\geq 0)$. It may be possible to establish some general formula, expressing I_n in terms of some of the quantities $I_{n-1}, I_{n-2}, \ldots$. Such a formula is a **reduction formula** and can be used to evaluate I_n for a particular value of n. The method is useful in integration. For example, if

$$I_n = \int_0^{\pi/2} \sin^n x\,dx,$$

it can be shown, by integration by parts, that $I_n = ((n-1)/n)I_{n-2}$ $(n \geq 2)$. It is easy to see that $I_0 = \pi/2$, and then the reduction formula can be used, for example, to find that

$$I_6 = \frac{5}{6}I_4 = \frac{5}{6} \times \frac{3}{4}I_2 = \frac{5}{6} \times \frac{3}{4} \times \frac{1}{2}I_0 = \frac{5}{6} \times \frac{3}{4} \times \frac{1}{2} \times \frac{\pi}{2} = \frac{5\pi}{32}.$$

re-entrant An interior angle of a polygon is **re-entrant** if it is greater than two right angles.

reflection (of the plane) Let l be a line in the plane. Then the **mirror-image** of a point P is the point P' such that PP' is perpendicular to l and l cuts PP' at its midpoint. The **reflection** of the plane in the line l is the transformation of the plane that maps each point P to its mirror-image P'. Suppose that the line l passes through the origin O and makes an angle α with the x-axis. If P has polar coordinates (r, θ), its mirror-image P' has polar coordinates $(r, 2\alpha - \theta)$. In terms of Cartesian coordinates, reflection in the line l maps P with coordinates (x, y) to P' with coordinates (x', y'), where

$$x' = x\cos 2\alpha + y \sin 2\alpha,$$
$$y' = x\sin 2\alpha - y\cos 2\alpha.$$

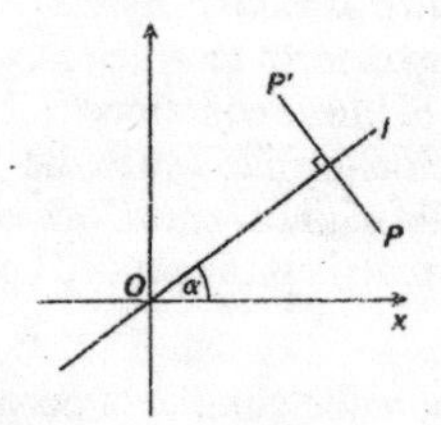

reflex angle An angle that is greater than 2 *right angles* and less than 4 right angles.

reflexive relation A *binary relation* ~ on a set S is **reflexive** if $a \sim a$ for all a in S.

Regiomontanus (1436–1476) A central figure in mathematics in the fifteenth century. Born Johann Müller, he took as his name the Latin form of Königsberg, his birthplace. His *De triangulis omnimodis* ('On All Classes of Triangles') was the first modern account of trigonometry and, even though it did not appear in print until 1533, it was influential in the revival of the subject in the West.

regression A statistical procedure to determine the relationship between a *dependent variable* and one or more *explanatory variables.* The purpose is normally to enable the value of the dependent variable to be predicted from given values of the explanatory variables. It is **multiple regression** if there are two or more explanatory variables. Usually the model supposes that $E(Y)$, where Y is the dependent variable, is given by some formula involving certain unknown parameters. In **simple linear regression,** $E(Y) = b_0 + b_1 X$. In multiple linear regression, with k explanatory variables $X_1, X_2, \ldots, X_k$, $E(Y) = b_0 + b_1 X_1 + b_2 X_2 + \cdots + b_k X_k$. Here $b_0, b_1, \ldots, b_k$ are the **regression coefficients.**

regular graph A *graph* in which all the vertices have the same *degree.* It is r-**regular** or **regular of degree** r if every vertex has degree r.

regular polygon See *polygon.*

regular polyhedron See *Platonic solid.*

regular tessellation See *tessellation.*

relation A **relation** on a set S is usually a *binary relation* on S, though the notion can be extended to involve more than two elements. An example of a **ternary relation,** involving three elements, is 'a lies between b and c', where a, b and c are real numbers.

relative complement If the set A is included in the set B, the *difference set* $B \setminus A$ is the **(relative) complement** of A in B, or the **complement of** A **relative to** B.

relative efficiency See *estimator.*

relative error Let x be an approximation to a value X and let $X = x + e$. The **relative error** is $|e/X|$. When 1.9 is used as an approximation for 1.875, the relative error equals $0.025/1.875 = 0.013$, to 3 decimal places. (This may be expressed as a *percentage error* of 1.3%.) Notice, in this example, that $0.025/1.9 = 0.013$, to 3 decimal places, too. In general, when e is small, it does not make much difference if the relative error is taken as $|e/x|$, instead of $|e/X|$; this has to be done if the exact value is not known but only the approximation. The relative error may be a more helpful figure than the absolute *error.* An absolute error of 0.025 in a value of 1.9, as above, may be acceptable. But the same absolute error in a value of 0.2, say, would give a relative error of $0.025/0.2 = 0.125$ (a percentage error of $12\frac{1}{2}$%), which would probably be considered quite serious.

relatively prime Integers a and b are **relatively prime** if their *greatest common divisor* (g.c.d.) is 1. Similarly, any number of integers $a_1, a_2, \ldots, a_n$ are relatively prime if their g.c.d. is 1.

relative position, relative velocity and **relative acceleration** Let $\mathbf{r}_P$ and $\mathbf{r}_Q$ be the position vectors of particles P and Q with respect to some *frame of reference* with origin O, as shown in the diagram. The position vector of P

relative to Q is $\mathbf{r}_P - \mathbf{r}_Q$. If $\mathbf{v}_P = \dot{\mathbf{r}}_P$ and $\mathbf{v}_Q = \dot{\mathbf{r}}_Q$, then $\mathbf{v}_P$ and $\mathbf{v}_Q$ are the velocities of P and Q relative to the frame of reference with origin O and $\mathbf{v}_P - \mathbf{v}_Q$ is the velocity of P relative to Q. If $\mathbf{a}_P = \dot{\mathbf{v}}_P$ and $\mathbf{a}_Q = \dot{\mathbf{v}}_Q$, then $\mathbf{a}_P$ and $\mathbf{a}_Q$ are the accelerations of P and Q relative to the frame of reference with origin O, and $\mathbf{a}_P - \mathbf{a}_Q$ is the acceleration of P relative to Q. These quantities may be called the **relative position**, the **relative velocity** and the **relative acceleration** of P with respect to Q.

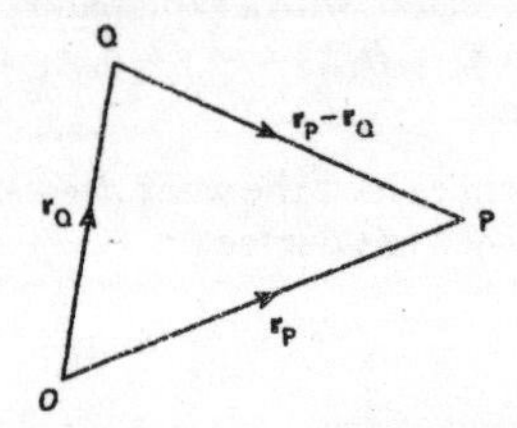

These notions are important when there are two or more frames of reference, each with an associated observer. For example, in a problem involving a ship and an aeroplane, the ship's captain and an observer on the land may both be viewing the aeroplane. The velocity of the aeroplane, for example, depends on which of them is making the measurement. The distinction must be made between the velocity of the aeroplane relative to the ship and the velocity of the aeroplane relative to the land.

remainder See *Division Algorithm* and *Taylor's Theorem.*

Remainder Theorem The following result about polynomials:

THEOREM: If a polynomial $f(x)$ is divided by $x - h$, then the remainder is equal to $f(h)$.

It is proved as follows. Divide the polynomial $f(x)$ by $x - h$ to get a quotient $q(x)$ and a remainder which will be a constant, r. This means that $f(x) = (x - h)q(x) + r$. Replacing x by h in this equation gives $r = f(h)$, thus proving the theorem. An important consequence of the Remainder Theorem is the *Factor Theorem.*

repeated root See *root.*

repeating decimal See *decimal representation.*

representation (of a vector) When the directed line-segment $\overrightarrow{AB}$ represents the *vector* **a**, then $\overrightarrow{AB}$ is a **representation** of **a**.

representative Given an *equivalence relation* on a set, any one of the *equivalence classes* can be specified by giving one of the elements in it. The particular element a used can be called a **representative** of the class, and the class can be denoted by $[a]$.

residual The difference between an observed value and the value predicted by some statistical model. The residuals may be checked to assess how well the model fits the data, perhaps by using a *chi-squared test*. A large residual may indicate an *outlier*.

residue class (modulo n) An *equivalence class* for the *equivalence relation* of *congruence* modulo n. So, two integers are in the same class if they have the same remainder upon division by n. If $[a]$ denotes the residue class modulo n containing a, the residue classes modulo n can be taken as $[0]$, $[1]$, $[2], \dots, [n-1]$. The sum and product of residue classes can be defined by

$$[a] + [b] = [a+b], \qquad [a][b] = [ab],$$

where it is necessary to show that the definitions here do not depend upon which *representatives* a and b are chosen for the two classes. With this addition and multiplication, the set, denoted by $\mathbf{Z}_n$, of residue classes modulo n forms a *ring* (in fact, a commutative ring with identity). If n is composite, the ring $\mathbf{Z}_n$ has *divisors of zero*, but when p is prime $\mathbf{Z}_p$ is a *field*.

resistive force A force that opposes the motion of a body. It acts on the body in a direction opposite to the direction of the velocity of the body. For example, a ball-bearing falling in a cylindrical glass jar containing oil experiences the downward pull of gravity and a resistive force upwards caused by the oil acting to oppose the motion of the falling ball-bearing. Other examples are *friction* and *aerodynamic drag*. When there is a resistive force, the principle of *conservation of energy* does not hold, but the *work–energy principle* does hold.

resolution The process of writing a vector in terms of its *components* in two or three mutually perpendicular directions. The process is used, for example, when a particle is acted on by two or more forces. It may then be useful to **resolve** all the forces in two or three mutually perpendicular directions. In one problem, it may be appropriate to resolve the forces horizontally and vertically; in another, along and perpendicular to an inclined plane. The equation of motion in vector form may then be replaced by two or three scalar equations of motion.

resolve See *resolution*.

resonance Suppose that a body capable of performing *oscillations* is subject to an applied force which is itself oscillatory, and not subject to any *resistive force*. In certain circumstances the body may oscillate with an amplitude that, in theory, increases indefinitely. This happens when the applied force has the same period as the period of the natural oscillations of the body, and then **resonance** is said to occur. For example, if the equation $m\ddot{x} + kx = F\cos\Omega t$ holds, there is resonance when $\Omega = \sqrt{k/m}$, in which case the solution of the equation involves the particular integral

$$\frac{Ft}{2\sqrt{km}} \sin\sqrt{\frac{k}{m}}\,t,$$

which gives oscillations of increasing amplitude.

When a resistive force is included, the amplitude of the forced oscillations may take a maximum value for a certain value of the *angular frequency* of the applied force. This example of resonance is important in the design of seismographs, instruments for measuring the strength of earthquakes.

response variable = *dependent variable.*

restriction (of a mapping) A mapping $g: S_1 \to T_1$ is a **restriction** of the mapping $f: S \to T$ if $S_1 \subseteq S$, $T_1 \subseteq T$ and $g(s_1) = f(s_1)$ for all s_1 in S_1. Thus a restriction is obtained by taking, perhaps, a subset of the domain or codomain or both, but otherwise following the same rule for defining the mapping.

resultant The sum of two or more vectors, particularly if they represent forces, may be called their **resultant**.

retardation = *deceleration.*

rhombohedron A *polyhedron* with six faces, each of which is a rhombus. It is thus a *parallelepiped* whose edges are all the same length.

rhombus (plural: rhombi) A quadrilateral all of whose sides have the same length. A rhombus is both a *kite* and a *parallelogram.*

Riemann, (Georg Friedrich) Bernhard (1826–1866) German mathematician who was a major figure in nineteenth-century mathematics. In many ways, he was the intellectual successor of Gauss. In geometry, he started the development of those tools which Einstein would eventually use to describe the universe and which in the twentieth century would be turned into the theory of manifolds. His basic geometrical ideas were presented in his famous inaugural lecture at Göttingen, to an audience including Gauss. He did much significant work in analysis, in which his name is preserved in the *Riemann integral,* the Cauchy–Riemann equations and Riemann surfaces. He also made connections between prime number theory and analysis: he formulated the Riemann hypothesis, a conjecture concerning the so-called zeta function, which if proved would give information about the distribution of prime numbers.

Riemann integral, Riemann sum See *integral.*

right angle A quarter of a complete revolution. It is thus equal to 90° or $\pi/2$ radians.

right-circular See *cone* and *cylinder.*

right derivative See *left and right derivative.*

right-handed system Let Ox, Oy and Oz be three mutually perpendicular directed lines, intersecting at the point O. In the order Ox, Oy, Oz, they form a **right-handed system** if a person standing with their head in the positive z-direction and facing the positive y-direction would have the positive x-direction to their right. Putting it another way, when seen from a position facing the positive z-direction, a rotation from the positive x-direction to the positive y-direction passes through a right angle clockwise. Following the normal practice, the figures in this book that use a Cartesian coordinate system for 3-dimensional space have Ox, Oy and Oz forming a right-handed system.

The three directed lines Ox, Oy and Oz (in that order) form a **left-handed system** if, taken in the order Oy, Ox, Oz, they form a right-handed system. If the direction of any one of three lines of a right-handed system is reversed, the three directed lines form a left-handed system.

Similarly, an ordered set of three oblique directed lines may be described as forming a right- or left-handed system. Three vectors, in a given order, form a right- or left-handed system if directed line-segments representing them define directed lines that do so.

right-regular See *prism* and *pyramid*.

rigid body An object with the property that it does not change shape whatever forces are applied to it. It is used in a *mathematical model* to represent an object in the real world. It may be a system of particles held in a rigid formation, or it may be a distribution of mass in the form of a rod, a lamina or some 3-dimensional shape. Problems on the motion of a rigid body are concerned with such matters as the position in space of the rigid body; the angular velocity, angular momentum and kinetic energy of the rigid body; and the position vector, velocity and acceleration of its centre of mass. In general, a rigid body has six *degrees of freedom*. One equation of motion in vector form governs the motion of the centre of mass, and another relates the rate of change of angular momentum to the moment of the forces acting on the rigid body.

ring Sets of entities with two operations, often called addition and multiplication, occur in different situations in mathematics and sometimes share many of the same properties. It is useful to recognize these similarities by identifying certain of the common characteristics. One such set of properties is specified in the definition of a ring: a **ring** is a set R, closed under two operations called addition and multiplication, denoted in the usual way, such that

1. for all a, b and c in R, $a + (b + c) = (a + b) + c$,
2. for all a and b in R, $a + b = b + a$,
3. there is an element 0 in R such that $a + 0 = a$ for all a in R,
4. for each a in R, there is an element $-a$ in R such that $a + (-a) = 0$,
5. for all a, b and c in R, $a(bc) = (ab)c$,
6. for all a, b and c in R, $a(b + c) = ab + ac$ and $(a + b)c = ac + bc$.

The element guaranteed by 3 is a *neutral element* for addition. It can be shown that in a ring this element is unique and has the extra property that

$a0 = 0$ for all a in R, so it is usually called the *zero element*. Also, for each a, the element $-a$ guaranteed by **4** is unique and is the *negative* of a.

The ring is a **commutative ring** if it is true that

7. for all a and b in R, $ab = ba$,

and it is a **commutative ring with identity** if also

8. there is an element $1 (\neq 0)$ such that $a1 = a$ for all a in R.

If certain further properties are added, the definitions of an *integral domain* and a *field* are obtained. So any integral domain and any field is a ring. Further examples of rings (which are not integral domains or fields) are the set of 2×2 real matrices and the set of all even integers, each with the appropriate addition and multiplication. Another example of a ring is $\mathbf{Z}_n$, the set $\{0, 1, 2, \ldots, n-1\}$ with addition and multiplication modulo n.

A ring may be denoted by $\langle R, +, \times \rangle$ and another ring by, say, $\langle R', \oplus, \otimes \rangle$ when it is necessary to distinguish the operations in one ring from the operations in the other. But it is sufficient to refer simply to the ring R when the operations intended are clear.

robust A test or *estimator* is **robust** if it is not sensitive to small discrepancies in assumptions, such as the assumption of the normality of the underlying distribution. A test is said to be robust to *outliers* if it is not unduly affected by their presence.

rod An object considered as being 1-dimensional, having length and density but having no width or thickness. It is used in a *mathematical model* to represent a thin straight object in the real world. It may be rigid or flexible, depending on the circumstances.

Rolle, Michel (1652–1719) French mathematician, remembered primarily for the theorem that bears his name, which appears in a book of his published in 1691.

Rolle's Theorem The following result concerning the existence of *stationary points* of a function f:

THEOREM: Let f be a function which is continuous on $[a, b]$ and differentiable in (a, b), such that $f(a) = f(b)$. (Some authors require that $f(a) = f(b) = 0$.) Then there is a number c with $a < c < b$ such that $f'(c) = 0$.

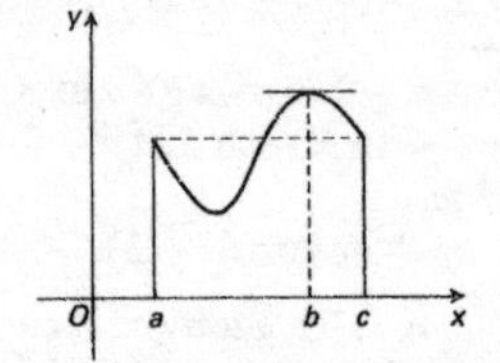

The result stated in the theorem can be expressed as a statement about the graph of *f*: with appropriate conditions on *f*, between any two points on the graph $y = f(x)$ that are level with each other, there must be a stationary point; that is, a point at which the tangent is horizontal. The theorem is, in fact, a special case of the *Mean Value Theorem*; however, it is normal to establish Rolle's Theorem first and deduce the Mean Value Theorem from it. A rigorous proof relies on the non-elementary result that a *continuous function* on a closed interval attains its bounds.

Roman numeral See *numeral*.

root (of an equation) Let $f(x) = 0$ be an equation that involves the indeterminate x. A **root** of the equation is a value h such that $f(h) = 0$. Such a value is also called a **zero** of the function *f*. Some authors use 'root' and 'zero' interchangeably.

If $f(x)$ is a polynomial, then $f(x) = 0$ is a polynomial equation. By the *Factor Theorem*, h is a root of this equation if and only if $x - h$ is a factor of $f(x)$. The value h is a **simple root** if $x - h$ is a factor but $(x - h)^2$ is not a factor of $f(x)$; and h is a root of **order** (or **multiplicity**) n if $(x - h)^n$ is a factor but $(x - h)^{n+1}$ is not. A root of order 2 is a **double root**; a root of order 3 is a **triple root**. A root of order n, where $n \geq 2$, is a **multiple** (or **repeated**) **root**.

If h is a double root of the polynomial equation $f(x) = 0$, then close to $x = h$ the graph $y = f(x)$ looks something like one of the diagrams in the first row of the figure. If h is a triple root, then the graph looks like one of the diagrams in the second row of the figure. The value h is a root of order at least n if and only if $f(h) = 0, f'(h) = 0, \ldots, f^{(n-1)}(h) = 0$.

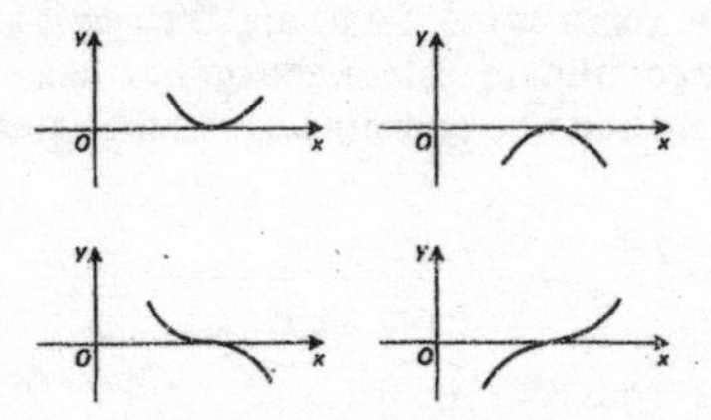

If α and β are the roots of the quadratic equation $ax^2 + bx + c = 0$, with $a \neq 0$, then $\alpha + \beta = -b/a$ and $\alpha\beta = c/a$. If α, β and γ are the roots of the cubic equation $ax^3 + bx^2 + cx + d = 0$, with $a \neq 0$, then $\alpha + \beta + \gamma = -b/a$, $\beta\gamma + \gamma\alpha + \alpha\beta = c/a$ and $\alpha\beta\gamma = -d/a$. Similar results hold for polynomial equations of higher degree.

root (of a tree) See *tree*.

root mean squared deviation The positive square root of the *mean squared deviation*.

root of unity See *n-th root of unity*.

rotation (of the plane) A **rotation** of the plane about the origin O through an angle α is the *transformation* of the plane in which O is mapped to itself, and a point P with polar coordinates (r, θ) is mapped to the point P' with polar coordinates $(r, \theta + \alpha)$. In terms of Cartesian coordinates, P with coordinates (x, y) is mapped to P' with coordinates (x', y'), where

$$x' = x\cos\alpha - y\sin\alpha,$$
$$y' = x\sin\alpha + y\cos\alpha.$$

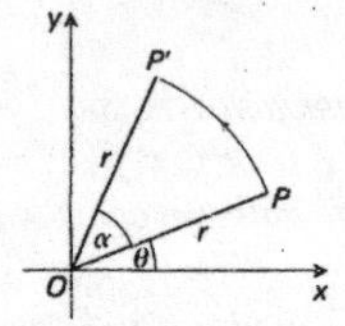

rotational symmetry A plane figure has **rotational symmetry** about a point O if the figure appears the same when it is rotated about O through some positive angle less than one complete revolution. For example, an equilateral triangle (and, indeed, any regular polygon) has rotational symmetry about its centre.

rotation of axes (in the plane) Suppose that a Cartesian coordinate system has a given x-axis and y-axis with origin O and given unit length, so that a typical point P has coordinates (x, y). Consider taking a new coordinate system with the same origin O and the same unit length, with X-axis and Y-axis, such that a rotation through an angle α (with the positive direction taken anticlockwise) carries the x-axis to the X-axis and the y-axis to the Y-axis. With respect to the new coordinate system, the point P has coordinates (X, Y). Then the old and new coordinates in such a **rotation of axes** are related by

$$x = X\cos\alpha - Y\sin\alpha,$$
$$y = X\sin\alpha + Y\cos\alpha.$$

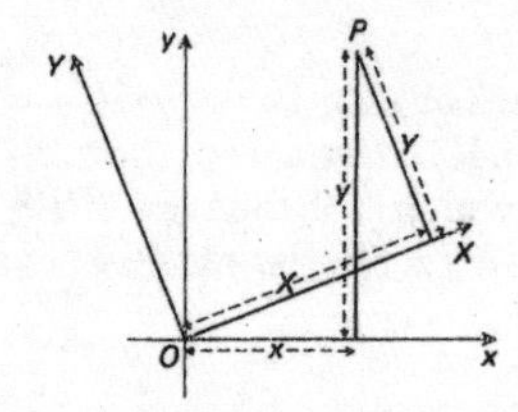

In matrix notation, these equations become

$$\begin{bmatrix} x \\ y \end{bmatrix} = \begin{bmatrix} \cos\alpha & -\sin\alpha \\ \sin\alpha & \cos\alpha \end{bmatrix} \begin{bmatrix} X \\ Y \end{bmatrix}$$

and, conversely,

$$\begin{bmatrix} X \\ Y \end{bmatrix} = \begin{bmatrix} \cos\alpha & \sin\alpha \\ -\sin\alpha & \cos\alpha \end{bmatrix} \begin{bmatrix} x \\ y \end{bmatrix}.$$

For example, in a rotation of axes through an angle of $-\pi/4$ radians, the coordinates are related by

$$x = \frac{1}{\sqrt{2}}(X + Y), \qquad y = \frac{1}{\sqrt{2}}(-X + Y),$$

and the curve with equation $x^2 - y^2 = 2$ has equation $XY = 1$ in the new coordinate system.

rough surface See *contact force*.

rounding Suppose that a number has more digits than can be conveniently handled or stored. In **rounding** (as opposed to *truncation*), the original number is replaced by the number with the required number of digits that is closest to it. Thus, when rounding to 1 decimal place, the number 1.875 becomes 1.9 and the number 1.845 becomes 1.8. It is said that the number is accordingly **rounded up** or **rounded down**. When the original is precisely at a halfway point (for example, if 1.85 is to be rounded to 1 decimal place), it may be rounded up (to 1.9) or rounded down (to 1.8). Some authors like to recommend a particular way of deciding which to do. See also *decimal places* and *significant figures*.

rounding error = *round-off error*.

round-off error When a number X is rounded to a certain number of digits to obtain an approximation x, the *error* is called the **round-off error**. For some authors this is $X - x$, and for others it is $x - X$. For example, if 1.875 is rounded to 1 decimal place, or to 2 significant figures, to give 1.9, the round-off error is either 0.025 or -0.025. When a number is rounded to k decimal places, the round-off error lies between $\pm 5 \times 10^{-(k+1)}$; for example, when rounding to 3 decimal places, the round-off error lies between ± 0.0005. For some authors, the error is $|X - x|$ (see *absolute value*) and so, for them, it is always greater than or equal to zero.

row matrix A *matrix* with exactly one row, that is, a $1 \times n$ matrix of the form $[a_1 \quad a_2 \quad \ldots \quad a_n]$. Given an $m \times n$ matrix, it may be useful to treat its rows as individual row matrices.

row operation See *elementary row operation*.

row rank See *rank*.

row stochastic matrix See *stochastic matrix*.

row vector = *row matrix*.

ruled surface A surface that can be traced out by a moving straight line; in other words, every point of the surface lies on a straight line lying wholly in the surface. Examples are a *cone*, a *cylinder*, a *hyperboloid of one sheet* and a *hyperbolic paraboloid*.

run A sequence of consecutive observations from a sample that share a specified property. Often the observations have either one or other of two properties A and B. For example, A may be 'above the median' and B 'below the median'. If the sequence of observations then gives the sequence $AABAAABBAB$, for example, the number of runs equals 6. This statistic is used in some *non-parametric methods*.

Russell, Bertrand Arthur William (1872–1970) British philosopher, logician and writer on many subjects. He is remembered in mathematics as the author, with A. N. Whitehead, of *Principia mathematica*, published between 1910 and 1913 in three volumes, which set out to show that pure mathematics could all be derived from certain fundamental logical axioms. Although the attempt was not completely successful, the work was highly influential. He was also responsible for the discovery of *Russell's paradox*.

Russell's paradox By using the notation of set theory, a set can be defined as the set of all x that satisfy some property. Now it is clearly possible for a set not to belong to itself: any set of numbers, say, does not belong to itself because to belong to itself it would have to be a number. But it is also possible to have a set that does belong to itself: for example, the set of all sets belongs to itself. In 1901, Bertrand Russell drew attention to what has become known as **Russell's paradox**, by considering the set R, defined by $R = \{ x \mid x \notin x \}$. If $R \in R$ then $R \notin R$; and if $R \notin R$ then $R \in R$. The paradox points out the danger of the unrestricted use of abstraction, and various solutions have been proposed to avoid the paradox.

S

saddle-point Suppose that a surface has equation $z = f(x, y)$, with, as usual, the z-axis vertically upwards. A point P on the surface is a **saddle-point** if the *tangent plane* at P is horizontal and if P is a local minimum on the curve obtained by one vertical cross-section and a local maximum on the curve obtained by another vertical cross-section. It is so called because the central point on the seat of a horse's saddle has this property. The *hyperbolic paraboloid*, for example, has a saddle-point at the origin. See also *stationary point* (in two variables).

sample A subset of a *population* selected in order to make *inferences* about the population. (Strictly speaking, it is not a subset because elements may be repeated.) It is a **random sample** if it is chosen in such a way that every sample of the same size has an equal chance of being selected. If the sample is chosen in such a way that no member of the population can be selected more than once, this is **sampling without replacement**. In **sampling with replacement**, an element has a chance of being selected more than once. A **quota sample** is a sample in which a predetermined number of elements have to be selected from each category in a specified list.

sample space The set of all possible outcomes of an experiment. For example, suppose that the aim of the experiment is to toss a coin three times and record the results. Then the sample space could be expressed as {HHH, HHT, HTH, HTT, THH, THT, TTH, TTT} where, for example, 'HTT' indicates that the coin came up 'heads' on the first toss and 'tails' on the second and third tosses. If the aim of the experiment is to toss a coin three times and count the numbers of heads obtained, the sample space could be taken as $\{0, 1, 2, 3\}$.

sampling distribution Every *statistic* is a random variable because its value varies from one sample to another. The *distribution* of this random variable is a **sampling distribution**.

scalar (for matrices) In work with *matrices*, the entries must belong to some particular set S. An element of S may be called a **scalar** to emphasize that it is not a matrix. For example, the occasion may arise when a matrix $\mathbf{A}$ is to be multiplied by a scalar k to form $k\mathbf{A}$, or when a certain row of a matrix is to be multiplied by a scalar.

scalar (for vectors) In work with *vectors*, a quantity that is a real number, in other words, *not* a vector, is called a **scalar**.

scalar multiple (of a matrix) Let $\mathbf{A}$ be an $m \times n$ matrix, with $\mathbf{A} = [a_{ij}]$, and k a scalar. The **scalar multiple** $k\mathbf{A}$ is the $m \times n$ matrix $\mathbf{C}$, where $\mathbf{C} = [c_{ij}]$ and $c_{ij} = ka_{ij}$. Multiplication by scalars has the following properties:

(i) $(h + k)\mathbf{A} = h\mathbf{A} + k\mathbf{A}$.

(ii) $k(\mathbf{A}+\mathbf{B}) = k\mathbf{A}+k\mathbf{B}$.
(iii) $h(k\mathbf{A}) = (hk)\mathbf{A}$.
(iv) $0\mathbf{A} = \mathbf{O}$, the zero matrix.
(v) $(-1)\mathbf{A} = -\mathbf{A}$, the negative of $\mathbf{A}$.

scalar multiple (of a vector) Let $\mathbf{a}$ be a non-zero vector and k a non-zero scalar. The **scalar multiple** of $\mathbf{a}$ by k, denoted by $k\mathbf{a}$, is the vector whose magnitude is $|k||\mathbf{a}|$ and whose direction is that of $\mathbf{a}$, if $k > 0$, and that of $-\mathbf{a}$, if $k < 0$. Also, $k\mathbf{0}$ and $0\mathbf{a}$ are defined to be $\mathbf{0}$, for all k and $\mathbf{a}$. Multiplication by scalars has the following properties:

(i) $(h+k)\mathbf{a} = h\mathbf{a}+k\mathbf{a}$.
(ii) $k(\mathbf{a}+\mathbf{b}) = k\mathbf{a}+k\mathbf{b}$.
(iii) $h(k\mathbf{a}) = (hk)\mathbf{a}$.
(iv) $1\mathbf{a} = \mathbf{a}$.
(v) $(-1)\mathbf{a} = -\mathbf{a}$, the negative of $\mathbf{a}$.

scalar product For vectors $\mathbf{a}$ and $\mathbf{b}$, the **scalar product** $\mathbf{a}\,.\,\mathbf{b}$ is defined by $\mathbf{a}\,.\,\mathbf{b} = |\mathbf{a}||\mathbf{b}|\cos\theta$, where θ is the angle, in radians with $0 \le \theta \le \pi$, between $\mathbf{a}$ and $\mathbf{b}$. This is a scalar quantity; that is, a real number, not a vector. The scalar product has the following properties:

(i) $\mathbf{a}\,.\,\mathbf{b} = \mathbf{b}\,.\,\mathbf{a}$.
(ii) For non-zero vectors $\mathbf{a}$ and $\mathbf{b}$, $\mathbf{a}\,.\,\mathbf{b} = 0$ if and only if $\mathbf{a}$ is perpendicular to $\mathbf{b}$.
(iii) $\mathbf{a}\,.\,\mathbf{a} = |\mathbf{a}|^2$; the scalar product $\mathbf{a}\,.\,\mathbf{a}$ may be written $\mathbf{a}^2$.
(iv) $\mathbf{a}\,.\,(\mathbf{b}+\mathbf{c}) = \mathbf{a}\,.\,\mathbf{b}+\mathbf{a}\,.\,\mathbf{c}$, the distributive law.
(v) $\mathbf{a}\,.\,(k\mathbf{b}) = (k\mathbf{a})\,.\,\mathbf{b} = k(\mathbf{a}\,.\,\mathbf{b})$.
(vi) If vectors $\mathbf{a}$ and $\mathbf{b}$ are given in terms of their components (with respect to the standard vectors $\mathbf{i}$, $\mathbf{j}$ and $\mathbf{k}$) as $\mathbf{a} = a_1\mathbf{i}+a_2\mathbf{j}+a_3\mathbf{k}$, $\mathbf{b} = b_1\mathbf{i}+b_2\mathbf{j}+b_3\mathbf{k}$, then $\mathbf{a}\,.\,\mathbf{b} = a_1b_1+a_2b_2+a_3b_3$.

scalar projection (of a vector on a vector) See *vector projection* (of a vector on a vector).

scalar triple product For vectors $\mathbf{a}$, $\mathbf{b}$ and $\mathbf{c}$, the *scalar product*, $\mathbf{a}\,.\,(\mathbf{b}\times\mathbf{c})$, of $\mathbf{a}$ with the vector $\mathbf{b}\times\mathbf{c}$ (see *vector product*), is called a **scalar triple product**. It is a scalar quantity and is denoted by $[\mathbf{a},\mathbf{b},\mathbf{c}]$. It has the following properties:

(i) $[\mathbf{a},\mathbf{b},\mathbf{c}] = -[\mathbf{a},\mathbf{c},\mathbf{b}]$.
(ii) $[\mathbf{a},\mathbf{b},\mathbf{c}] = [\mathbf{b},\mathbf{c},\mathbf{a}] = [\mathbf{c},\mathbf{a},\mathbf{b}]$.
(iii) The vectors $\mathbf{a}$, $\mathbf{b}$ and $\mathbf{c}$ are coplanar if and only if $[\mathbf{a},\mathbf{b},\mathbf{c}] = 0$.
(iv) If the vectors are given in terms of their components (with respect to the standard vectors $\mathbf{i}$, $\mathbf{j}$ and $\mathbf{k}$) as $\mathbf{a} = a_1\mathbf{i}+a_2\mathbf{j}+a_3\mathbf{k}$, $\mathbf{b} = b_1\mathbf{i}+b_2\mathbf{j}+b_3\mathbf{k}$, and $\mathbf{c} = c_1\mathbf{i}+c_2\mathbf{j}+c_3\mathbf{k}$, then

$$[\mathbf{a},\mathbf{b},\mathbf{c}] = a_1(b_2c_3-b_3c_2)+a_2(b_3c_1-b_1c_3)+a_3(b_1c_2-b_2c_1)$$
$$= \begin{bmatrix} a_1 & a_2 & a_3 \\ b_1 & b_2 & b_3 \\ c_1 & c_2 & c_3 \end{bmatrix}.$$

(v) Let $\overrightarrow{OA}$, $\overrightarrow{OB}$ and $\overrightarrow{OC}$ represent **a**, **b** and **c** and let P be the *parallelepiped* with OA, OB and OC as three of its edges. The *absolute value* of [**a**, **b**, **c**] then gives the volume of the parallelepiped P. (If **a**, **b** and **c** form a right-handed system, then [**a**, **b**, **c**] is positive; and if **a**, **b** and **c** form a left-handed system, then [**a**, **b**, **c**] is negative.)

scalene triangle A triangle in which all three sides have different lengths.

scatter diagram A two-dimensional diagram showing the points corresponding to n paired sample observations $(x_1, y_1), (x_2, y_2), \ldots, (x_n, y_n)$, where $x_1, x_2, \ldots, x_n$ are the observed values of the *explanatory variable* and $y_1, y_2, \ldots, y_n$ are the observed values of the *dependent variable*.

scatter plot = *scatter diagram*.

scientific notation A number is said to be in **scientific notation** when it is written as a number between 1 and 10, times a power of 10; that is to say, as $a \times 10^n$, where $1 \leq a < 10$ and n is an integer. Thus 634.8 and 0.002 34 are written in scientific notation as 6.348×10^2 and 2.34×10^{-3}. The notation is particularly useful for very large and very small numbers.

seasonal variation See *time series*.

secant See *trigonometric function*.

sech See *hyperbolic function*.

second (angular measure) See *degree* (angular measure).

second (time) In *SI units*, the base unit used for measuring time, abbreviated to 's'. It was once defined as 1/86 400 of the mean solar day, which is the average time it takes for the Earth to rotate relative to the Sun. Now it is defined in terms of the radiation emitted by the caesium-133 atom.

second derivative See *higher derivative*.

second-order partial derivative See *higher-order partial derivative*.

section formula (in vectors) Let A and B be two points with *position vectors* **a** and **b**, and P a point on the line through A and B, such that $AP : PB = m : n$. Then P has position vector **p**, given by the **section formula**

$$\mathbf{p} = \frac{1}{m+n}(m\mathbf{b} + n\mathbf{a}).$$

It is possible to choose m and n such that $m + n = 1$ and thus to suppose, changing notation, that $AP : PB = k : 1 - k$. Then $\mathbf{p} = (1 - k)\mathbf{a} + k\mathbf{b}$.

section formulae (in the plane) Let A and B be two points with Cartesian coordinates (x_1, y_1) and (x_2, y_2), and P a point on the line through A and B

such that $AP : PB = m : n$. Then the **section formulae** give the coordinates of P as

$$\left(\frac{mx_2 + nx_1}{m+n}, \frac{my_2 + ny_1}{m+n}\right).$$

If, instead, P is a point on the line such that $AP : PB = k : 1 - k$, its coordinates are $((1-k)x_1 + kx_2, (1-k)y_1 + ky_2)$.

section formulae (in 3-dimensional space) Let A and B be two points with Cartesian coordinates (x_1, y_1, z_1) and (x_2, y_2, z_2), and P a point on the line through A and B. If $AP : PB = m : n$, the **section formulae** give the coordinates of P as

$$\left(\frac{mx_2 + nx_1}{m+n}, \frac{my_2 + ny_1}{m+n}, \frac{mz_2 + nz_1}{m+n}\right),$$

and, if P is a point on the line such that $AP : PB = k : 1 - k$, its coordinates are $((1-k)x_1 + kx_2,\ (1-k)y_1 + ky_2, (1-k)z_1 + kz_2)$.

sector A **sector** of a circle, with centre *O*, is the region bounded by an arc *AB* of the circle and the two radii *OA* and *OB*. The area of a sector is equal to $\frac{1}{2}r^2\theta$, where r is the radius and θ is the angle in radians.

segment A **segment** of a circle is the region bounded by an arc *AB* of the circle and the chord *AB*.

selection The number of **selections** of n objects taken r at a time (that is, the number of ways of selecting r objects out of n) is denoted by ${}^nC_r$ and is equal to

$$\frac{n!}{r!(n-r)!}.$$

(See *binomial coefficient*, where the alternative notation $\binom{n}{r}$ is defined.)

For example, from four objects *A*, *B*, *C* and *D*, there are six ways of selecting two: *AB*, *AC*, *AD*, *BC*, *BD*, *CD*. The property that ${}^{n+1}C_r = {}^nC_{r-1} + {}^nC_r$ can be seen displayed in *Pascal's triangle*.

semicircle One half of a circle cut off by a diameter.

semi-interquartile range A measure of *dispersion* equal to half the difference between the first and third *quartiles* in a set of numerical data.

semi-regular polyhedron See *Archimedean solid*.

semi-regular tessellation See *tessellation*.

semi-vertical angle See *cone*.

sensitivity analysis The varying of *parameters* in a *simulation* to discover which parameters have the greatest influence on the features of interest.

separable first-order differential equation A first-order differential equation $dy/dx = f(x, y)$ in which the function f can be expressed as the product of a function of x and a function of y. The differential equation then has the form $dy/dx = g(x)h(y)$, and its solution is given by the equation

$$\int \frac{1}{h(y)}\, dy = \int g(x)\, dx + c,$$

where c is an arbitrary constant.

sequence A **finite sequence** consists of n terms $a_1, a_2, \dots, a_n$, one corresponding to each of the integers $1, 2, \dots, n$, where n, some positive integer, is the **length** of the sequence. An **infinite sequence** consists of terms $a_1, a_2, a_3, \dots$, one corresponding to each positive integer. Sometimes, it is more convenient to denote the terms of a sequence by $a_0, a_1, a_2, \dots$. See also *limit* (of a sequence).

sequential sampling A statistical process to decide which of two hypotheses to accept as true. Observations are made one at a time, and a test is carried out to decide whether one of the hypotheses is to be accepted or whether more observations should be made. The sampling stops when it has been decided which hypothesis to accept.

series A **finite series** is written as $a_1 + a_2 + \cdots + a_n$, where $a_1, a_2, \dots, a_n$ are n numbers called the **terms** in the series, and n, some positive integer, is the **length** of the series. The **sum** of the series is simply the sum of the n terms. For certain finite series, such as *arithmetic series* and *geometric series*, the sum of the series is given by a known formula. The following can also be established:

$$\sum_{r=1}^{n} r^2 = 1^2 + 2^2 + \cdots + n^2 = \tfrac{1}{6}n(n+1)(2n+1),$$

$$\sum_{r=1}^{n} r^3 = 1^3 + 2^3 + \cdots + n^3 = \tfrac{1}{4}n^2(n+1)^2.$$

An **infinite series** is written as $a_1 + a_2 + a_3 + \cdots$, with terms $a_1, a_2, a_3, \dots$, one corresponding to each positive integer. Let s_n be the sum of the first n terms of such a series. If the sequence $s_1, s_2, s_3, \dots$ has a limit s, then the value s is called the **sum** (or **sum to infinity**) of the infinite series. Otherwise, the infinite series has no sum. See also *arithmetic series, geometric series, binomial series, Taylor series* and *Maclaurin series*.

set A well-defined collection of objects. It may be possible to define a set by listing the elements: $\{a, e, i, o, u\}$ is the set consisting of the vowels of the alphabet, $\{1, 2, \dots, 100\}$ is the set of the first 100 positive integers. The meaning of $\{1, 2, 3, \dots\}$ is also clear: it is the set of all positive integers. It may be possible to define a set as consisting of all elements, from some universal set, that satisfy some property. Thus the set of all real numbers that are greater than 1 can be written as either $\{ x \mid x \in \mathbf{R} \text{ and } x > 1 \}$ or $\{ x : x \in \mathbf{R} \text{ and } x > 1 \}$, both of which are read as 'the set of *x* such that *x* belongs to **R** and *x* is greater than 1'. The same set is sometimes written $\{ x \in \mathbf{R} \mid x > 1 \}$.

sgn See *signum function.*

sheet See *hyperboloid of one sheet* and *hyperboloid of two sheets.*

SHM = *simple harmonic motion.*

sieve of Eratosthenes The following method of finding all the *primes* up to some given number *N*. List all the positive integers from 2 up to *N*. Leave the first number, 2, but delete all its multiples; leave the next remaining number, 3, but delete all its multiples; leave the next remaining number, 5, but delete all its multiples, and so on. The integers not deleted when the process ends are the primes.

significance level See *hypothesis testing.*

significance test See *hypothesis testing.*

significant figures To count the number of **significant figures** in a given number, start with the first non-zero digit from the left and, moving to the right, count all the digits thereafter, counting final zeros if they are to the right of the decimal point. For example, 1.2048, 1.2040, 0.012 048, 0.001 2040 and 1204.0 all have 5 significant figures. In *rounding* or *truncation* of a number to *n* significant figures, the original is replaced by a number with *n* significant figures.

Note that final zeros to the left of the decimal point may or may not be significant: the number 1 204 000 has at least 4 significant figures, but without more information there is no way of knowing whether or not any more figures are significant. When 1 203 960 is rounded to 5 significant figures to give 1 204 000, an explanation that this has 5 significant figures is required. This could be made clear by writing it in *scientific notation*: 1.2040×10^6.

To say that $a = 1.2048$ to 5 significant figures means that the exact value of *a* becomes 1.2048 after rounding to 5 significant figures; that is to say, $1.204\,75 \leq a \leq 1.204\,85$.

sign test A non-parametric test (see *non-parametric methods*) to test the *null hypothesis* that a sample is selected from a population with median *m*. If the null hypothesis is true, the sample of size *n* is expected to have an equal number of observations above and below *m*, and the probability that *r* of them are greater than *m* has the *binomial distribution* $B(n, 0.5)$. The null hypothesis is rejected if the value lies in the critical region determined by the chosen significance level.

The test may also be used to compare the medians of two populations from paired data.

signum function The real function, denoted by sgn, defined by

$$\operatorname{sgn} x = \begin{cases} -1, & \text{if } x < 0, \\ 0, & \text{if } x = 0, \\ 1, & \text{if } x > 0. \end{cases}$$

The name and notation come from the fact that the value of the function depends on the sign of x.

similar (of figures) Two geometrical figures are **similar** if they are of the same shape but not necessarily of the same size. This includes the case when one is a mirror-image of the other, so the three triangles shown in the figure are all similar. For two similar triangles, there is a correspondence between their vertices such that corresponding angles are equal and the ratios of corresponding sides are equal. In the figure, $\angle A = \angle P, \angle B = \angle Q, \angle C = \angle R$ and $QR/BC = RP/CA = PQ/AB = 3/2$.

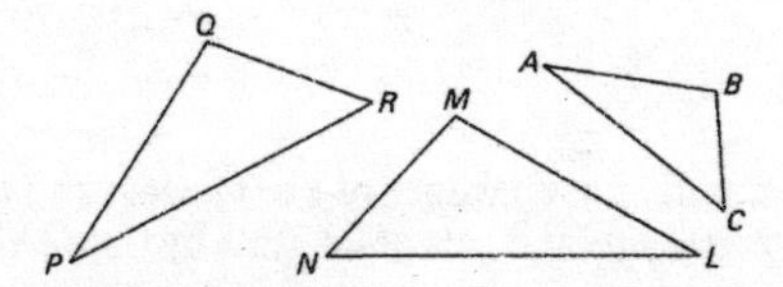

simple curve A continuous plane curve that does not intersect itself.

simple fraction A *fraction* in which the numerator and denominator are positive integers, as opposed to a *compound fraction*.

simple graph A *graph* with no loops or multiple edges.

simple harmonic motion Suppose that a particle is moving in a straight line so that its displacement x at time t is given by $x = A\sin(\omega t + \alpha)$, where $A\ (> 0)$, ω and α are constants. Then the particle is executing **simple harmonic motion**, with *amplitude* A, *period* $2\pi/\omega$ and *phase* α. This equation gives the general solution of the differential equation $\ddot{x} + \omega^2 x = 0$.

An example of simple harmonic motion is the motion of a particle suspended from a fixed support by a spring (see *Hooke's law*). Also, the motion of a *simple pendulum* or a *compound pendulum* performing oscillations of small amplitude is approximately simple harmonic motion.

simple interest Suppose that a sum of money P is invested, attracting interest at i per cent a year. When simple interest is given, the interest due each year is $(i/100)P$ and so, after n years, the amount becomes

$$P\left(1 + \frac{ni}{100}\right).$$

When points are plotted on graph paper to show how the amount increases, they lie on a straight line. Most banks and building societies in fact do not operate in this way but use the method of *compound interest*.

simple linear regression See *regression*.

simple pendulum In a *mathematical model*, a **simple pendulum** is represented by a particle of mass m, suspended by a string of constant length l and negligible mass from a fixed point. The particle, representing the bob of the pendulum, is free to move in a specified vertical plane through the point of suspension. The forces on the particle are the uniform gravitational force and the tension in the string. Suppose the string makes an angle θ with the vertical at time t. The equation of motion can be shown to give $\ddot{\theta} + (g/l)\sin\theta = 0$. When θ is small for all time, $\sin\theta \approx \theta$ and the equation becomes $\ddot{\theta} + \omega^2\theta = 0$, where $\omega^2 = g/l$. It follows that the pendulum performs approximately *simple harmonic motion* with period $2\pi\sqrt{l/g}$.

simple root See *root*.

Simpson's rule An approximate value can be found for the definite integral

$$\int_a^b f(x)\,dx,$$

using the values of $f(x)$ at equally spaced values of x between a and b, as follows. Divide $[a, b]$ into n equal subintervals of length h by the *partition*

$$a = x_0 < x_1 < x_2 < \cdots < x_{n-1} < x_n = b,$$

where $x_{i+1} - x_i = h = (b-a)/n$. Denote $f(x_i)$ by f_i, and let P_i be the point (x_i, f_i). The *trapezium rule* uses the line segment P_iP_{i+1} as an approximation to the curve. Instead, take an arc of a certain parabola (in fact, the graph of a polynomial function of degree two) through the points P_0, P_1 and P_2, an arc of a parabola through P_2, P_3 and P_4, similarly, and so on. Thus n must be even. The resulting **Simpson's rule** gives

$$\tfrac{1}{3}h(f_0 + 4f_1 + 2f_2 + 4f_3 + 2f_4 + \cdots + 2f_{n-2} + 4f_{n-1} + f_n)$$

as an approximation to the value of the integral. This approximation has an *error* that is roughly proportional to $1/n^4$. In general, Simpson's rule can be expected to be much more accurate than the trapezium rule, for a given value of n. It was devised by the British mathematician Thomas Simpson (1710–1761).

simulation An attempt to replicate a physical procedure mathematically, where the system being studied is too complicated for explicit analytic methods to be used. Because of the complexity, simulation is often carried out by computer, perhaps using *Monte Carlo methods*. As simulation is often only an approximation to the physical procedure, the use of *sensitivity analysis* is crucial.

simultaneous linear equations The solution of a set of m linear equations in n unknowns can be investigated by the method of *Gaussian elimination* that transforms the augmented matrix to *echelon form*. The number of non-zero rows in the echelon form cannot be greater than the number of unknowns, and three cases can be distinguished:

(i) If the echelon form has a row with all its entries zero except for a non-zero entry in the last place, then the set of equations is inconsistent.
(ii) If case (i) does not occur and, in the echelon form, the number of non-zero rows is equal to the number of unknowns, then the set of equations has a unique solution.
(iii) If case (i) does not occur and, in the echelon form, the number of non-zero rows is less than the number of unknowns, then the set of equations has infinitely many solutions.

When the set of equations is consistent, that is, in cases (ii) and (iii), the solution or solutions can be found either from the echelon form using *back-substitution* or by using *Gauss–Jordan elimination* to find the *reduced echelon form*. When there are infinitely many solutions, they can be expressed in terms of *parameters* that replace those unknowns free to take arbitrary values.

sine See *trigonometric function.*

sine rule See *triangle.*

singleton A set containing just one element.

singular A square matrix **A** is **singular** if det **A** $= 0$, where det **A** is the determinant of **A**. A singular matrix is not invertible (see *inverse matrix*).

sinh See *hyperbolic function.*

SI units The units used for measuring physical quantities in the internationally agreed system **Système International d'Unités** are known as **SI units**. There are seven **base units** of which the *metre*, for measuring length, the *kilogram*, for measuring mass, and the *second*, for measuring time, are the commonest in mathematics. The so-called **supplementary units** are the *radian*, for measuring angle, and the *steradian*, for measuring solid angle. From these base and supplementary units, further **derived units** are defined, such as the 'square metre' for measuring area and the 'metre per second' for measuring velocity. Some derived units have special names, such as the *newton, joule, watt, pascal* and *hertz*.

Each base unit has an agreed abbreviation, and in the abbreviations of derived units positive and negative indices are used. There should be a small space between different units involved. For example, the abbreviations for 'square metre' and 'metre per second' are 'm^2' and '$\mathrm{m\,s}^{-1}$'. Derived units with special names have their own abbreviations.

Prefixes are used to define multiples of units by powers of 10. Powers in which the index is a multiple of 3 are preferred. In the list below, the abbreviation for the prefix is given in brackets. For example, 1 megawatt equals 10^6 watts, and 1 milligram equals 10^{-3} grams; the abbreviation for megawatt is 'MW' and the abbreviation for milligram is 'mg'.

10^3	kilo-	(k)	10^{-3}	milli-	(m)
10^6	mega-	(M)	10^{-6}	micro-	(μ)
10^9	giga-	(G)	10^{-9}	nano-	(n)
10^{12}	tera-	(T)	10^{-12}	pico-	(p)
10^{15}	peta-	(P)	10^{-15}	femto-	(f)
10^{18}	exa-	(E)	10^{-18}	atto-	(a)

The following additional prefixes sometimes occur:

10	deka-	(da)	10^{-1}	deci-	(d)
10^2	hecto-	(h)	10^{-2}	centi-	(c)

skew lines Two straight lines in 3-dimensional space that do not intersect and are not parallel.

skewness The amount of asymmetry of a *distribution*. One measure of skewness is the **coefficient of skewness**, which is defined to be equal to $\mu_3/(\mu_2)^{3/2}$, where μ_2 and μ_3 are the second and third *moments* about the mean. This measure is zero if the distribution is symmetrical about the mean. If the distribution has a long tail to the left, as in the left-hand figure, it is said to be **skewed to the left** and to have negative skewness, because the coefficient of skewness is negative. If the distribution has a long tail to the right, as in the right-hand figure, it is said to be **skewed to the right** and to have positive skewness, because the coefficient of skewness is positive. A **skew distribution** is one that is skewed to the left or skewed to the right.

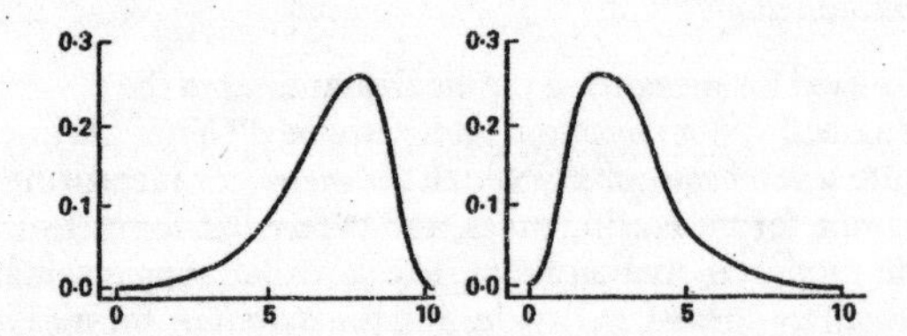

skew-symmetric matrix Let **A** be the square matrix $[a_{ij}]$. Then the matrix **A** is **skew-symmetric** if $\mathbf{A}^T = -\mathbf{A}$ (see *transpose*); that is to say, if $a_{ij} = -a_{ji}$ for all i and j. It follows that in a skew-symmetric matrix the entries in the main diagonal are all zero: $a_{ii} = 0$ for all i.

slant asymptote See *asymptote*.

slant height See *cone*.

slide rule A mechanical device used for mathematical calculations such as multiplication and division. In the simplest form, one piece slides alongside another piece, each piece being marked with a *logarithmic scale*. To multiply x and y, the reading 'x' on one scale is placed opposite the '1' on the other scale, and the required product 'xy' then appears opposite the 'y'. More elaborate slide rules have log–log scales or scales giving trigonometric functions. Slide rules have now been superseded by electronic calculators.

slope = *gradient*.

small circle A circle on the surface of a sphere that is not a *great circle*. It is the curve of intersection obtained when a sphere is cut by a plane not through the centre of the sphere.

smooth surface See *contact force*.

snowflake curve The *Koch curve*, or any similar curve constructed in a similar way.

solid angle The 3-dimensional analogue of the 2-dimensional concept of angle. Just as an angle is bounded by two lines, a solid angle is bounded by the generators of a cone.

A solid angle is measured in **steradians**: this is defined to be the area of the intersection of the solid angle with a sphere of unit radius. Thus the 'complete' solid angle at a point measures 4π steradians (comparable with one complete revolution measuring 2π radians). The steradian is the SI unit for measuring solid angle and is abbreviated to 'sr'.

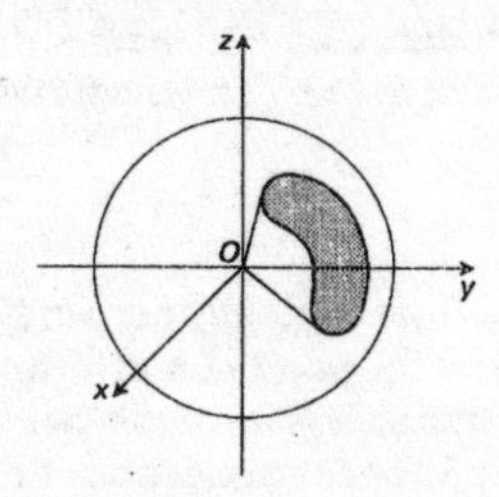

solid of revolution Suppose that a plane region is rotated through one revolution about a line in the plane that does not cut the region. The 3-dimensional region thus obtained is a **solid of revolution**. See also *volume of a solid of revolution*.

solution A **solution** of a set of equations is an element, belonging to some appropriate *universal set*, specified or understood, that satisfies the equations. For a set of equations in n unknowns, a solution may be considered to be an n-tuple $(x_1, x_2, \ldots, x_n)$, or a column matrix

$$\begin{bmatrix} x_1 \\ x_2 \\ \vdots \\ x_n \end{bmatrix},$$

such that $x_1, x_2, \ldots, x_n$ satisfy the equations. See also *inequality*.

solution set The **solution set** of a set of equations is the set consisting of all the solutions. It may be considered as a subset of some appropriate *universal set*, specified or understood. See also *inequality*.

space-filling curve See *Peano curve*.

spanning set See *basis*.

speed In mathematics, it is useful to distinguish between *velocity* and speed. First, when considering motion of a particle in a straight line, specify a positive direction so that it is a *directed line*. Then the velocity of the particle is positive if it is moving in the positive direction and negative if it is moving in the negative direction. The **speed** of the particle is the *absolute value* of its velocity. In more advanced work, when the velocity is a vector **v**, the **speed** is the magnitude $|\mathbf{v}|$ of the velocity.

sphere The **sphere** with centre C and radius r is the locus of all points (in 3-dimensional space) whose distance from C is equal to r. If C has Cartesian coordinates (a, b, c), this sphere has equation

$$(x-a)^2+(y-b)^2+(z-c)^2=r^2.$$

The equation $x^2+y^2+z^2+2ux+2vy+2wz+d=0$ represents a sphere, provided that $u^2+v^2+w^2-d>0$, and it is then an equation of the sphere with centre $(-u, -v, -w)$ and radius $\sqrt{u^2+v^2+w^2-d}$.

The volume of a sphere of radius r is equal to $\frac{4}{3}\pi r^3$, and the surface area equals $4\pi r^2$.

spherical cap See *zone*.

spherical polar coordinates Suppose that three mutually perpendicular directed lines Ox, Oy and Oz, intersecting at the point O, and forming a right-handed system, are taken as coordinate axes (see *coordinates* (in 3-dimensional space)). For any point P, let M be the projection of P on the xy-plane. Let $r=|OP|$, let θ be the angle $\angle zOP$ in radians $(0 \leq \theta \leq \pi)$ and let ϕ be the angle $\angle xOM$ in radians $(0 \leq \phi < 2\pi)$. Then (r, θ, ϕ) are the **spherical polar coordinates** of P. (The point O gives no value for θ or ϕ, but is simply said to correspond to $r=0$.) The value $\phi+2k\pi$, where k is an

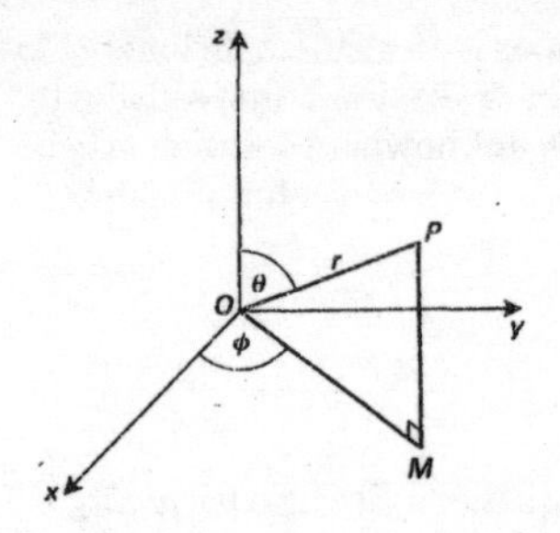

integer, may be allowed in place of ϕ. The Cartesian coordinates (x, y, z) of P can be found from r, θ and ϕ by $x = r \sin\theta \cos\phi$, $y = r \sin\theta \sin\phi$, $z = r\cos\theta$. Spherical polar coordinates may be useful in treating problems involving spheres, for a sphere with centre at the origin then has equation $r =$ constant.

spherical triangle A triangle on a sphere, with three vertices and three sides that are arcs of *great circles*. The angles of a spherical triangle do not add up to 180°. In fact, the sum of the angles can be anything between 180° and 540°. Consider, for example, a spherical triangle with one vertex at the North Pole and the other two vertices on the equator of the Earth.

spherical trigonometry The application of the methods of trigonometry to the study of such matters as the angles, sides and areas of *spherical triangles* and other figures on a sphere.

spiral See *Archimedean spiral* and *equiangular spiral*.

spread = *dispersion*.

spring A device, usually made out of wire in the form of a *helix*, that can be extended and compressed. It is elastic and so resumes its natural length when the forces applied to extend or compress it are removed. How the *tension* in the spring varies with the *extension* may be complicated. In the simplest mathematical model, it is assumed that *Hooke's law* holds, so that the tension is proportional to the extension.

square matrix A *matrix* with the same number of rows as columns.

square root A **square root** of a real number a is a number x such that $x^2 = a$. If a is negative, there is no such real number. If a is positive, there are two such numbers, one positive and one negative. For $a \geq 0$, the notation $\sqrt{a}$ is used to denote quite specifically the non-negative square root of a.

squaring the circle One of the problems that the Greek geometers attempted (like the *duplication of the cube* and the *trisection of an angle*) was to find a construction, with ruler and pair of compasses, to obtain a square whose area was equal to that of a given circle. This is equivalent to a geometrical construction to obtain a length of $\sqrt{\pi}$ from a given unit length. Now constructions of the kind envisaged can give only lengths that are algebraic numbers (and not even all algebraic numbers at that: for instance, $\sqrt[3]{2}$ cannot be obtained). So the proof by Lindemann in 1882 that π is *transcendental* established the impossibility of squaring the circle.

stability The nature of the *equilibrium* of a particle or rigid body in an equilibrium position, whether stable, unstable or neutral.

stable equilibrium See *equilibrium*.

standard deviation The positive square root of the *variance*, a commonly used measure of the *dispersion* of observations in a sample. For a *normal* *distribution* $N(\mu, \sigma^2)$, with mean μ and standard deviation σ, approximately 95% of the distribution lies in the interval between $\mu - 2\sigma$ and $\mu + 2\sigma$.

The standard deviation of an *estimator* of a population parameter is the *standard error.*

standard error The standard deviation of an *estimator* of a population parameter. The standard error of the sample mean, from a sample of size n, is $\sigma/\sqrt{n}$, where σ^2 is the population variance.

standard form (of a number) = *scientific notation.*

standard normal distribution See *normal distribution.*

state, state space See *stochastic process.*

statement In mathematical logic, the fundamental property of a **statement** is that it makes sense and is either true or false. For example, 'there is a real number x such that $x^2 = 2$' makes sense and is true; the statement 'if x and y are positive integers then $x - y$ is a positive integer' makes sense and is false. In contrast, '$x = 2$' is not a statement.

stationary point (in one variable) A point on the graph $y = f(x)$ at which f is *differentiable* and $f'(x) = 0$. The term is also used for the number c such that $f'(c) = 0$. The corresponding value $f(c)$ is a **stationary value.** A stationary point c can be classified as one of the following, depending upon the behaviour of f in the neighbourhood of c:

(i) a *local maximum,* if $f'(x) > 0$ to the left of c and $f'(x) < 0$ to the right of c,
(ii) a *local minimum,* if $f'(x) < 0$ to the left of c and $f'(x) > 0$ to the right of c,
(iii) neither local maximum nor local minimum.

The case (iii) can be subdivided to distinguish between the following two cases. It may be that $f'(x)$ has the same sign to the left and to the right of c, in which case c is a horizontal *point of inflexion*; or it may be that there is an interval at every point of which $f'(x)$ equals zero and c is an end-point or interior point of this interval.

stationary point (in two variables) A point P on the surface $z = f(x, y)$ is a **stationary point** if the *tangent plane* at P is horizontal. This is so if $\partial f/\partial x = 0$ and $\partial f/\partial y = 0$. Now let

$$r = \frac{\partial^2 f}{\partial x^2}, \qquad s = \frac{\partial^2 f}{\partial x \partial y}, \qquad t = \frac{\partial^2 f}{\partial y^2}.$$

If $rt > s^2$ and $r < 0$, the stationary point P is a local maximum (all the vertical cross-sections through P have a local maximum at P). If $rt > s^2$ and $r > 0$, the stationary point is a local minimum (all the vertical cross-sections through P have a local minimum at P). If $rt < s^2$, the stationary point is a *saddle-point.*

stationary value See *stationary point*.

statistic A function of a sample; in other words, a quantity calculated from a set of observations. Often a statistic is an *estimator* for a population parameter. For example, the sample mean, sample variance and sample median are each a statistic. The sum of the observations in a sample is also a statistic, but this is not an estimator.

statistical model A statistical description of an underlying system, intended to match a real situation as nearly as possible. The model for a population is fitted to a sample by estimating the *parameters* in the model. It is then possible to perform *hypothesis testing*, construct *confidence intervals* and draw *inferences* about the population.

stem-and-leaf plot A method of displaying *grouped data*, by listing the observations in each group, resulting in something like a histogram on its side. For the observations 45, 25, 67, 49, 12, 9, 45, 34, 37, 61, 23, grouped using class intervals 0–9, 10–19, 20–29, 30–39, 40–49, 50–59 and 60–69, a stem-and-leaf plot is shown on the left below. If, as in this case, the class intervals are defined by the digit occurring in the 'tens' position, the diagram can also be written as shown on the right.

0–9	9		
10–19	12		
20–29	23	25	
30–39	34	37	
40–49	45	45	49
50–59			
60–69	61	67	

0	9		
1	2		
2	3	5	
3	4	7	
4	5	5	9
5			
6	1	7	

step function A real function is a **step function** if its domain can be partitioned into a number of intervals on each of which the function takes a constant value. An example is the function f defined by $f(x) = [x]$, where this denotes the *integer part* of x.

steradian See *solid angle*.

stereographic projection Suppose that a sphere, centre O, touches a plane at the point S, and let N be the opposite end of the diameter through S. If P

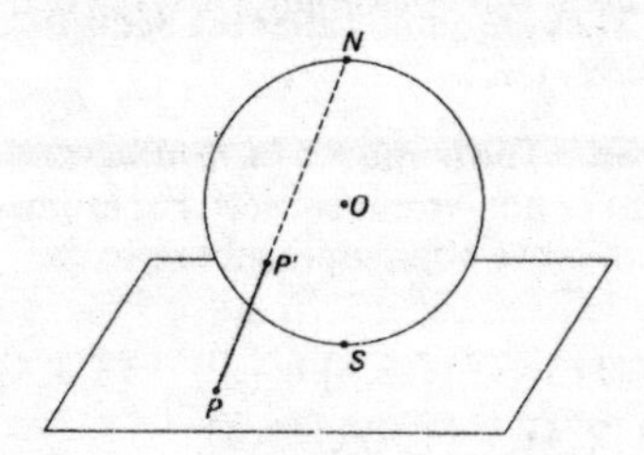

is any point (except N) on the sphere, the line NP meets the plane in a corresponding point P'. Conversely, each point in the plane determines a point on the sphere, so there is a *one-to-one correspondence* between the points of the sphere (except N) and the points of the plane. This means of mapping sphere to plane is called **stereographic projection**. Circles (great or small) not through N on the sphere's surface are mapped to circles in the plane; circles (great or small) through N are mapped to straight lines. The angles at which curves intersect are preserved by the projection.

Stevin, Simon (1548–1620) Flemish engineer and mathematician, the author of a popular pamphlet that explained decimal fractions simply and was influential in bringing about their everyday use, replacing sexagesimal fractions. He made discoveries in statics and hydrostatics and was also responsible for an experiment in 1586, in which two lead spheres, one ten times the weight of the other, took the same time to fall 30 feet. This probably preceded any similar experiment by Galileo.

stiffness A parameter that describes the ability of a spring to be extended or compressed. It occurs as a constant of proportionality in *Hooke's law*. The stiffness can be determined as follows. Plot the magnitude of an applied force against the extension that it produces. If Hooke's law holds, the graph produced should be a straight line through the origin and its gradient is the stiffness.

Stirling number of the first kind The number $s(n, r)$ of ways of partitioning a set of n elements into r *cycles*. For example, the set $\{1, 2, 3, 4\}$ can be partitioned into two cycles in the following ways:

$$[1,2,3][4],\quad [1,3,2][4],\quad [1,2,4][3],\quad [1,4,2][3],$$
$$[1,3,4][2],\quad [1,4,3][2],\quad [2,3,4][1],\quad [2,4,3][1],$$
$$[1,2][3,4],\quad [1,3][2,4],\quad [1,4][2,3].$$

So $s(4, 2) = 11$. Clearly $s(n, 1) = (n-1)!$ and $s(n, n) = 1$. It can be shown that

$$s(n+1, r) = s(n, r-1) + n\,s(n, r).$$

Some authors define these numbers differently so that they satisfy $s(n+1, r) = s(n, r-1) - n\,s(n, r)$. The result is that the values are the same except that some of them occur with a negative sign.

Rather like the *binomial coefficients*, the Stirling numbers occur as coefficients in certain identities. They are named after the Scottish mathematician James Stirling (1692–1770).

Stirling number of the second kind The number $S(n, r)$ of ways of partitioning a set of n elements into r non-empty subsets. For example, the set $\{1, 2, 3, 4\}$ can be partitioned into two non-empty subsets in the following ways:

$$\{1,2,3\} \cup \{4\},\quad \{1,2,4\} \cup \{3\},\quad \{1,3,4\} \cup \{2\},\quad \{2,3,4\} \cup \{1\},$$
$$\{1,2\} \cup \{3,4\},\quad \{1,3\} \cup \{2,4\},\quad \{1,4\} \cup \{2,3\}.$$

So $S(4,2) = 7$. Clearly, $S(n,1) = 1$ and $S(n,n) = 1$. It can be shown that

$$S(n+1,r) = S(n,r-1) + r\,S(n,r).$$

Rather like the *binomial coefficients*, the Stirling numbers occur as coefficients in certain identities. They are named after the Scottish mathematician James Stirling (1692–1770).

Stirling's formula The formula

$$\frac{\sqrt{2\pi n}(n/e)^n}{n!} \to 1.$$

Named after the Scottish mathematician James Stirling (1692–1770), it was known earlier by De Moivre. It gives the approximation $n! \approx \sqrt{2\pi n}(n/e)^n$ for large values of n.

stochastic matrix A square matrix is **row stochastic** if all the entries are non-negative and the entries in each row add up to 1. It is **column stochastic** if all the entries are non-negative and the entries in each column add up to 1. A square matrix is **stochastic** if it is either row stochastic or column stochastic. It is **doubly stochastic** if it is both row stochastic and column stochastic.

stochastic process A family $\{X(t), t \in T\}$ of *random variables*, where T is some *index set*. Often the index set is the set **N** of natural numbers, and the stochastic process is a sequence $X_1, X_2, X_3, \ldots$ of random variables. For example, X_n may be the outcome of the n-th trial of some experiment, or the n-th in a set of observations. The possible values taken by the random variables are often called **states**, and these form the **state space**. The state space is said to be discrete if it is finite or countably infinite, and continuous if it is an interval, finite or infinite. If there are countably many states, they may be denoted by $0, 1, 2, \ldots$, and if $X_n = i$ then X_n is said to be in state i.

straight line (in the plane) A **straight line** in the plane is represented in Cartesian coordinates by a linear equation; that is, an equation of the form $ax + by + c = 0$, where the constants a and b are not both zero. A number of different forms are useful for obtaining an equation of a given line.

(i) The equation $y = mx + c$ represents the line with gradient m that cuts the y-axis at the point $(0, c)$. The value c may be called the **intercept**.

(ii) The line through a given point (x_1, y_1) with gradient m has equation $y - y_1 = m(x - x_1)$.

(iii) The line through the two points (x_1, y_1) and (x_2, y_2) has, if $x_2 \neq x_1$, equation

$$y - y_1 = \frac{y_2 - y_1}{x_2 - x_1}(x - x_1),$$

and has equation $x = x_1$, if $x_2 = x_1$.

(iv) The line that meets the coordinate axes at the points $(p, 0)$ and $(0, q)$, where $p \neq 0$ and $q \neq 0$, has equation $x/p + y/q = 1$.

straight line (in 3-dimensional space) A **straight line** in 3-dimensional space can be specified as the intersection of two planes. Thus a straight line is given, in general, by two *linear equations*, $a_1x + b_1y + c_1z + d_1 = 0$ and $a_2x + b_2y + c_2z + d_2 = 0$. (If these equations represent identical or parallel planes, they do not define a straight line.) Often it is more convenient to obtain *parametric equations* for the line, which can also be written in 'symmetric form'. See also *vector equation* (of a line).

strictly decreasing See *decreasing function* and *decreasing sequence.*

strictly determined game A *matrix game* is **strictly determined** if there is an entry in the matrix that is the smallest in its row and the largest in its column. If the game is strictly determined then, when the two players R and C use *conservative strategies*, the pay-off is always the same and is the value of the game.

The game given by the matrix on the left below is strictly determined. The conservative strategy for R is to choose Row 3, and the conservative strategy for C is to choose Column 1. The value of the game is 5. The game on the right is not strictly determined.

$$\begin{bmatrix} 2 & 8 & 4 \\ 3 & 1 & 4 \\ 5 & 6 & 7 \end{bmatrix}, \qquad \begin{bmatrix} 2 & 8 & 4 \\ 6 & 1 & 3 \\ 5 & 4 & 7 \end{bmatrix}.$$

strictly included See *proper subset.*

strictly increasing See *increasing function* and *increasing sequence.*

strictly monotonic See *monotonic function* and *monotonic sequence.*

string See *elastic string* and *inextensible string.*

Student See *Gosset, William Sealy.*

Student's t-distribution = *t-distribution.*

subdivision (of an interval) = *partition* (of an interval).

subfield Let *F* be a *field* with operations of addition and multiplication. If *S* is a subset of *F* that forms a field with the same operations, then *S* is a **subfield** of *F*. For example, the set **Q** of rational numbers forms a subfield of the field **R** of real numbers.

subgroup Let *G* be a *group* with a given operation. If *H* is a subset of *G* that forms a group with the same operation, then *H* is a **subgroup** of *G*. For example, $\{1, i, -1, -i\}$ forms a subgroup of the group of all non-zero complex numbers with multiplication.

submatrix A **submatrix** of a *matrix* **A** is obtained from **A** by deleting from **A** some number of rows and some number of columns. For example, suppose that **A** is a 4×4 matrix and that $\mathbf{A} = [a_{ij}]$. Deleting the first and third rows and the second column gives the submatrix

$$\begin{bmatrix} a_{21} & a_{23} & a_{24} \\ a_{41} & a_{43} & a_{44} \end{bmatrix}.$$

subring Let R be a *ring* with operations of addition and multiplication. If S is a subset of R that forms a ring with the same operations, then S is a **subring** of R. For example, the set $\mathbf{Z}$ of all integers forms a subring of the ring $\mathbf{R}$ of all real numbers, and the set of all even integers forms a subring of $\mathbf{Z}$.

subset The set A is a **subset** of the set B if every element of A is an element of B. When this is so, A is **included** in B, written $A \subseteq B$, and B **includes** A, written $B \supseteq A$. The following properties hold:

(i) For all sets A, $\emptyset \subseteq A$ and $A \subseteq A$.
(ii) For all sets A and B, $A = B$ if and only if $A \subseteq B$ and $B \subseteq A$.
(iii) For all sets A, B and C, if $A \subseteq B$ and $B \subseteq C$, then $A \subseteq C$.

See also *proper subset*.

subspace See *n-dimensional space*.

substitution See *integration*.

subtend The angle that a line segment or arc with end-points A and B **subtends** at a point P is the angle APB. For example, one of the circle theorems may be stated as follows: the angle subtended by a diameter of a circle at any point on the circumference is a right angle.

subtraction (of matrices) For matrices **A** and **B** of the same order, the operation of **subtraction** is defined by taking $\mathbf{A} - \mathbf{B}$ to mean $\mathbf{A} + (-\mathbf{B})$, where $-\mathbf{B}$ is the *negative* of **B**. Thus if $\mathbf{A} = [a_{ij}]$ and $\mathbf{B} = [b_{ij}]$, then $\mathbf{A} - \mathbf{B} = \mathbf{C}$, where $\mathbf{C} = [c_{ij}]$ and $c_{ij} = a_{ij} - b_{ij}$.

sufficient estimator An *estimator* of a parameter θ that gives as much information about θ as is possible from the sample. The sample mean is a sufficient estimator of the population mean of a normal distribution.

sum (of matrices) See *addition* (of matrices).

sum (of a series) See *series*.

summation notation The finite series $a_1 + a_2 + \cdots + a_n$ can be written, using the capital Greek letter sigma, as

$$\sum_{r=1}^{n} a_r.$$

Similarly, for example,

$$1^2 + 2^2 + \cdots + 10^2 = \sum_{r=1}^{10} r^2, \qquad 1 + x + x^2 + \cdots + x^{n-1} = \sum_{r=0}^{n-1} x^r.$$

(The letter '*r*' used here could equally well be replaced by any other letter.) In a similar way, the infinite series $a_1 + a_2 + \cdots$ can be written as

$$\sum_{r=1}^{\infty} a_r.$$

The infinite series may not have a sum to infinity (see *series*), but if it does the same abbreviation is also used to denote the sum to infinity. For example, the harmonic series

$$\sum_{r=1}^{\infty} \frac{1}{r}$$

is the infinite series $1 + \frac{1}{2} + \frac{1}{3} + \frac{1}{4} + \cdots$ and has no sum to infinity. But, for $-1 < x < 1$, the geometric series $1 + x + x^2 + x^3 + \cdots$ has sum $1/(1 - x)$, and this can be written

$$\sum_{r=0}^{\infty} x^r = \frac{1}{1-x} \qquad (-1 < x < 1).$$

sum to infinity See *series* and *geometric series.*

supplement See *supplementary angles.*

supplementary angles Two angles that add up to two right angles. Each angle is the **supplement** of the other.

supplementary unit See *SI units.*

supremum (plural: suprema) See *bound.*

surd An *irrational number* in the form of a root of some number (such as $\sqrt{5}$, $\sqrt[3]{2}$ or $4^{2/3}$) or a numerical expression involving such numbers. The term is rarely used nowadays.

surface of revolution Suppose that an arc of a plane curve is rotated through one revolution about a line in the plane that does not cut the arc. The surface of the 3-dimensional figure obtained is a **surface of revolution.** See also *area of a surface of revolution.*

surjection = *onto mapping.*

surjective mapping = *onto mapping.*

symmetrical about a line A plane figure is **symmetrical about the line** l if, whenever P is a point of the figure, so too is P', where P' is the *mirror-image* of P in the line l. The line l is called a **line of symmetry**; and the figure is said to have **bilateral symmetry** or to be symmetrical by reflection in the line l. The letter A, for example, is symmetrical about the vertical line down the middle.

symmetrical about a point A plane figure is **symmetrical about the point** O if, whenever P is a point of the figure, so too is P', where O is the midpoint of $P'P$. The point O is called a **centre of symmetry**; and the figure is said to have **half-turn symmetry** about O because the figure appears the same when rotated through half a revolution about O. The letter S, for example, is symmetrical about the point at its centre.

symmetric difference For sets A and B (subsets of some *universal set*), the **symmetric difference**, denoted by $A + B$, is the set $(A \setminus B) \cup (B \setminus A)$. The notation $A \Delta B$ is also used. The set is represented by the shaded regions of the *Venn diagram* shown below. The following properties hold, for all A, B and C (subsets of some universal set E):

(i) $A + A = \varnothing, A + \varnothing = A, A + A' = E, A + E = A'$.
(ii) $A + B = (A \cup B) \setminus (A \cap B) = (A \cup B) \cap (A' \cup B')$.
(iii) $A + B = B + A$, the commutative law.
(iv) $(A + B) + C = A + (B + C)$, the associative law.
(v) $A \cap (B + C) = (A \cap B) + (A \cap C)$, the operation $\cap$ is distributive over the operation $+$.

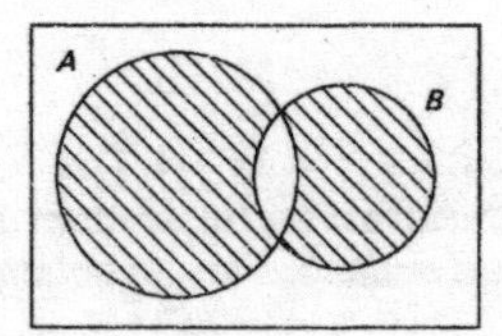

symmetric group For any set X, a permutation of X is a *one-to-one onto mapping* from X to X. If X has n elements, there are $n!$ permutations of X and the set of all of these, with *composition* of mappings as the operation, forms a *group* called the **symmetric group** of degree n, denoted by S_n.

symmetric matrix Let **A** be the square matrix $[a_{ij}]$. Then **A** is **symmetric** if $\mathbf{A}^T = \mathbf{A}$ (see *transpose*); that is, if $a_{ij} = a_{ji}$ for all i and j.

symmetric relation A *binary relation* $\sim$ on a set S is **symmetric** if, for all a and b in S, whenever $a \sim b$ then $b \sim a$.

symmetry (of a graph) Two particular **symmetries** that a given graph $y = f(x)$ may have are symmetry about the y-axis, if f is an *even function*, and symmetry about the origin, if f is an *odd function*.

synthetic division A method of finding the quotient and remainder when a polynomial $f(x)$ is divided by a factor $x - h$, in which the numbers are laid out in the form of a table. The rule to remember is that, at each step, 'you multiply and then add'. To divide $ax^3 + bx^2 + cx + d$ by $x - h$, set up a table as follows:

h	a	b	c	d
	0	...	...	...
	...	...	...	...

Working from the left, each total, written below the line, is multiplied by h and entered above the line in the next column. The two numbers in that column are added to form the next total:

h	a	b	c	d
	0	ah	...	...
	a	$ah + b$	...	...

The last total is equal to the remainder, and the other numbers below the line give the coefficients of the quotient. By the *Remainder Theorem*, the remainder equals $f(h)$.

For example, suppose that $2x^3 - 7x^2 + 5x + 11$ is to be divided by $x - 2$. The resulting table is:

2	2	−7	5	11
	0	4	−6	−2
	2	−3	−1	9

So the remainder equals 9, and the quotient is $2x^2 - 3x - 1$. The calculations here correspond exactly to those that are made when the polynomial is evaluated for $x = 2$ by *nested multiplication*, to obtain $f(2) = 9$.

Système International d'Unités See *SI units*.

system of forces The collection of forces acting on a particle, a system of particles or a rigid body.

system of particles A collection of particles whose motion is under investigation. In a given mathematical model, the particles may be able to move freely relative to each other, or they may be connected by light rods, springs or strings.

T

tangent See *trigonometric function.*

tangent (to a curve) Let P be a point on a (plane) curve. Then the **tangent** to the curve at P is the line through P that touches the curve at P. The gradient of the tangent at P is equal to the gradient of the curve at P. See also *gradient* (of a curve).

tangent line See *tangent plane.*

tangent plane Let P be a point on a smooth surface. The tangent at P to any curve through P on the surface is called a **tangent line** at P, and the tangent lines are all perpendicular to a line through P called the **normal** to the surface at P. The tangent lines all lie in the plane through P perpendicular to this normal, and this plane is called the **tangent plane** at P. (A precise definition of 'smooth' cannot be given here. A tangent plane does *not* exist, however, at a point on the edge of a cube or at the vertex of a (double) cone, for example.)

At a point P on the surface, it may be that all the points near P (apart from P itself) lie on one side of the tangent plane at P. On the other hand, this may not be so—there may be some points close to P on one side of the tangent plane and some on the other. In this case, the tangent plane cuts the surface in two curves that intersect at P. This is what happens at a *saddle-point*.

tanh See *hyperbolic function.*

Tartaglia, Niccolò (1499–1557) Italian mathematician remembered as the discoverer of a method of solving cubic equations, though he may not have been the first to find a solution. His method was published by Cardano, to whom he had communicated it in confidence. Born Niccolò Fontana, his adopted nickname means 'Stammerer'.

tautochrone As well as being the solution to the *brachistochrone* problem, the *cycloid* has another property. Suppose that a cycloid is positioned as in the diagram for the brachistochrone problem. If a particle starts from rest at any point of the cycloid and travels along the curve under the force of gravity, the time it takes to reach the lowest point is independent of its starting point. So the cycloid is also called the **tautochrone** (from the Greek for 'same time').

tautology A *compound statement* that is true for all possible truth values of its components. For example, $p \Rightarrow (q \Rightarrow p)$ is a tautology, as can be seen by calculating its *truth table.*

Taylor, Brook (1685–1731) English mathematician who contributed to the development of calculus. His text of 1715 contains what has become known as the *Taylor series*. Its importance was not appreciated until much later, but it had in fact been discovered earlier by James Gregory and others.

Taylor polynomial, Taylor series (or **expansion**) See *Taylor's Theorem*.

Taylor's Theorem Applied to a suitable function f, **Taylor's Theorem** gives a polynomial, called a **Taylor polynomial**, of any required degree, that is an approximation to $f(x)$.

THEOREM: Let f be a function such that, in an interval I, the derived functions $f^{(r)}$ $(r = 1, \ldots, n)$ are continuous, and suppose that $a \in I$. Then, for all x in I,

$$f(x) = f(a) + \frac{f'(a)}{1!}(x-a) + \frac{f''(a)}{2!}(x-a)^2 + \cdots + \frac{f^{(n-1)}(a)}{(n-1)!}(x-a)^{n-1} + R_n,$$

where various forms for the **remainder** R_n are available.

Two possible forms for R_n are

$$R_n = \frac{1}{(n-1)!}\int_a^x f^{(n)}(t)(x-t)^{n-1}\,dt \quad \text{and}$$

$$R_n = \frac{f^{(n)}(c)}{n!}(x-a)^n,$$

where c lies between a and x. By taking $x = a + h$, where h is small (positive or negative), the formula

$$f(a+h) = f(a) + \frac{f'(a)}{1!}h + \frac{f''(a)}{2!}h^2 + \cdots + \frac{f^{(n-1)}(a)}{(n-1)!}h^{n-1} + R_n$$

is obtained, where the second form of the remainder now becomes

$$R_n = \frac{f^{(n)}(a+k)}{n!}h^n,$$

and k lies between 0 and h. This enables $f(a+h)$ to be determined up to a certain degree of accuracy, the remainder R_n giving the *error*. Suppose now that, for the function f, Taylor's Theorem holds for all values of n, and that $R_n \to 0$ as $n \to \infty$; then an infinite series can be obtained whose sum is $f(x)$. In such a case, it is customary to write

$$f(x) = f(a) + \frac{f'(a)}{1!}(x-a) + \frac{f''(a)}{2!}(x-a)^2 + \cdots.$$

This is the **Taylor series** (or **expansion**) for f at (or about) a. The special case with $a = 0$ is the *Maclaurin series* for f.

t-distribution The continuous probability *distribution* of a random variable formed from the ratio of a random variable with a standard *normal distribution* and the square root of a random variable with a *chi-squared distribution* divided by its degrees of freedom. Alternatively, it is formed from the square root of a random variable with an *F-distribution* in which the numerator has one degree of freedom. The shape, which depends on the degrees of freedom, is similar in shape to a standard normal distribution, but is less peaked and has fatter tails. The distribution is used to test for significant differences between a sample mean and the population mean, and can be used to test for differences between two sample means. The table gives, for different degrees of freedom, the values corresponding to certain values of α, to be used in a *one-tailed* or two-tailed *t-test*, as explained below.

α	2α	$\nu = 1$	2	3	4	6	8	10	15	20	30	60	∞
0.05	0.10	6.34	2.92	2.35	2.13	1.94	1.86	1.81	1.75	1.72	1.70	1.67	1.64
0.025	0.05	12.71	4.30	3.18	2.78	2.45	2.31	2.23	2.13	2.09	2.04	2.00	1.96
0.01	0.02	31.82	6.96	4.54	3.75	3.14	2.90	2.76	2.60	2.53	2.46	2.39	2.33
0.005	0.01	63.66	9.92	5.84	4.60	3.71	3.36	3.17	2.95	2.84	2.75	2.66	2.58

Corresponding to the first column value α, the table gives the one-tailed value $t_{\alpha,\nu}$ such that $\Pr(t > t_{\alpha,\nu}) = \alpha$, for the *t*-distribution with ν degrees of freedom. Corresponding to the second column value 2α, the table gives the two-tailed value $t_{\alpha,\nu}$ such that $\Pr(|t| > t_{\alpha,\nu}) = 2\alpha$. Interpolation may be used for values of ν not included.

techniques of integration See *integration.*

tension An internal force that exists at each point of a string or spring. If the string or spring were cut at any particular point, then equal and opposite forces applied to the two cut ends would be needed to maintain the illusion that it remains uncut. The magnitude of these applied forces is the magnitude of the tension at that point. When a particle is suspended from a fixed point by a light string, the tension in the string exerts a force vertically upwards on the particle, and it exerts an equal and opposite force on the point of suspension.

tera- Prefix used with *SI units* to denote multiplication by 10^{12}.

term See *sequence* and *series.*

terminal speed When an object falls to Earth from a great height, its speed, in certain *mathematical models*, tends to a value called its **terminal speed.** One possible mathematical model gives the equation $m\ddot{\mathbf{r}} = -mg\mathbf{j} - c\dot{\mathbf{r}}$, where m is the mass of the particle and $\mathbf{j}$ the unit vector in the direction vertically upwards. The second term on the right-hand side, in which c is a positive constant, arises from the *air resistance.* The velocity corresponding to $\ddot{\mathbf{r}} = 0$ is equal to $(-mg/c)\mathbf{j}$, which is called the **terminal velocity.** The terminal speed is the magnitude mg/c of this velocity. If, instead, the equation $m\ddot{\mathbf{r}} = -mg\mathbf{j} - c|\dot{\mathbf{r}}|\dot{\mathbf{r}}$ is used, the terminal speed is $\sqrt{mg/c}$ and the terminal velocity is $-\sqrt{mg/c}\mathbf{j}$. As the particle falls, its velocity tends to the terminal velocity as t increases, irrespective of the initial conditions.

terminating decimal See *decimal representation.*

ternary relation See *relation.*

ternary representation The representation of a number to *base* 3.

tessellation In its most general form, a **tessellation** is a covering of the plane with shapes. Often the shapes are polygons and the pattern is in some sense repetitive. A tessellation is **regular** if it consists of congruent regular polygons. There are just the three possibilities shown here: the polygon is either an equilateral triangle, a square or a regular hexagon.

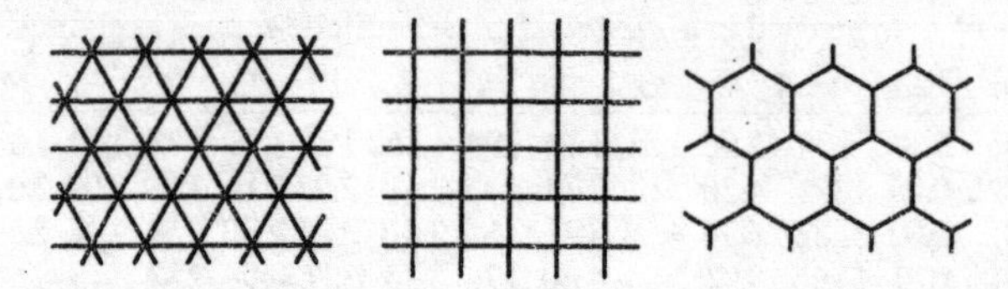

A tessellation is **semi-regular** if it consists of regular polygons, not all congruent. It can be shown that there are just eight of these, one of which has two forms that are mirror-images of each other. They use triangles, squares, hexagons, octagons and dodecagons. For example, one consisting of octagons and squares and another consisting of hexagons and triangles, shown here, may be familiar as patterns of floor-coverings.

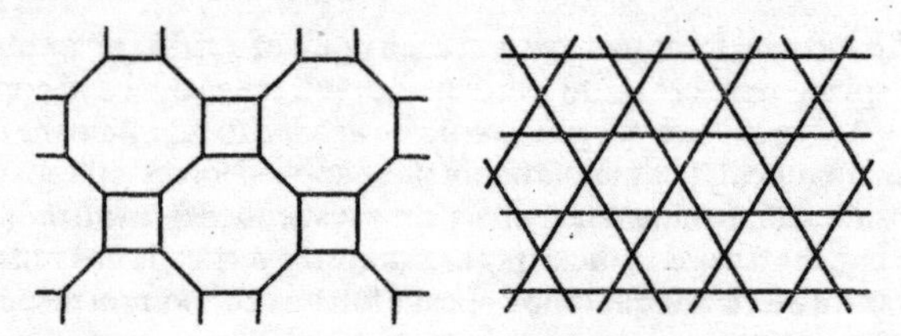

The possibilities are unlimited if tessellations in which the polygons are not regular are considered. For example, there are many interesting tessellations using congruent non-regular pentagons.

test statistic A *statistic* used in *hypothesis testing* that has a known distribution if the null hypothesis is true.

tetrahedral number An integer of the form $\frac{1}{6}n(n+1)(n+2)$, where n is a positive integer. This number equals the sum of the first n *triangular numbers*. The first few tetrahedral numbers are 1, 4, 10 and 20, and the reason for the name can be seen from the figure.

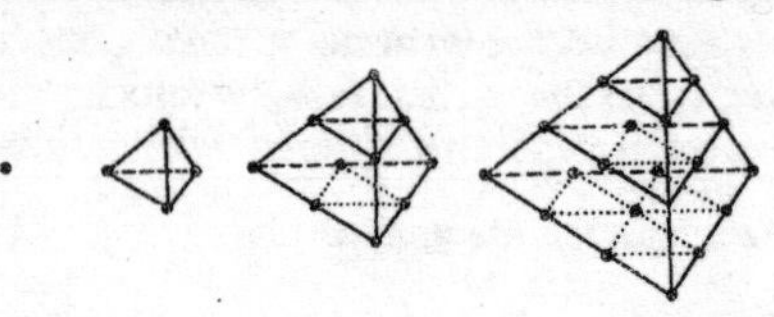

tetrahedron (plural: tetrahedra) A solid figure bounded by four triangular faces, with four vertices and six edges. A regular tetrahedron has equilateral triangles as its faces, and so all its edges have the same length.

Thales of Miletus (about 585 BC) Greek philosopher frequently regarded as the first mathematician in the sense of being one to whom particular discoveries have been attributed. These include a number of geometrical propositions, including the theorem that the angle in a semicircle is a right angle, and methods of measuring heights by means of shadows and calculating the distances of ships at sea.

theorem A mathematical statement established by means of a proof.

third derivative See *higher derivative*.

Thomson, William See *Kelvin, Lord*.

tiling = *tessellation*.

time In the real world, the passage of **time** is a universal experience which clocks of various kinds have been designed to measure. In a *mathematical model*, time is represented by a real variable, usually denoted by t, with $t = 0$ corresponding to some suitable starting-point. An *observer* associated with a *frame of reference* has the capability to measure the duration of time intervals between events occurring in the problem being investigated.

Time has the dimension T, and the SI unit of measurement is the *second*.

time series A sequence of observations taken over a period of time, usually at equal intervals. Time series analysis is concerned with identifying the factors that influence the variation in a time series, perhaps with a view to predicting what will happen in the future. For many time series, two important components are **seasonal variation**, which is periodic change usually in yearly cycles, and **trend**, which is the long-term change in the seasonally adjusted figures.

torque = *moment*.

torus (plural: tori) Suppose that a circle of radius a is rotated through one revolution about a line, in the plane of the circle, a distance b from the centre of the circle, where $b > a$. The resulting surface or solid is called a **torus**, the shape of a 'doughnut' or 'anchor ring'. The surface area of such a torus is equal to $4\pi^2 ab$ and its volume equals $2\pi^2 a^2 b$.

total force The (vector) sum of all the forces acting on a particle, a system of particles or a rigid body.

total probability law Let $A_1, A_2, \ldots, A_n$ be mutually exclusive events whose union is the whole sample space of an experiment. Then the **total probability law** is the following formula for the probability of an event B:

$$\Pr(B) = \Pr(B \mid A_1)\Pr(A_1) + \Pr(B \mid A_2)\Pr(A_2) + \cdots + \Pr(B \mid A_n)\Pr(A_n).$$

Tower of Brahma See *Tower of Hanoi*.

Tower of Hanoi Imagine three poles with a number of discs of different sizes initially all on one of the poles in decreasing order. The problem is to transfer all the discs to one of the other poles, moving the discs individually from pole to pole, so that one disc is never placed above another of smaller diameter. A version with 8 discs, known as the **Tower of Hanoi**, was invented by Edouard Lucas and sold as a toy in 1883, but the idea may be much older. The toy referred to a legend about the **Tower of Brahma** which has 64 discs being moved from pole to pole in the same way by a group of priests, the prediction being that the world will end when they have finished their task.

If t_n is the number of moves it takes to transfer n discs from one pole to another, then $t_1 = 1$ and $t_{n+1} = 2t_n + 1$. It can be shown that the solution of this *difference equation* is $t_n = 2^n - 1$, so the original Tower of Hanoi puzzle takes 255 moves, and the Tower of Brahma takes $2^{64} - 1$.

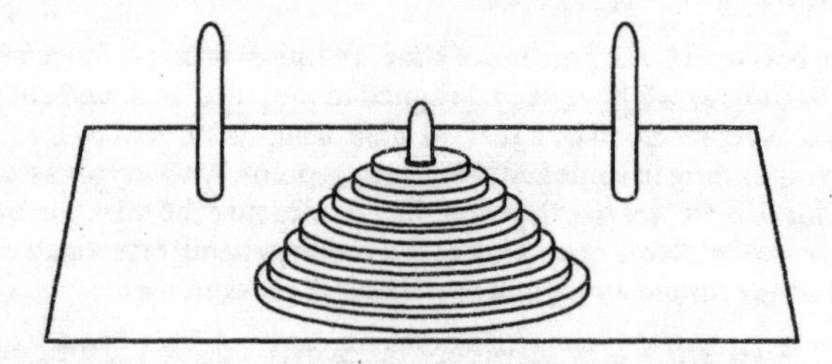

trace The **trace** of a square *matrix* is the sum of the entries in the *main diagonal*.

trajectory The path traced out by a *projectile*.

transcendental number A real number that is not a root of a polynomial equation with integer coefficients. In other words, a number is transcendental if it is not an *algebraic number*. In 1873, Hermite showed that e is transcendental; and it was shown by Lindemann, in 1882, that π is transcendental.

transformation (of the plane) Let S be the set of points in the plane. A **transformation** of the plane is a *one-to-one mapping* from S to S. The most important transformations of the plane are the **linear transformations**, which are those that, in terms of Cartesian coordinates, can be represented by linear equations. For a linear transformation T, there are constants a, b, c, d, h and k such that T maps the point P with coordinates (x, y) to the point P' with coordinates (x', y'), where

$$x' = ax + by + h,$$
$$y' = cx + dy + k.$$

When $h = k = 0$, the origin O is a **fixed point**, since T maps O to itself, and then the transformation can be written $\mathbf{x}' = \mathbf{A}\mathbf{x}$, where

$$\mathbf{x} = \begin{bmatrix} x \\ y \end{bmatrix}, \quad \mathbf{x}' = \begin{bmatrix} x' \\ y' \end{bmatrix}, \quad \mathbf{A} = \begin{bmatrix} a & b \\ c & d \end{bmatrix}.$$

Examples of such transformations are *rotations* about O, *reflections* in lines through O and *dilatations* from O. *Translations* are examples of linear transformations in which O is not a fixed point.

transition matrix, transition probability See *Markov chain*.

transitive relation A *binary relation* $\sim$ on a set S is **transitive** if, for all a, b and c in S, whenever $a \sim b$ and $b \sim c$ then $a \sim c$.

translation (of the plane) *A transformation* of the plane in which a point P with coordinates (x, y) is mapped to the point P' with coordinates (x', y'), where $x' = x + h$, $y' = y + k$. Thus the origin O is mapped to the point O' with coordinates (h, k), and the point P is mapped to the point P', where the *directed line-segment* $\overrightarrow{PP'}$ has the same direction and length as $\overrightarrow{OO'}$.

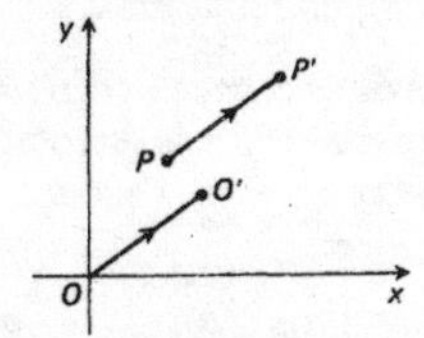

translation of axes (in the plane) Suppose that a Cartesian coordinate system has a given x-axis and y-axis with origin O and given unit length, so that a typical point P has coordinates (x, y). Let O' be the point with coordinates (h, k), and consider a new coordinate system with X-axis and Y-axis parallel and similarly directed to the x-axis and y-axis, with the same unit length, with origin at O'. With respect to the new coordinate system, the point P has coordinates (X, Y). The old and new coordinates in such a **translation of axes** are related by $x = X + h$, $y = Y + k$, or, put another way, $X = x - h$, $Y = y - k$.

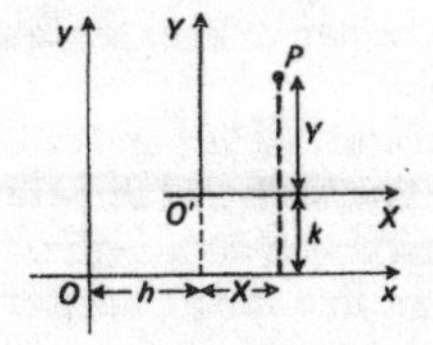

This procedure is useful for investigating, for example, a curve with equation $9x^2 + 4y^2 - 18x + 16y - 11 = 0$. Completing the square in x and y gives $9(x-1)^2 + 4(y+2)^2 = 36$, and so, with respect to a new coordinate system with origin at the point $(1, -2)$, the curve has the simple equation $9X^2 + 4Y^2 = 36$.

translation of axes (in 3-dimensional space) Just like a translation of axes in the plane, a **translation of axes** in 3-dimensional space can be made, to a new origin with coordinates (h, k, l), with new axes parallel and similarly directed to the old. The old coordinates (x, y, z) and new coordinates (X, Y, Z) are related by $x = X + h$, $y = Y + k$, $z = Z + l$, or, to put it another way, $X = x - h$, $Y = y - k$, $Z = z - l$.

transportation problem A problem in which units of a certain product are to be transported from a number of factories to a number of retail outlets in a way that minimizes the total cost. For example, suppose that there are m factories and n retail outlets, and that the transportation costs are specified by an $m \times n$ matrix $[c_{ij}]$, where c_{ij} is the cost, in suitable units, of transporting one unit of the product from the i-th factory to the j-th retail outlet. Suppose also that the maximum number of units that each factory can supply and the minimum number of units that each outlet requires are specified. By introducing suitable variables, the problem of minimizing the total cost can be formulated as a *linear programming* problem.

transpose The **transpose** of an $m \times n$ matrix is the $n \times m$ matrix obtained by interchanging the rows and columns. The transpose of $\mathbf{A}$ is denoted by $\mathbf{A}^T$, $\mathbf{A}^t$ or $\mathbf{A}'$. Thus if $\mathbf{A} = [a_{ij}]$, then $\mathbf{A}^T = [a'_{ij}]$, where $a'_{ij} = a_{ji}$; that is,

$$\mathbf{A} = \begin{bmatrix} a_{11} & a_{12} & \cdots & a_{1n} \\ a_{21} & a_{22} & \cdots & a_{2n} \\ \vdots & \vdots & \ddots & \vdots \\ a_{m1} & a_{m2} & \cdots & a_{mn} \end{bmatrix}, \qquad \mathbf{A}^T = \begin{bmatrix} a_{11} & a_{21} & \cdots & a_{m1} \\ a_{12} & a_{22} & \cdots & a_{m2} \\ \vdots & \vdots & \ddots & \vdots \\ a_{1n} & a_{2n} & \cdots & a_{mn} \end{bmatrix}.$$

The following properties hold, for matrices $\mathbf{A}$ and $\mathbf{B}$ of appropriate orders:

(i) $(\mathbf{A}^T)^T = \mathbf{A}$.
(ii) $(\mathbf{A} + \mathbf{B})^T = \mathbf{A}^T + \mathbf{B}^T$.
(iii) $(k\mathbf{A})^T = k\mathbf{A}^T$.
(iv) $(\mathbf{AB})^T = \mathbf{B}^T\mathbf{A}^T$.

transversal A straight line that intersects a given set of two or more straight lines in the plane. When a transversal intersects a given pair of lines, eight angles are formed; the four angles between the pair of lines are **interior angles** and the four others are **exterior angles**.

An interior angle between the transversal and one line and an exterior angle between the transversal and the other line, such that these two angles are on the same side of the transversal, are **corresponding angles**. An interior angle between the transversal and one line and an interior angle between the transversal and the other line, such that these two angles are on opposite sides of the transversal, are **alternate angles**. The given two lines are parallel if and only if corresponding angles are equal: they are also parallel if and only if alternate angles are equal.

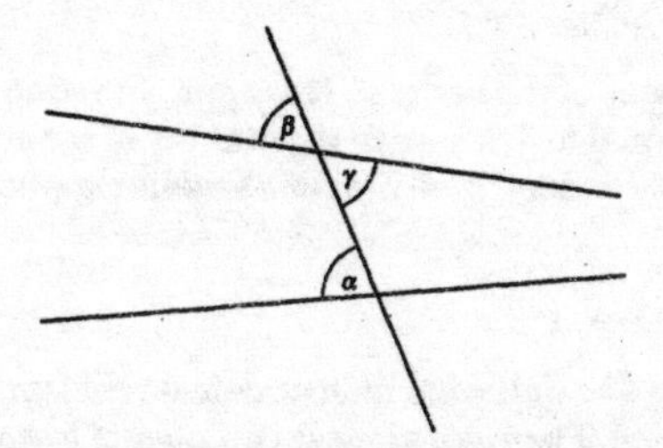

The figure shows a transversal intersecting two lines. The angles α and β are corresponding angles and α and γ are alternate angles.

transverse axis See *hyperbola.*

transverse component See *radial and transverse components.*

trapezium (plural: trapezia) A quadrilateral with two parallel sides. If the parallel sides have lengths a and b, and the distance between them is h, the area of the trapezium equals $\frac{1}{2}h(a+b)$.

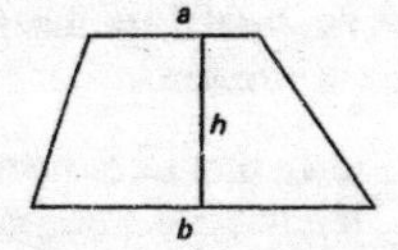

trapezium rule An approximate value can be found for the definite integral

$$\int_a^b f(x)\,dx,$$

using the values of $f(x)$ at equally spaced values of x between a and b, as follows. Divide $[a, b]$ into n equal subintervals of length h by the partition

$$a = x_0 < x_1 < x_2 < \cdots < x_{n-1} < x_n = b,$$

where $x_{i-1} - x_i = h = (b-a)/n$. Denote $f(x_i)$ by f_i, and let P_i be the point (x_i, f_i). If the line segment P_iP_{i+1} is used as an approximation to the curve $y = f(x)$ between x_i and x_{i+1}, the area under that part of the curve is approximately the area of the trapezium shown in the figure, which equals $\frac{1}{2}h(f_i + f_{i+1})$. By adding up the areas of such trapezia between a and b, the resulting **trapezium rule** gives

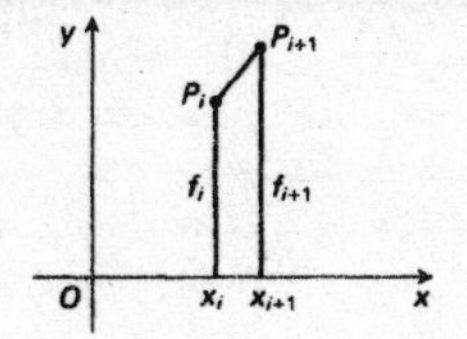

$$\tfrac{1}{2}h(f_0 + 2f_1 + 2f_2 + \cdots + 2f_{n-1} + f_n)$$

as an approximation to the value of the integral. This approximation has an *error* that is roughly proportional to $1/n^2$; when the number of subintervals is doubled, the error is roughly divided by 4. *Simpson's rule* is significantly more accurate.

trapezoidal rule = *trapezium rule.*

travelling salesman problem The following mathematical problem, derived from a real-life situation. There are a certain number of towns, with routes between them, each route, from town i to town j, being assigned a value c_{ij} giving the distance from town i to town j. The problem is to find how to visit all the towns, returning to the starting point, in a way that minimizes the total distance travelled. (Alternatively, c_{ij} can be the cost of travelling, or the time taken to travel, from town i to town j, and the total cost or time is to be minimized.) Theoretically, the problem can be solved by considering all the possible routes. But if there are n towns, the number to be considered is $(n-1)!/2$ and, for all but the smallest values of n, this is impossible even for a large computer. What is sought is a manageable algorithm that works in what is considered to be a reasonable time. Such algorithms are known that find a way for which the total distance travelled is close to, but not necessarily quite equal to, the minimum.

tree A *connected graph* with no *cycles.* It can be shown that a connected simple graph with n vertices is a tree if and only if it has $n-1$ edges. The figure shows all the trees with up to five vertices.

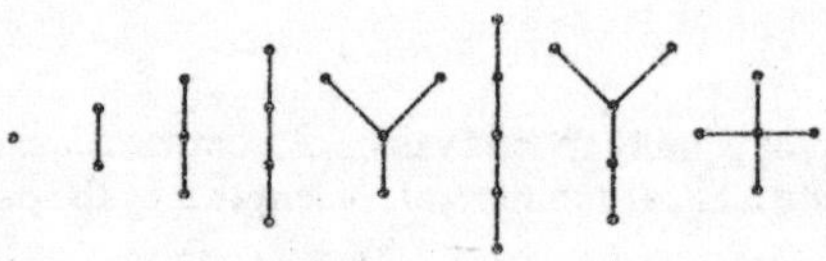

Particularly in applications, one of the vertices (which may be called **nodes**) of a tree may be designated as the **root**, and the tree may be drawn with the vertices at different levels indicating their distance from the root. A rooted tree in which every vertex (except the root, of degree 2) has degree either 1 or 3, such as the one shown below, is called a **binary tree**.

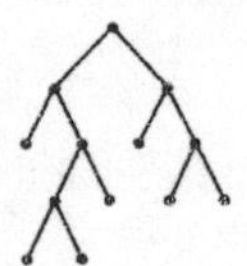

trend See *time series.*

triangle Using the properties of angles made when one line cuts a pair of parallel lines, it is proved, as illustrated on the left, that the angles of a triangle ABC add up to 180°. By considering separate areas, illustrated on the right, it can be shown that the area of the triangle is half that of the rectangle shown, and so the area of a triangle is 'half base times height'.

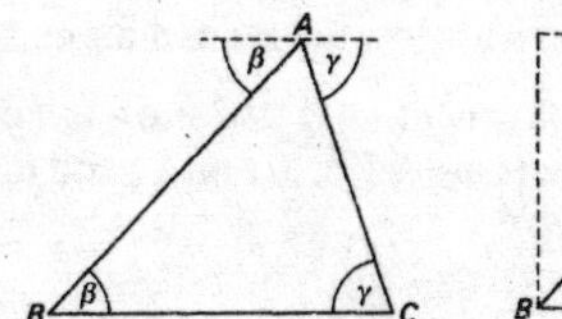

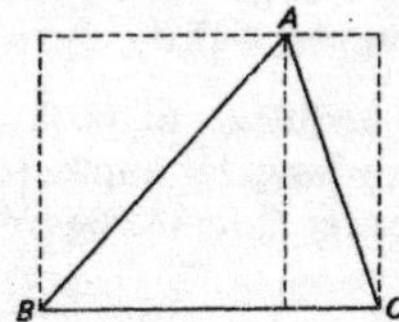

If now A, B and C denote the angles of the triangle, and a, b and c the lengths of the sides opposite them, the following results hold:

(i) The area of the triangle equals $\frac{1}{2}bc \sin A$.

(ii) The **sine rule**:

$$\frac{a}{\sin A} = \frac{b}{\sin B} = \frac{c}{\sin C} = 2R,$$

where R is the radius of the *circumcircle*.

(iii) The **cosine rule**: $a^2 = b^2 + c^2 - 2bc \cos A$, or, in another form,

$$\cos A = \frac{b^2 + c^2 - a^2}{2bc}.$$

(iv) **Hero's formula**: Let $s = \frac{1}{2}(a + b + c)$. Then the area of the triangle equals $\sqrt{s(s-a)(s-b)(s-c)}$.

triangle inequality (for complex numbers) If z_1 and z_2 are complex numbers, then $|z_1 + z_2| \leq |z_1| + |z_2|$. This result is known as the **triangle inequality** because it follows from the fact that $|OQ| \leq |OP_1| + |P_1Q|$, where P_1, P_2 and Q represent z_1, z_2 and $z_1 + z_2$ in the *complex plane*.

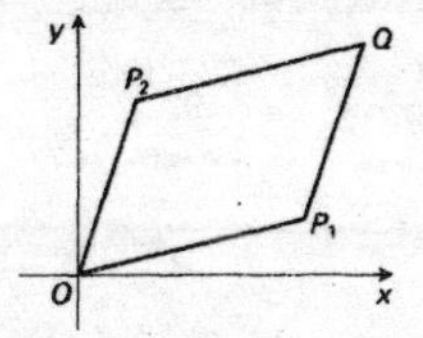

triangle inequality (for points in the plane) For points A, B and C in the plane, $|AC| \leq |AB| + |BC|$. This result, the **triangle inequality**, says that the length of one side of a triangle is less than or equal to the sum of the lengths of the other two sides.

triangle inequality (for vectors) Let $|\mathbf{a}|$ denote the length of the vector **a**. For vectors **a** and **b**, $|\mathbf{a}+\mathbf{b}| \leq |\mathbf{a}| + |\mathbf{b}|$. This result is known as the **triangle inequality**, since it is equivalent to saying that the length of one side of a triangle is less than or equal to the sum of the lengths of the other two sides.

triangular matrix A square matrix that is either lower triangular or upper triangular. It is **lower triangular** if all the entries above the main diagonal are zero, and **upper triangular** if all the entries below the main diagonal are zero.

triangular number An integer of the form $\frac{1}{2}n(n+1)$, where n is a positive integer. The first few triangular numbers are 1, 3, 6, 10 and 15, and the reason for the name can be seen from the figure.

trigonometric function Though the distinction tends to be overlooked, each of the trigonometric functions has two forms, depending upon whether degrees or radians are used.

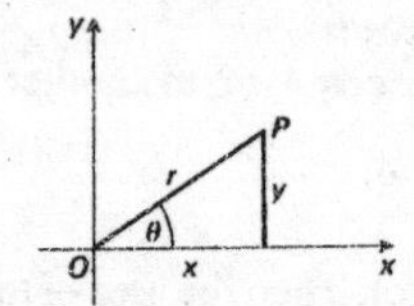

USING DEGREES: The basic trigonometric functions, **cosine**, **sine** and **tangent**, are first introduced by using a right-angled triangle, but $\cos\theta^\circ$, $\sin\theta^\circ$ and $\sin\theta^\circ$ can also be defined when θ is larger than 90 and when θ is negative. Let P be a point (not at O) with Cartesian coordinates (x, y). Suppose that OP makes an angle of θ° with the positive x-axis and that $|OP| = r$. Then the following are the definitions: $\cos\theta^\circ = x/r$, $\sin\theta^\circ = y/r$, and (when $x \neq 0$) $\tan\theta^\circ = y/x$. It follows that $\tan\theta^\circ = \sin\theta^\circ/\cos\theta^\circ$, and that $\cos^2\theta^\circ + \sin^2\theta^\circ = 1$. Some of the most frequently required values are given in the table.

θ	0	30	45	60	90
$\cos\theta^\circ$	1	$\frac{\sqrt{3}}{2}$	$\frac{1}{\sqrt{2}}$	$\frac{1}{2}$	0
$\sin\theta^\circ$	0	$\frac{1}{2}$	$\frac{1}{\sqrt{2}}$	$\frac{\sqrt{3}}{2}$	1
$\tan\theta^\circ$	0	$\frac{1}{\sqrt{3}}$	1	$\sqrt{3}$	not defined

2 sin positive	1 all positive
3 tan positive	4 cos positive

The point P may be in any of the four quadrants. By considering the signs of x and y, the quadrants in which the different functions take positive values can be found, and are shown in the figure above. The following are useful for calculating values, when P is in quadrant 2, 3 or 4:

$$\cos(180-\theta)^\circ = -\cos\theta^\circ, \qquad \sin(180-\theta)^\circ = \sin\theta^\circ,$$
$$\cos(180+\theta)^\circ = -\cos\theta^\circ, \qquad \sin(180+\theta)^\circ = -\sin\theta^\circ,$$
$$\cos(-\theta)^\circ = \cos\theta^\circ, \qquad \sin(-\theta)^\circ = -\sin\theta^\circ.$$

The functions cosine and sine are periodic, of period 360; that is to say, $\cos(360+\theta)^\circ = \cos\theta^\circ$ and $\sin(360+\theta)^\circ = \sin\theta^\circ$. The function tangent is periodic, of period 180; that is, $\tan(180+\theta)^\circ = \tan\theta^\circ$.

USING RADIANS: In more advanced work, it is essential that angles are always measured in *radians*. It is necessary to introduce new trigonometric functions $\cos x$, $\sin x$ and $\tan x$, where now x is a real number. These agree with the former functions, if x is treated as being the measure in radians of the angle formerly measured in degrees. For example, since π radians $= 180^\circ$,

$$\cos\pi = \cos 180^\circ = -1, \qquad \sin\frac{\pi}{3} = \sin 60^\circ = \frac{\sqrt{3}}{2},$$
$$\tan\frac{\pi}{4} = \tan 45^\circ = \frac{1}{\sqrt{2}}.$$

The functions cos and sin are periodic with period 2π, and tan has period π.

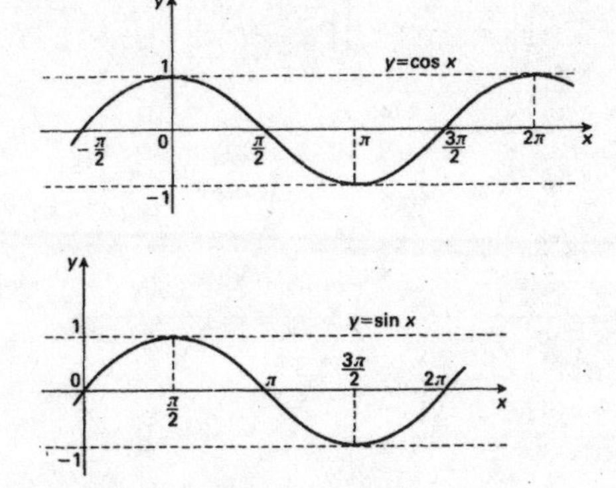

It is important to distinguish between the functions $\cos x$ and $\cos x°$. They are different functions, but are related since $\cos x° = \cos(\pi x/180)$. The same applies to sin and tan. Sometimes authors do not make the distinction and use the notation $\cos A$, for example, where A is an angle measured in degrees or radians. This is what has been done in the Table of Trigonometric Formulae (Appendix 5).

The other trigonometric functions, **cotangent**, **secant** and **cosecant**, are defined as follows:

$$\cot x = \frac{\cos x}{\sin x}, \qquad \sec x = \frac{1}{\cos x}, \qquad \operatorname{cosec} x = \frac{1}{\sin x},$$

where, in each case, values of x that make the denominator zero must be excluded from the domain. The basic identities satisfied by the trigonometric functions will be found in the Table of Trigonometric Formulae (Appendix 5).

In order to find the derivatives of the trigonometric functions, it is first necessary to show that

$$\lim_{x\to 0} \frac{\sin x}{x} = 1.$$

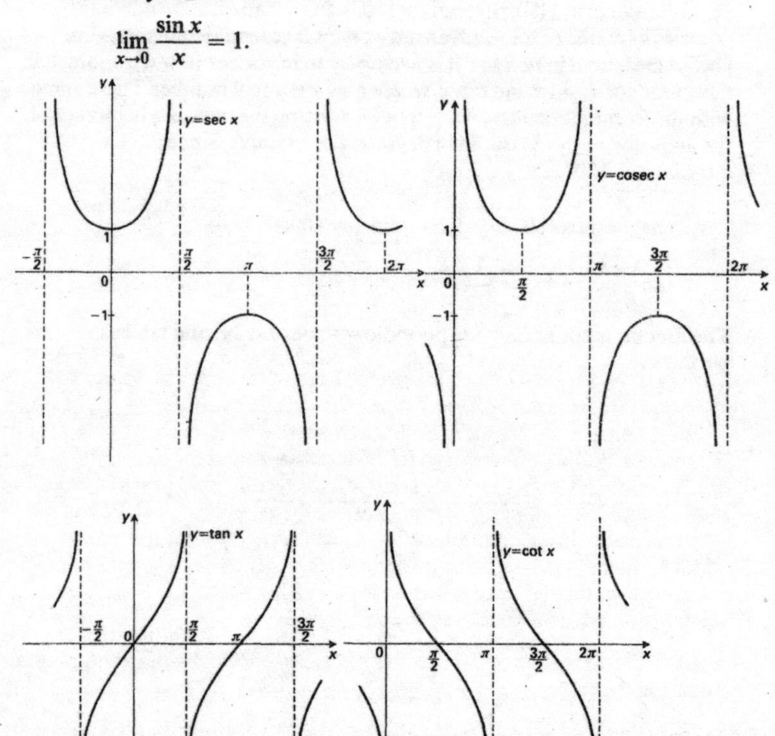

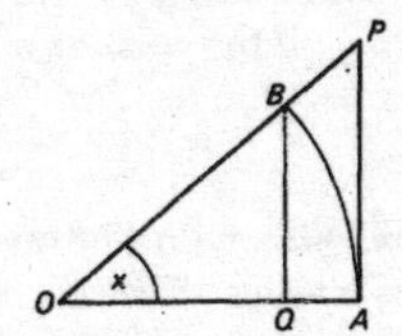

In view of the essentially geometric definition of sin x, a geometric method has to be used. By considering (when $0 < x < \pi/2$) the areas of $\triangle OBQ$, sector OAB and $\triangle OAP$, where $\angle OAB$ measures x radians, illustrated above, it can be shown that

$$\cos x < \frac{\sin x}{x} < \frac{1}{\cos x}.$$

After dealing also with $x < 0$, the required limit can be deduced. Hence it can be shown that

$$\frac{d}{dx}(\sin x) = \cos x, \qquad \frac{d}{dx}(\cos x) = -\sin x.$$

The derivatives of the other trigonometric functions are found from these, by using the rules for differentiation, and are given in the Table of Derivatives (Appendix 2).

trillion In Britain, a million cubed (10^{18}); in the United States, the number 10^{12}. Consequently, as with *billion*, use of the word can lead to ambiguity.

triple A **triple** consists of 3 objects normally taken in a particular order, denoted, for example, by (x_1, x_2, x_3).

triple product (of vectors) See *scalar triple product* and *vector triple product*.

triple root See *root*.

trisection of an angle One of the problems that the Greek geometers attempted (like the *duplication of the cube* and the *squaring of the circle*) was to find a construction, with ruler and compasses, to trisect any angle; that is, to divide it into three equal parts. (The construction for bisecting an angle is probably familiar.) Now constructions of the kind envisaged can give only lengths belonging to a class of numbers obtained, essentially, by addition, subtraction, multiplication, division and the taking of square roots. It can be shown that the trisection of certain angles is equivalent to the construction of numbers that do not belong to this class. So, in general, the trisection of an angle is impossible.

trisector The **trisectors** of an angle are the two lines that divide the angle into three equal angles.

trivial solution The solution of a *homogeneous set of linear equations* in which all the unknowns are equal to zero.

truncated cube One of the *Archimedean solids*, with 6 octagonal faces and 8 triangular faces. It can be formed by cutting off the corners of a cube in such a way that the original square faces become regular octagons.

truncated tetrahedron One of the *Archimedean solids*, with 4 hexagonal faces and 4 triangular faces. It can be formed by cutting off the corners of a (regular) *tetrahedron* in such a way that the original triangular faces become regular hexagons.

truncation Suppose that a number has more digits than can be conveniently handled or stored. In **truncation** (as opposed to *rounding*), the extra digits are simply dropped; for example, when truncated to 1 decimal place, the numbers 1.875 and 1.845 both become 1.8. See also *decimal places* and *significant figures*.

truth table The *truth value* of a *compound statement* can be determined from the truth values of its components. A table that gives, for all possible truth values of the components, the resulting truth values of the compound statement is a **truth table**. The truth table for $\neg p$ is

p	$\neg p$
T	F
F	T

and combined truth tables for $p \wedge q$, $p \vee q$ and $p \Rightarrow q$ are as follows:

p	q	$p \wedge q$	$p \vee q$	$p \Rightarrow q$
T	T	T	T	T
T	F	F	T	F
F	T	F	T	T
F	F	F	F	T

From these, any other truth table can be completed. For example, the final column below, giving the truth table for the compound statement $(p \wedge q) \vee (\neg r)$, is found by first completing columns for $p \wedge q$ and $\neg r$:

p	q	r	$p \wedge q$	$\neg r$	$(p \wedge q) \vee (\neg r)$
T	T	T	T	F	T
T	T	F	T	T	T
T	F	T	F	F	F
T	F	F	F	T	T
F	T	T	F	F	F
F	T	F	F	T	T
F	F	T	F	F	F
F	F	F	F	T	T

truth value A term whose meaning is apparent from the following usage: if a statement is true, its truth value is *T* (or TRUE); if the statement is false, its truth value is *F* (or FALSE).

***t*-test** A test to determine whether or not a sample of size *n* with mean $\bar{x}$ comes from a *normal distribution* with mean μ. The *statistic t* given by

$$t = \frac{\sqrt{n}(\bar{x} - \mu)}{s}$$

has a *t*-distribution with $n - 1$ degrees of freedom, where s^2 is the (unbiased) sample *variance*. The *t*-test can also be used to test whether a sample mean differs significantly from a population mean, and to test whether two samples have been drawn from the same population.

See *t-distribution* for a table for use in the *t*-test.

Turing, Alan Mathison (1912–1954) British mathematician and logician who conceived the notion of the *Turing machine*. During the Second World War, he was involved in cryptanalysis, the breaking of codes, and afterwards worked on the construction of some of the early digital computers and the development of their programming systems.

Turing machine A theoretical machine which operates according to extremely simple rules, invented by Turing with the aim of obtaining a mathematically precise definition of what is 'computable'. It has been generally agreed that the machine can calculate or compute anything for which there is an 'effective' *algorithm*. The resulting understanding of computability has been shown to be equivalent to other attempts at defining the concept.

turning point A point on the graph $y = f(x)$ at which $f'(x) = 0$ and $f'(x)$ changes sign. A turning point is either a *local maximum* or a *local minimum*. Some authors use 'turning point' as equivalent to *stationary point*.

twin primes A pair of prime numbers that differ by 2. For example, 29 and 31, 71 and 73, and 10 006 427 and 10 006 429, are twin primes. A conjecture that there are infinitely many such pairs has been neither proved nor disproved.

two-sided test See *hypothesis testing*.

two-tailed test See *hypothesis testing*.

Type I and **Type II errors** See *hypothesis testing*.

U

unary operation A **unary operation** on a set S is a rule that associates with any element of S a resulting element. If this resulting element is always also in S, then it is said that S is **closed under** the operation. The following are examples of unary operations: the rule that associates with each integer a its negative $-a$; the rule that associates with each non-zero real number a its inverse $1/a$; and the rule that associates with any subset A of a universal set E its complement A'.

unbiased estimator See *estimator.*

uniform A quantity in mechanics may be called **uniform** when it is, in some sense, constant. For example, a particle may be moving with uniform velocity or with uniform acceleration. A uniform rod, lamina or rigid body is one whose density is uniform; that is, whose density is the same at every point.

uniform distribution The **uniform distribution** on the interval $[a, b]$ is the continuous probability *distribution* whose *probability density function* f is given by $f(x) = 1/(b-a)$, where $a \leq x \leq b$. It has mean $(a+b)/2$ and variance $(b-a)^2/12$. There is also a discrete form: on the set $0, 1, 2, \ldots, n$, it is the probability distribution whose *probability mass function* is given by $\Pr(X = r) = 1/(n+1)$, for $r = 0, 1, 2, \ldots, n$. For example, the random variable for the winning number in a lottery has a uniform distribution on the set of all the numbers entered in the lottery.

uniform gravitational force A *gravitational force*, acting on a particular body, that is independent of the position of the body. The gravitational force on a body in a limited region near the surface of a planet is approximately uniform.

For example, the gravitational force acting on a particle of mass m near the Earth's surface, assumed to be a horizontal plane, can be taken to be $-mg\mathbf{k}$, where $\mathbf{k}$ is a unit vector directed vertically upwards and g is the magnitude of the acceleration due to gravity.

union The **union** of sets A and B (subsets of a *universal set*) is the set consisting of all objects that belong to A or B (or both), and it is denoted by $A \cup B$ (read as 'A **union** B'). Thus the term 'union' is used for both the resulting set and the operation, a *binary operation* on the set of all subsets of a universal set. The following properties hold:

(i) For all A, $A \cup A = A$ and $A \cup \emptyset = A$.
(ii) For all A and B, $A \cup B = B \cup A$; that is, $\cup$ is commutative.
(iii) For all A, B and C, $(A \cup B) \cup C = A \cup (B \cup C)$; that is, $\cup$ is associative.

In view of (iii), the union $A_1 \cup A_2 \cup \cdots \cup A_n$ of more than two sets can be written without brackets, and it may also be denoted by

$$\bigcup_{i=1}^{n} A_i.$$

For the union of two events, see *event*.

Unique Factorization Theorem The process of writing any positive integer as the product of its prime factors is probably familiar; it may be taken as self-evident that this can be done in only one way. Known as the **Unique Factorization Theorem**, this result of elementary number theory can be proved from basic axioms about the integers.

THEOREM: Any positive integer ($\neq 1$) can be expressed as a product of primes. This expression is unique except for the order in which the primes occur.

Thus, any positive integer n ($\neq 1$) can be written as $p_1^{\alpha_1} p_2^{\alpha_2} \dots p_r^{\alpha_r}$, where $p_1, p_2, \dots, p_r$ are primes, satisfying $p_1 < p_2 < \dots < p_r$, and $\alpha_1, \alpha_2, \dots, \alpha_r$ are positive integers. This is the **prime decomposition** of n. For example, writing $360 = 2^3 \times 3^2 \times 5$ shows the prime decomposition of 360.

unit See *SI units*.

unit circle In the plane, the **unit circle** is the circle of radius 1 with its centre at the origin. In Cartesian coordinates, it has equation $x^2 + y^2 = 1$. In the *complex plane*, it represents those complex numbers z such that $|z| = 1$.

unit matrix = *identity matrix*.

unit vector A vector with magnitude, or length, equal to 1. For any non-zero vector $\mathbf{a}$, a unit vector in the direction of $\mathbf{a}$ is $\mathbf{a}/|\mathbf{a}|$. The row vector $[a_1 \quad a_2 \quad \dots \quad a_n]$ or the column vector

$$\begin{bmatrix} a_1 \\ a_2 \\ \vdots \\ a_n \end{bmatrix}$$

may be called a unit vector if $\sqrt{a_1^2 + a_2^2 + \dots + a_n^2} = 1$.

universal gravitational constant = *gravitational constant*.

universal quantifier See *quantifier*.

universal set In a particular piece of work, it may be convenient to fix the **universal set** E, a set to which all the objects to be discussed belong. Then all the sets considered are subsets of E.

unstable equilibrium See *equilibrium*.

upper bound See *bound*.

upper limit See *limit of integration*.

upper triangular matrix See *triangular matrix*.

valency = *degree* (of a vertex of a graph).

Vallée-Poussin, Charles-Jean De La (1866–1962) Belgian mathematician who in 1896 proved the *Prime Number Theorem* independently of Hadamard.

value See *constant function, function* and *infinite product.*

value (of a matrix game) See *Fundamental Theorem of Game Theory.*

Vandermonde's convolution formula The following relationship between *binomial coefficients*:

$$\binom{m+n}{r} = \binom{m}{0}\binom{n}{r} + \binom{m}{1}\binom{n}{r-1} + \cdots + \binom{m}{r}\binom{n}{0}.$$

The formula may be proved by equating the coefficients of x^r on both sides of the identity $(1+x)^{m+n} = (1+x)^m(1+x)^n$.

variance A measure of the *dispersion* of a random variable or of a sample. For a random variable X, the **population variance** is the second moment about the population mean μ and is equal to $\mathrm{E}((X-\mu)^2)$ (see *expected value*). It is usually denoted by σ^2 or $\mathrm{Var}(X)$. For a sample, the **sample variance**, denoted by s^2, is the second moment of the data about the sample mean $\bar{x}$ but the denominator is usually taken as $n-1$ rather than n in order to make it an *unbiased estimator* of the population variance. So

$$s^2 = \frac{\sum(x_i - \bar{x})^2}{n-1}.$$

For computational purposes, notice that $\sum(x_i-\bar{x})^2 = \sum x_i^2 - n\bar{x}^2$.

variate = *random variable.*

varies directly, varies inversely See *proportion.*

vector In physics or engineering, the term 'vector' is used to describe a physical quantity like velocity or force that has a magnitude and a direction. Sometimes there may also be a specified point of application, but generally in mathematics that is not of concern. Thus a vector is to be 'something' that has magnitude and direction.

One approach is to define a **vector** to be an *ordered pair* consisting of a positive real number, the **magnitude** or **length**, and a direction in space. The vectors **a** and **b** are said to be equal if they have the same magnitude and the same direction. The **zero vector 0** that has magnitude 0 and no direction is also allowed.

Another way is to make use of the well-defined notion of *directed line-segment* and define a **vector** to be the collection of all directed line-segments with a given length and a given direction. If $\overrightarrow{AB}$ is a directed line-segment in the collection that is the vector **a**, it is said that $\overrightarrow{AB}$ **represents a**. If $\overrightarrow{AB}$ and $\overrightarrow{CD}$ represent the same vector, then $\overrightarrow{AB}$ and $\overrightarrow{CD}$ are parallel and have the same length. The **magnitude**, or **length**, $|\mathbf{a}|$ of the vector **a** is the length of any of the directed line-segments that represent **a**. According to this definition, two vectors are **equal** if they are the same collection of directed line-segments. It is also necessary to permit directed line-segments of length zero and define the **zero vector 0** to be the collection of all these.

In the first approach, the connection with directed line-segments is made by saying that a directed line-segment $\overrightarrow{AB}$ **represents** the vector **a** if the length and direction of $\overrightarrow{AB}$ are equal to the magnitude and direction of **a**. Some authors write $\overrightarrow{AB} = \mathbf{a}$ if $\overrightarrow{AB}$ represents the vector **a**, and then, if $\overrightarrow{AB} = \mathbf{a}$ and $\overrightarrow{CD} = \mathbf{a}$, they go on to write $\overrightarrow{AB} = \overrightarrow{CD}$. Some authors actually use the word 'vector' for what we prefer to call a directed line-segment. They then write $\overrightarrow{AB} = \overrightarrow{CD}$ if $\overrightarrow{AB}$ and $\overrightarrow{CD}$ have the same length and the same direction.

vector equation (of a line) Given a line in space, let **a** be the *position vector* of a point A on the line, and **u** any vector with direction along the line. Then the line consists of all points P whose position vector **p** is given by $\mathbf{p} = \mathbf{a} + t\mathbf{u}$ for some value of t. This is a **vector equation** of the line. It is established by noting that P lies on the line if and only if $\mathbf{p} - \mathbf{a}$ has the direction of **u** or its negative, which is equivalent to $\mathbf{p} - \mathbf{a}$ being a scalar multiple of **u**. If, instead, the line is specified by two points A and B on it, with position vectors **a** and **b**, then the line has vector equation $\mathbf{p} = (1 - t)\mathbf{a} + t\mathbf{b}$. This is obtained by setting $\mathbf{u} = \mathbf{b} - \mathbf{a}$ in the previous form.

vector equation (of a plane) Given a plane in 3-dimensional space, let **a** be the *position vector* of a point A in the plane, and **n** a *normal vector* to the plane. Then the plane consists of all points P whose position vector **p** satisfies $(\mathbf{p} - \mathbf{a}) \,.\, \mathbf{n} = 0$. This is a **vector equation** of the plane. It may also be written $\mathbf{p} \,.\, \mathbf{n} =$ constant. By supposing that **p** has components x, y, z, that **a** has components x_1, y_1, z_1, and that **n** has components l, m, n, the first form of the equation becomes $l(x - x_1) + m(y - y_1) + n(z - z_1) = 0$, and the second form becomes the standard linear equation $lx + my + nz =$ constant.

vector product Let **a** and **b** be non-zero non-parallel vectors, and let θ be the angle between them (θ in radians, with $0 < \theta < \pi$). The **vector product** $\mathbf{a} \times \mathbf{b}$ of **a** and **b** is defined as follows. Its magnitude equals $|\mathbf{a}||\mathbf{b}| \sin\theta$, and its direction is perpendicular to **a** and **b** such that **a**, **b** and $\mathbf{a} \times \mathbf{b}$ form a right-handed system. So, viewed from a position facing the direction of $\mathbf{a} \times \mathbf{b}$, the vector **a** has to be rotated clockwise through the angle θ to have the direction of **b**. If **a** is parallel to **b** or if one of them is the zero vector **0**, then $\mathbf{a} \times \mathbf{b}$ is defined to be $\mathbf{0}$. The notation $\mathbf{a} \wedge \mathbf{b}$ is also used for $\mathbf{a} \times \mathbf{b}$. The following properties hold, for all vectors $\mathbf{a}$, $\mathbf{b}$ and $\mathbf{c}$:

(i) $\mathbf{b} \times \mathbf{a} = -(\mathbf{a} \times \mathbf{b})$.
(ii) $\mathbf{a} \times (k\mathbf{b}) = (k\mathbf{a}) \times \mathbf{b} = k(\mathbf{a} \times \mathbf{b})$.
(iii) The magnitude $|\mathbf{a} \times \mathbf{b}|$ is equal to the area of the parallelogram with sides determined by $\mathbf{a}$ and $\mathbf{b}$.
(iv) $\mathbf{a} \times (\mathbf{b} + \mathbf{c}) = \mathbf{a} \times \mathbf{b} + \mathbf{a} \times \mathbf{c}$, the distributive law.
(v) If the vectors, in terms of components (with respect to standard vectors $\mathbf{i}, \mathbf{j}, \mathbf{k}$), are $\mathbf{a} = a_1\mathbf{i} + a_2\mathbf{j} + a_3\mathbf{k}$, $\mathbf{b} = b_1\mathbf{i} + b_2\mathbf{j} + b_3\mathbf{k}$, then $\mathbf{a} \times \mathbf{b} = (a_2b_3 - a_3b_2)\mathbf{i} + (a_3b_1 - a_1b_3)\mathbf{j} + (a_1b_2 - a_2b_1)\mathbf{k}$. This can be written, with an abuse of 3×3 determinant notation, as

$$\mathbf{a} \times \mathbf{b} = \begin{vmatrix} \mathbf{i} & \mathbf{j} & \mathbf{k} \\ a_1 & a_2 & a_3 \\ b_1 & b_2 & b_3 \end{vmatrix}.$$

vector projection (of a vector on a vector) Given non-zero vectors $\mathbf{a}$ and $\mathbf{b}$, let $\overrightarrow{OA}$ and $\overrightarrow{OB}$ be *directed line-segments* representing $\mathbf{a}$ and $\mathbf{b}$, and let θ be the angle between them (θ in radians, with $0 \leq \theta \leq \pi$). Let C be the *projection* of B on the line OA. The **vector projection** of $\mathbf{b}$ on $\mathbf{a}$ is the vector represented by $\overrightarrow{OC}$. Since $|OC| = |OB| \cos\theta$, this vector projection is equal to $|\mathbf{b}| \cos\theta$ times the unit vector $\mathbf{a}/|\mathbf{a}|$. Thus the vector projection of $\mathbf{b}$ on $\mathbf{a}$ equals

$$\left(\frac{\mathbf{a} \cdot \mathbf{b}}{\mathbf{a} \cdot \mathbf{a}}\right)\mathbf{a}.$$

The **scalar projection** of $\mathbf{b}$ on $\mathbf{a}$ is equal to $(\mathbf{a} \cdot \mathbf{b})/|\mathbf{a}|$, which equals $|\mathbf{b}| \cos\theta$. It is positive when the vector projection of $\mathbf{b}$ on $\mathbf{a}$ is in the same direction as $\mathbf{a}$, and negative when the vector projection is in the opposite direction to $\mathbf{a}$; its absolute value gives the length of the vector projection of $\mathbf{b}$ on $\mathbf{a}$.

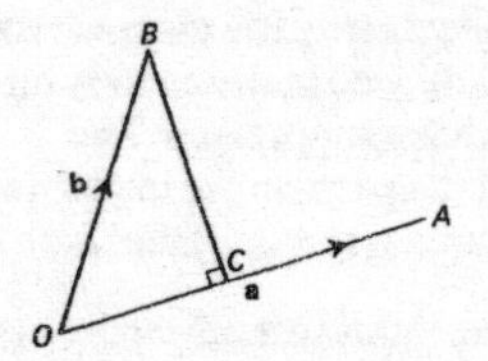

vector triple product For vectors $\mathbf{a}$, $\mathbf{b}$ and $\mathbf{c}$, the vector $\mathbf{a} \times (\mathbf{b} \times \mathbf{c})$, being the *vector product* of $\mathbf{a}$ with the vector $\mathbf{b} \times \mathbf{c}$, is called a **vector triple product**. The use of brackets here is essential, since $(\mathbf{a} \times \mathbf{b}) \times \mathbf{c}$, another vector triple product, gives, in general, quite a different result. The vector $\mathbf{a} \times (\mathbf{b} \times \mathbf{c})$ is perpendicular to $\mathbf{b} \times \mathbf{c}$ and so lies in the plane determined by $\mathbf{b}$ and $\mathbf{c}$. In fact, $\mathbf{a} \times (\mathbf{b} \times \mathbf{c}) = (\mathbf{a} \cdot \mathbf{c})\mathbf{b} - (\mathbf{a} \cdot \mathbf{b})\mathbf{c}$.

velocity Suppose that a particle is moving in a straight line, with a point O on the line taken as origin and one direction taken as positive. Let x be the *displacement* of the particle at time t. The **velocity** of the particle is equal to $\dot{x}$ or dx/dt, the *rate of change* of x with respect to t. The velocity is positive when the particle is moving in the positive direction and negative when it is moving in the negative direction.

In the preceding paragraph, a common convention has been followed in which the unit vector $\mathbf{i}$ in the positive direction along the line has been suppressed. Velocity is in fact a vector quantity, and in the one-dimensional case above is equal to $\dot{x}\mathbf{i}$.

When the motion is in 2 or 3 dimensions, vectors are used explicitly. The velocity $\mathbf{v}$ of a particle P with position vector $\mathbf{r}$ is given by $\mathbf{v} = d\mathbf{r}/dt = \dot{\mathbf{r}}$. When Cartesian coordinates are used, $\mathbf{r} = x\mathbf{i} + y\mathbf{j} + z\mathbf{k}$, and then $\dot{\mathbf{r}} = \dot{x}\mathbf{i} + \dot{y}\mathbf{j} + \dot{z}\mathbf{k}$.

Velocity has the dimensions LT^{-1}, and the SI unit of measurement is the metre per second, abbreviated to '$m\ s^{-1}$'.

velocity ratio See *machine*.

velocity–time graph A graph that shows velocity plotted against time for a particle moving in a straight line. Let $x(t)$ and $v(t)$ be the displacement and velocity, respectively, of the particle at time t. The velocity–time graph is the graph $y = v(t)$, where the t-axis is horizontal and the y-axis is vertical with the positive direction upwards. When the graph is above the horizontal axis, the particle is moving in the positive direction, when it is below the horizontal axis, the particle is moving in the negative direction.

The gradient of the velocity–time graph at any point is equal to the acceleration of the particle at that time. Also,

$$\int_{t_1}^{t_2} v(t)\,dt = x(t_2) - x(t_1).$$

So, with the convention that any area below the horizontal axis is negative, the area under the graph gives the change in position during the time interval concerned. Note that this is not necessarily the distance travelled.

(Here a common convention has been followed in which the unit vector $\mathbf{i}$ in the positive direction along the line has been suppressed. The displacement, velocity and acceleration of the particle are in fact vector quantities equal to $x(t)\mathbf{i}$, $v(t)\mathbf{i}$ and $a(t)\mathbf{i}$, where $v(t) = \dot{x}(t)$ and $a(t) = \dot{v}(t)$.)

Venn, John (1834–1923) British logician who, in his work *Symbolic Logic* of 1881, introduced what are now called *Venn diagrams*.

Venn diagram A method of displaying relations between subsets of some *universal set*. The universal set E is represented by the interior of a rectangle, say, and subsets of E are represented by regions inside this, bounded by simple closed curves. For instance, two sets A and B can be represented by the interiors of overlapping circles and then the sets $A \cup B$, $A \cap B$ and $A \setminus B$, for example, are represented by the shaded regions shown in the figures.

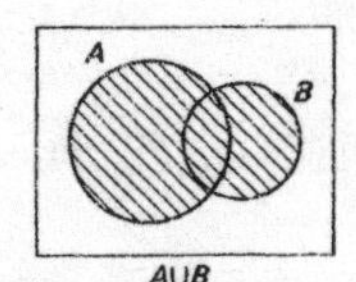

$A \cup B$

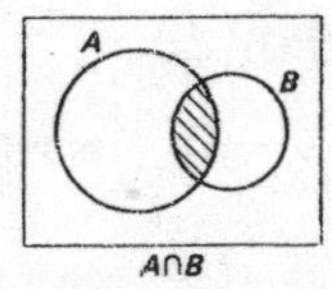

$A \cap B$

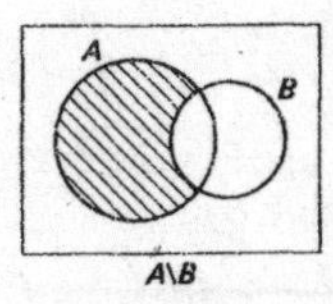

$A \backslash B$

Given one set, A, the universal set is divided into two disjoint subsets A and A', which can be clearly seen in a simple Venn diagram. Given two sets A and B, the universal set E is divided into four disjoint subsets $A \cap B$, $A' \cap B$, and $A \cap B'$ and $A' \cap B'$. A Venn diagram drawn with two overlapping circles for A and B clearly shows the four corresponding regions. Given three sets A, B and C, the universal set E is divided into eight disjoint subsets $A \cap B \cap C$, $A' \cap B \cap C$, $A \cap B' \cap C$, $A \cap B \cap C'$, $A \cap B' \cap C'$, $A' \cap B \cap C'$, $A' \cap B' \cap C$ and $A' \cap B' \cap C'$, and these can be illustrated in a Venn diagram as shown here.

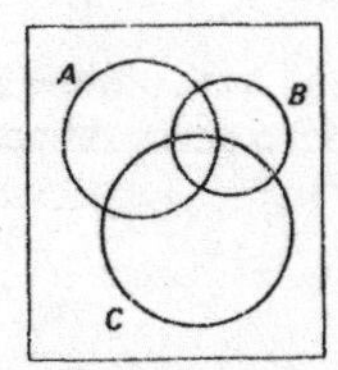

Venn diagrams can be used with care to prove properties such as *De Morgan's laws*, but some authors prefer other proofs because a diagram may only illustrate a special case. Four general sets, for example, should not be represented by four overlapping circles because they cannot be drawn in such a way as to make apparent the 16 disjoint subsets into which E should be divided.

vertex See *ellipse*, *hyperbola* and *parabola*; and also *cone*.

vertex (of a graph), **vertex-set** See *graph*.

Viète, François (1540–1603) French mathematician who moved a step towards modern algebraic notation by using vowels to represent unknowns and consonants to represent known numbers. This practice, together with other improvements in notation, facilitated the handling of equations and enabled him to make considerable advances in algebra. He also developed the subject of trigonometry.

volume of a solid of revolution Let $y = f(x)$ be the graph of a function f, a *continuous function* on $[a, b]$, and such that $f(x) \geq 0$ for all x in $[a, b]$. The **volume V of the solid of revolution** obtained by rotating, through one revolution about the x-axis, the region bounded by the curve $y = f(x)$, the x-axis and the lines $x = a$ and $x = b$, is given by

$$V = \int_a^b \pi y^2\,dx = \int_a^b \pi(f(x))^2\,dx.$$

PARAMETRIC FORM: For the curve $x = x(t), y = y(t)$ $(t \in [\alpha, \beta])$, the volume V is given by

$$V = \int_\alpha^\beta \pi y^2 \frac{dx}{dt}\,dt = \int_\alpha^\beta \pi(y(t))^2 x'(t)\,dt.$$

Von Neumann, John (1903–1957) Mathematician who made important contributions to a wide range of areas of pure and applied mathematics. He was born in Budapest but lived in the United States from 1930. In applied mathematics, he was one of the founders of optimization theory and the theory of games, with applications to economics. Within pure mathematics, his work in functional analysis is important. His lifelong interest in mechanical devices led to his being involved crucially in the initial development of the modern electronic computer and the important concept of the stored program.

vulgar fraction A *fraction* written as a/b where a and b are positive integers, as opposed to a *decimal fraction*. For example, 0.75 is a decimal fraction; $\frac{3}{4}$ is a vulgar fraction.

Wallis, John (1616–1703) The leading English mathematician before Newton. His most important contribution was in the results he obtained by using infinitesimals, developing Cavalieri's method of indivisibles. He also wrote on mechanics. Newton built on Wallis's work in his development of calculus and his laws of motion. It was Wallis who first made use of negative and fractional indices. He was instrumental in the founding of the Royal Society in 1662.

Wallis's Product See *pi*.

watt The SI unit of *power*, abbreviated to 'W'. One watt is equal to one *joule* per second.

Weierstrass, Karl (Theodor Wilhelm) (1815–1897) German mathematician who was a leading figure in the field of mathematical analysis. His work was concerned with providing the subject with the necessary rigour as it developed out of eighteenth-century calculus. One particular area in which he made significant contributions was in the expansion of functions in *power series*. To show that intuition is not always reliable, he gave an example of a function that is continuous at every point but not differentiable at any point. Some of his work was done while he was a provincial school teacher, having little contact with the world of professional mathematicians. He was promoted directly to professor of mathematics in Berlin at the age of 40.

weight The magnitude of the force acting on a body due to gravity. It is generally assumed that when a body of mass m is near the Earth's surface it is acted upon by a *uniform gravitational force* equal to $-mg\mathbf{k}$, where $\mathbf{k}$ is a unit vector directed vertically upwards and g is the magnitude of the acceleration due to gravity. Thus the weight of the body equals mg.

Weight has the dimensions MLT^{-2}, the same as force, and the SI unit of measurement is the *newton*. In common usage, when an object is weighed the measurement is recorded in kilograms. In fact, the weighing machine measures the weight, but the reading gives the corresponding mass.

weighted mean See *mean*.

Wilcoxon rank-sum test A non-parametric test (see *non-parametric methods*) for testing the null hypothesis that two independent samples of size n and m are from the same population. The observations are combined and ranked. If the null hypothesis is true, the sum U of the ranks of the observations in the sample of size n has mean $n(n+m+1)/2$ and variance $nm(n+m+1)/12$. For reasonably large values of n and m, the value of U can be tested against a *normal distribution*.

work done The **work done** by a force $\mathbf{F}$ during the time interval from $t = t_1$ to $t = t_2$ is equal to

$$\int_{t_1}^{t_2} \mathbf{F} \cdot \mathbf{v}\, dt,$$

where $\mathbf{v}$ is the velocity of the point of application of $\mathbf{F}$.

From the equation of motion $m\mathbf{a} = \mathbf{F}$ for a particle of mass m moving with acceleration $\mathbf{a}$, it follows that $m\mathbf{a} \cdot \mathbf{v} = \mathbf{F} \cdot \mathbf{v}$, and this then gives $(d/dt)(\frac{1}{2}m\mathbf{v} \cdot \mathbf{v}) = \mathbf{F} \cdot \mathbf{v}$. By integration, it follows that the change in *kinetic energy* is equal to the work done by the force.

Suppose that a particle, moving along the x-axis in the positive direction, with displacement $x(t)\mathbf{i}$ and velocity $v(t)\mathbf{i}$ at time t, is acted on by a constant force $F\mathbf{i}$ in the same direction. Then the work done by this constant force during the time interval from $t = t_1$ to $t = t_2$ is equal to

$$\int_{t_1}^{t_2} F\, v(t)\, dt,$$

which equals $F(x(t_2) - x(t_1))$. This is usually interpreted as 'work = force × distance'.

The work done against a force $\mathbf{F}$ should be interpreted as the work done by an applied force equal and opposite to $\mathbf{F}$. When the force is *conservative*, this is equal to the change in *potential energy*. When a person lifts an object of mass m from ground level to a height z above the ground, work is done against the *uniform gravitational force* and the work done equals the increase mgz in potential energy.

Work has the dimensions $\mathrm{ML^2T^{-2}}$, the same as energy, and the SI unit of measurement is the *joule*.

work–energy principle The principle that the change in *kinetic energy* of a particle during some time interval is equal to the *work done* by the total force acting on the particle during the time interval.

Z

Z See *integer*.

Z_n See *residue class* (modulo n).

Zeno of Elea (5th century BC) Greek philosopher whose paradoxes, known through the writings of Aristotle, may have significantly influenced the Greeks' perception of magnitude and number. The four paradoxes of motion are concerned with whether or not space and time are fundamentally made up of minute indivisible parts. If not, the first two paradoxes, the Dichotomy and the Paradox of Achilles and the Tortoise, appear to lead to absurdities. If so, the third and fourth, the Paradox of the Arrow and the Paradox of the Stadium, apparently give contradictions.

Zermelo, Ernst (Friedrich Ferdinand) (1871–1953) German mathematician considered to be the founder of axiomatic set theory. In 1908, he formulated a set of axioms for set theory, which attempted to overcome problems such as that posed by *Russell's paradox*. These, with modifications, have been the foundation on which much subsequent work in the subject has been built.

zero (of a function) See *root*.

zero divisor = *divisor of zero*.

zero element An element z is a **zero element** for a *binary operation* $\circ$ on a set S if, for all a in S, $a \circ z = z \circ a = z$. Thus the real number 0 is a zero element for multiplication since, for all a, $a0 = 0a = 0$. The term 'zero element', also denoted by 0, may be used for an element such that $a + 0 = 0 + a = a$ for all a in S, when S is a set with a binary operation $+$ called addition. Strictly speaking, this is a *neutral element* for the operation $+$.

zero function In real analysis, the **zero function** is the *real function* f such that $f(x) = 0$ for all x in **R**.

zero matrix The $m \times n$ **zero matrix O** is the $m \times n$ matrix with all its entries zero. A zero column matrix or row matrix may be denoted by **0**.

zero-sum game See *matrix game*.

zero vector See *vector*.

zone A **zone** of a sphere is the part between two parallel planes. If the sphere has radius r and the distance between the planes is h, the area of the curved surface of the zone equals $2\pi rh$.

A **circumscribing cylinder** of a sphere is a right-circular cylinder with the same radius as the sphere and with its axis passing through the centre of the

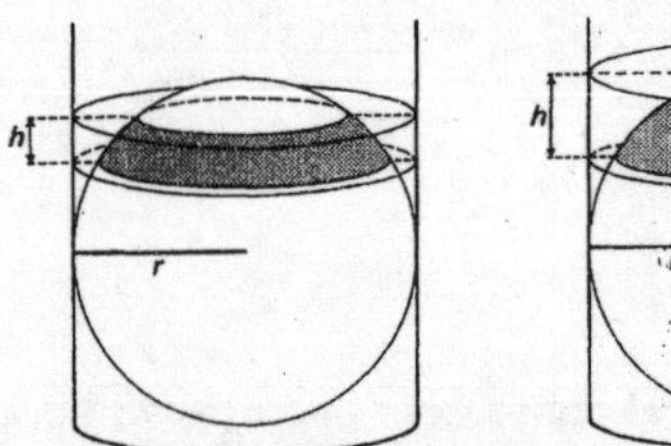

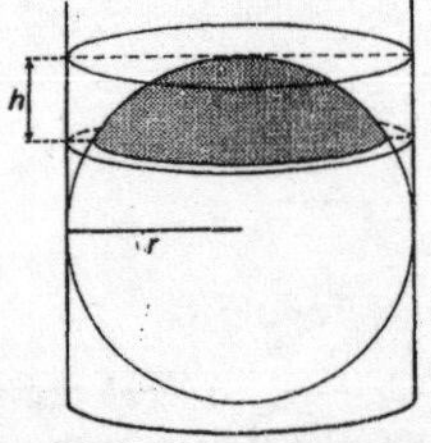

sphere. Suppose that a zone is formed by two parallel planes. Take the circumscribing cylinder with its axis perpendicular to the two planes as shown. It is an interesting fact that the area of the curved surface of the zone is equal to the area of the part of the circumscribing cylinder between the same two planes. The special case when one of the planes touches the sphere gives the area of a **spherical cap** as $2\pi rh$.

Appendix 1. Table of areas and volumes

(For unexplained notation, see under the relevant reference.)

Rectangle, length a, width b:
Area $= ab$.

Parallelogram:
Area $= bh = ab\sin\theta$.

Triangle:
Area $= \frac{1}{2}$base × height $= \frac{1}{2}bc\sin A$.

Trapezium:
Area $= \frac{1}{2}h(a+b)$.

Circle, radius r:
Area $= \pi r^2$,
Length of circumference $= 2\pi r$.

Right-circular cylinder, radius r, height h:
Volume $= \pi r^2 h$,
Curved surface area $= 2\pi rh$.

Right-circular cone:
Volume $= \frac{1}{3}\pi r^2 h$,
Curved surface area $= \pi rl$.

Frustum of a cone:
Volume $= \frac{1}{3}\pi h(a^2 + ab + b^2)$,
Curved surface area $= \pi(a+b)l$.

Sphere, radius r:
Volume $= \frac{4}{3}\pi r^3$,
Surface area $= 4\pi r^2$.

Zone of a sphere:
Curved surface area $= 2\pi rh$.

Appendix 2. Table of derivatives

$f(x)$	$f'(x)$
k (constant)	0
x	1
x^k	kx^{k-1}
$\sin x$	$\cos x$
$\cos x$	$-\sin x$
$\tan x$	$\sec^2 x$
$\sec x$	$\sec x \tan x$
$\operatorname{cosec} x$	$-\operatorname{cosec} x \cot x$
$\cot x$	$-\operatorname{cosec}^2 x$
e^{kx}	ke^{kx}
$\ln x$	$1/x$
$a^x (a > 0)$	$a^x \ln a$
$\sin^{-1} x$	$\frac{1}{\sqrt{1-x^2}}$
$\cos^{-1} x$	$-\frac{1}{\sqrt{1-x^2}}$
$\tan^{-1} x$	$\frac{1}{1+x^2}$
$\sinh x$	$\cosh x$
$\cosh x$	$\sinh x$
$\tanh x$	$\operatorname{sech}^2 x$
$\coth x$	$-\operatorname{cosech}^2 x$
$\sinh^{-1} x$	$\frac{1}{\sqrt{x^2+1}}$
$\cosh^{-1} x$	$\frac{1}{\sqrt{x^2-1}}$
$\tanh^{-1} x$	$\frac{1}{1-x^2}$

Appendix 3. Table of integrals

Notes:

(i) The table gives, for each function f, an antiderivative ϕ. The function ϕ_1 given by $\phi_1(x) = \phi(x) + c$, where c is an arbitrary constant, is also an antiderivative of f.

(ii) In certain cases, for example when $f(x) = 1/x$, $\tan x$, $\cot x$, $\sec x$, $\operatorname{cosec} x$ and $\sqrt{x^2 - a^2}$, the function f is not continuous for all x. In these cases, a definite integral $\int_a^b f(x)\,dx$ can be evaluated as $\phi(b) - \phi(a)$ only if a and b both belong to an interval in which f is continuous.

$f(x)$	$\phi(x) = \int f(x)\,dx$
$x^k \ (k \neq -1)$	$\dfrac{x^{k+1}}{k+1}$
$1/x$	$\ln\lvert x\rvert$
$\sin x$	$-\cos x$
$\cos x$	$\sin x$
$\tan x$	$-\ln\lvert\cos x\rvert = \ln\lvert\sec x\rvert$
$\sec x$	$\ln\lvert\sec x + \tan x\rvert$
$\operatorname{cosec} x$	$\ln\lvert\operatorname{cosec} x - \cot x\rvert = \ln\lvert\tan \frac{1}{2}x\rvert$
$\cot x$	$\ln\lvert\sin x\rvert$
$\sin^2 x$	$\frac{1}{2}(x - \frac{1}{2}\sin 2x)$
$\cos^2 x$	$\frac{1}{2}(x + \frac{1}{2}\sin 2x)$
$\sinh x$	$\cosh x$
$\cosh x$	$\sinh x$
$e^{kx} \ (k \neq 0)$	e^{kx}/k
$e^{ax}\sin bx$	$\dfrac{e^{ax}}{a^2+b^2}(a\sin bx - b\cos bx)$
$e^{ax}\cos bx$	$\dfrac{e^{ax}}{a^2+b^2}(a\cos bx + b\sin bx)$
$a^x \ (a > 0,\ a \neq 1)$	$a^x/\ln a$
$\ln x$	$x\ln x - x$
$\dfrac{1}{\sqrt{a^2 - x^2}} \ (a > 0)$	$\sin^{-1}\dfrac{x}{a}$
$\dfrac{1}{a^2 + x^2} \ (a > 0)$	$\dfrac{1}{a}\tan^{-1}\dfrac{x}{a}$

$f(x)$	$\phi(x) = \int f(x)\,dx$
$\dfrac{1}{x^2 - a^2}\ (a > 0)$	$\dfrac{1}{2a}\ln\left\|\dfrac{x-a}{x+a}\right\|$
$\dfrac{1}{\sqrt{x^2 + a^2}}\ (a > 0)$	$\sinh^{-1}\dfrac{x}{a}$ or $\ln(x + \sqrt{x^2 + a^2})$
$\dfrac{1}{\sqrt{x^2 - a^2}}\ (a > 0)$	$\cosh^{-1}\dfrac{x}{a}$ or $\ln\left\|x + \sqrt{x^2 - a^2}\right\|$
$\sqrt{x^2 + a^2}$	$\frac{1}{2}x\sqrt{x^2 + a^2} + \frac{1}{2}a^2\ln(x + \sqrt{x^2 + a^2})$
$\sqrt{x^2 - a^2}$	$\frac{1}{2}x\sqrt{x^2 - a^2} - \frac{1}{2}a^2\ln\left\|x + \sqrt{x^2 - a^2}\right\|$
$\sqrt{a^2 - x^2}\ (a > 0)$	$\frac{1}{2}x\sqrt{a^2 - x^2} + \frac{1}{2}a^2\sin^{-1}\dfrac{x}{a}$

Appendix 4. Table of series

$$e^x = 1 + \frac{x}{1!} + \frac{x^2}{2!} + \cdots + \frac{x^n}{n!} + \cdots \quad \text{(for all } x\text{)}$$

$$\sin x = x - \frac{x^3}{3!} + \frac{x^5}{5!} - \frac{x^7}{7!} + \cdots + (-1)^n \frac{x^{2n+1}}{(2n+1)!} + \cdots \quad \text{(for all } x\text{)}$$

$$\cos x = 1 - \frac{x^2}{2!} + \frac{x^4}{4!} - \frac{x^6}{6!} + \cdots + (-1)^n \frac{x^{2n}}{(2n)!} + \cdots \quad \text{(for all } x\text{)}$$

$$\sinh x = x + \frac{x^3}{3!} + \frac{x^5}{5!} + \frac{x^7}{7!} + \cdots + \frac{x^{2n+1}}{(2n+1)!} + \cdots \quad \text{(for all } x\text{)}$$

$$\cosh x = 1 + \frac{x^2}{2!} + \frac{x^4}{4!} + \frac{x^6}{6!} + \cdots + \frac{x^{2n}}{(2n)!} + \cdots \quad \text{(for all } x\text{)}$$

$$\ln(1+x) = x - \frac{x^2}{2} + \frac{x^3}{3} - \frac{x^4}{4} + \cdots + (-1)^{n-1} \frac{x^n}{n} + \cdots \quad (-1 < x \le 1)$$

$$\tan^{-1} x = x - \frac{x^3}{3} + \frac{x^5}{5} - \frac{x^7}{7} + \cdots + (-1)^n \frac{x^{2n+1}}{2n+1} + \cdots \quad (-1 \le x \le 1)$$

$$\sin^{-1} x = x + \frac{1}{2}\frac{x^3}{3} + \frac{1 \times 3}{2 \times 4}\frac{x^5}{5} + \frac{1 \times 3 \times 5}{2 \times 4 \times 6}\frac{x^7}{7} + \cdots + \binom{2n}{n}\frac{x^{2n+1}}{2^{2n}(2n+1)} + \cdots \quad (-1 \le x \le 1)$$

$$(1+x)^\alpha = 1 + \frac{\alpha}{1!}x + \frac{\alpha(\alpha-1)}{2!}x^2 + \cdots + \frac{\alpha(\alpha-1)\ldots(\alpha-n+1)}{n!}x^n + \cdots \quad (-1 < x < 1)$$

(This is a binomial expansion; when α is a non-negative integer, the expansion is a finite series and is then valid for all x.)

Appendix 5. Table of trigonometric formulae

$$\tan A = \frac{\sin A}{\cos A}, \qquad \cot A = \frac{\cos A}{\sin A} = \frac{1}{\tan A},$$

$$\sec A = \frac{1}{\cos A}, \qquad \operatorname{cosec} A = \frac{1}{\sin A},$$

$$\cos^2 A + \sin^2 A = 1, \qquad \sec^2 A = 1 + \tan^2 A, \qquad \operatorname{cosec}^2 A = 1 + \cot^2 A.$$

Addition formulae

$$\sin(A+B) = \sin A \cos B + \cos A \sin B,$$

$$\sin(A-B) = \sin A \cos B - \cos A \sin B,$$

$$\cos(A+B) = \cos A \cos B - \sin A \sin B,$$

$$\cos(A-B) = \cos A \cos B + \sin A \sin B,$$

$$\tan(A+B) = \frac{\tan A + \tan B}{1 - \tan A \tan B},$$

$$\tan(A-B) = \frac{\tan A - \tan B}{1 + \tan A \tan B}.$$

Double-angle formulae

$$\sin 2A = 2 \sin A \cos A,$$

$$\cos 2A = \cos^2 A - \sin^2 A,$$

$$\cos 2A = 1 - 2\sin^2 A, \qquad \sin^2 A = \tfrac{1}{2}(1 - \cos 2A),$$

$$\cos 2A = 2\cos^2 A - 1, \qquad \cos^2 A = \tfrac{1}{2}(1 + \cos 2A),$$

$$\tan 2A = \frac{2 \tan A}{1 - \tan^2 A}.$$

Tangent of half-angle formulae

Let $t = \tan \frac{1}{2}A$; then

$$\sin A = \frac{2t}{1+t^2}, \qquad \cos A = \frac{1-t^2}{1+t^2}, \qquad \tan A = \frac{2t}{1-t^2}.$$

Product formulae

$$\sin A \cos B = \tfrac{1}{2}(\sin(A+B) + \sin(A-B)),$$
$$\cos A \sin B = \tfrac{1}{2}(\sin(A+B) - \sin(A-B)),$$
$$\cos A \cos B = \tfrac{1}{2}(\cos(A+B) + \cos(A-B)),$$
$$\sin A \sin B = \tfrac{1}{2}(\cos(A-B) - \cos(A+B)).$$

Sums and differences

$$\sin C + \sin D = 2\sin\tfrac{1}{2}(C+D)\cos\tfrac{1}{2}(C-D),$$
$$\sin C - \sin D = 2\cos\tfrac{1}{2}(C+D)\sin\tfrac{1}{2}(C-D),$$
$$\cos C + \cos D = 2\cos\tfrac{1}{2}(C+D)\cos\tfrac{1}{2}(C-D),$$
$$\cos C - \cos D = -2\sin\tfrac{1}{2}(C+D)\sin\tfrac{1}{2}(C-D).$$

Appendix 6. Table of symbols

Symbol	Reference
$\neg$	negation
$\wedge$	conjunction
$\vee$	disjunction
$\Rightarrow, \Leftrightarrow$	implication
$\sim$	equivalence relation
$\exists, \forall$	quantifier
$\in, \notin$	belongs to
$\subseteq, \supseteq$	subset
$\subset, \supset$	proper subset
$\cup, \bigcup$	union
$\cap, \bigcap$	intersection
$A', \bar{A}$	complement
$\emptyset$	empty set
$A \times B$	Cartesian product
$A \setminus B, A - B$	difference set
$A + B, A \Delta B$	symmetric difference
$n(A), \#(A), \|A\|$	cardinality
$\mathcal{P}(A)$	power set
$n!$	factorial
$[a, b]$	closed interval, least common multiple
(a, b)	open interval, greatest common divisor
$[a, b), (a, b]$	interval
$\binom{n}{r}$	binomial coefficient
$[x]$	integer part
$\{x\}$	fractional part
$\|x\|$	absolute value
$\|z\|$	modulus
$\bar{z}$	conjugate
$\Re z, \mathrm{Re}\, z$	real part
$\Im z, \mathrm{Im}\, z$	imaginary part
$\overrightarrow{AB}$	directed line-segment
$\|AB\|, \|\overrightarrow{AB}\|$	length
$\\|P\\|$	norm

Symbol	Reference
$\sqrt{\ }$	square root
$\neq, <, \leq, >, \geq$	inequality
$\approx$	approximation
$\equiv$	congruence
$\propto$	proportion
$\Sigma, \sum$	summation notation
$\Pi, \prod$	product notation
π	pi
$f: x \mapsto y$	function, mapping
$f: S \to T$	function, mapping
$\to$	limit
$\nearrow, \searrow$	limit from the left and right
$f \circ g$	composition
f^{-1}	inverse function, inverse mapping
$f', \frac{df}{dx}, y', \frac{dy}{dx}$	derivative, derived function
$f'', f''', \ldots, f^{(n)}, \frac{d^2f}{dx^2}, \ldots, \frac{d^nf}{dx^n}$	higher derivative
$y'', y''', \ldots, y^{(n)}, \frac{d^2y}{dx^2}, \ldots, \frac{d^ny}{dx^n}$	higher derivative
$f_x, f_y, f_1, f_2, \frac{\partial f}{\partial x}, \frac{\partial f}{\partial y}$	partial derivative
$f_{xx}, f_{xy}, \ldots, f_{11}, f_{12}, \ldots$	higher-order partial derivative
$\frac{\partial^2 f}{\partial x^2}, \frac{\partial^2 f}{\partial x \partial y}, \ldots$	higher-order partial derivative
$\dot{x}, \ddot{x}$	rate of change
$\int$	integral, antiderivative
$\mathbf{a} . \mathbf{b}$	scalar product
$\mathbf{a} \times \mathbf{b}, \mathbf{a} \wedge \mathbf{b}$	vector product
$\mathbf{a} . (\mathbf{b} \times \mathbf{c}), [\mathbf{a}, \mathbf{b}, \mathbf{c}]$	scalar triple product
$\mathbf{a} \times (\mathbf{b} \times \mathbf{c})$	vector triple product
$\mathbf{A}^T, \mathbf{A}^t, \mathbf{A}'$	transpose
$\mathbf{A}^{-1}$	inverse matrix
$\lvert\mathbf{A}\rvert$	determinant

Symbol	**Reference**
$\langle G, \circ \rangle$	group
$\langle R, +, \times \rangle$	ring
$\mathrm{E}(X)$	expected value
$\mathrm{Var}(X)$	variance
$\mathrm{Cov}(X, Y)$	covariance
$\Pr(A)$	probability
$\Pr(A \mid B)$	conditional probability

Appendix 7. Table of Greek letters

Name	Lower case	Capital
Alpha	α	A
Beta	β	B
Gamma	γ	Γ
Delta	δ	Δ
Epsilon	ϵ	E
Zeta	ζ	Z
Eta	η	H
Theta	θ	Θ
Iota	ι	I
Kappa	κ	K
Lambda	λ	Λ
Mu	μ	M
Nu	ν	N
Xi	ξ	Ξ
Omicron	o	O
Pi	π	Π
Rho	ρ	P
Sigma	σ	Σ
Tau	τ	T
Upsilon	υ	Υ
Phi	ϕ	Φ
Chi	χ	X
Psi	ψ	Ψ
Omega	ω	Ω